INTERNATIONAL SEMINAR ON
NUCLEAR WAR AND PLANETARY EMERGENCIES
26th Session:

AIDS AND INFECTIOUS DISEASES — MEDICATION OR VACCINATION FOR DEVELOPING COUNTRIES; MISSILE PROLIFERATION AND DEFENSE; TCHERNOBYL — MATHEMATICS AND DEMOCRACY; TRANSMISSIBLE SPONGIFORM ENCEPHALOPATHY; FLOODS AND EXTREME WEATHER EVENTS — COASTAL ZONE PROBLEMS; SCIENCE AND TECHNOLOGY FOR DEVELOPING COUNTRIES; WATER — TRANSBOUNDARY WATER CONFLICTS; CLIMATIC CHANGES — GLOBAL MONITORING OF THE PLANET; INFORMATION SECURITY; POLLUTION IN THE CASPIAN SEA; PERMANENT MONITORING PANELS REPORTS; TRANSMISSIBLE SPONGIFORM ENCEPHALOPATHY WORKSHOP; AIDS AND INFECTIOUS DISEASES WORKSHOP; POLLUTION WORKSHOP

THE SCIENCE AND CULTURE SERIES
Nuclear Strategy and Peace Technology

Series Editor: Antonino Zichichi

1981 — International Seminar on Nuclear War — 1st Session: The World-wide Implications of Nuclear War

1982 — International Seminar on Nuclear War — 2nd Session: How to Avoid a Nuclear War

1983 — International Seminar on Nuclear War — 3rd Session: The Technical Basis for Peace

1984 — International Seminar on Nuclear War — 4th Session: The Nuclear Winter and the New Defence Systems: Problems and Perspectives

1985 — International Seminar on Nuclear War — 5th Session: SDI, Computer Simulation, New Proposals to Stop the Arms Race

1986 — International Seminar on Nuclear War — 6th Session: International Cooperation: The Alternatives

1987 — International Seminar on Nuclear War — 7th Session: The Great Projects for Scientific Collaboration East-West-North-South

1988 — International Seminar on Nuclear War — 8th Session: The New Threats: Space and Chemical Weapons — What Can be Done with the Retired I.N.F. Missiles-Laser Technology

1989 — International Seminar on Nuclear War — 9th Session: The New Emergencies

1990 — International Seminar on Nuclear War — 10th Session: The New Role of Science

1991 — International Seminar on Nuclear War — 11th Session: Planetary Emergencies

1991 — International Seminar on Nuclear War — 12th Session: Science Confronted with War (unpublished)

1991 — International Seminar on Nuclear War and Planetary Emergencies — 13th Session: Satellite Monitoring of the Global Environment (unpublished)

1992 — International Seminar on Nuclear War and Planetary Emergencies — 14th Session: Innovative Technologies for Cleaning the Environment

1992 — International Seminar on Nuclear War and Planetary Emergencies — 15th Session (1st Seminar after Rio): Science and Technology to Save the Earth (unpublished)

1992 — International Seminar on Nuclear War and Planetary Emergencies — 16th Session (2nd Seminar after Rio): Proliferation of Weapons for Mass Destruction and Cooperation on Defence Systems

1993 — International Seminar on Planetary Emergencies — 17th Workshop: The Collision of an Asteroid or Comet with the Earth (unpublished)

1993 — International Seminar on Nuclear War and Planetary Emergencies — 18th Session (4th Seminar after Rio): Global Stability Through Disarmament

1994 — International Seminar on Nuclear War and Planetary Emergencies — 19th Session (5th Seminar after Rio): Science after the Cold War

1995 — International Seminar on Nuclear War and Planetary Emergencies — 20th Session (6th Seminar after Rio): The Role of Science in the Third Millennium

1996 — International Seminar on Nuclear War and Planetary Emergencies — 21st Session (7th Seminar after Rio): New Epidemics, Second Cold War, Decommissioning, Terrorism and Proliferation

1997 — International Seminar on Nuclear War and Planetary Emergencies — 22nd Session (8th Seminar after Rio): Nuclear Submarine Decontamination, Chemical Stockpiled Weapons, New Epidemics, Cloning of Genes, New Military Threats, Global Planetary Changes, Cosmic Objects & Energy

1998 — International Seminar on Nuclear War and Planetary Emergencies — 23rd Session (9th Seminar after Rio): Medicine & Biotechnologies, Proliferation & Weapons of Mass Destruction, Climatology & El Nino, Desertification, Defence Against Cosmic Objects, Water & Pollution, Food, Energy, Limits of Development, The Role of Permanent Monitoring Panels

1999 — International Seminar on Nuclear War and Planetary Emergencies — 24th Session: HIV/AIDS Vaccine Needs, Biotechnology, Neuropathologies, Development Sustainability — Focus Africa, Climate and Weather Predictions, Energy, Water, Weapons of Mass Destruction, The Role of Permanent Monitoring Panels, HIV Think Tank Workshop, Fertility Problems Workshop

2000 — International Seminar on Nuclear War and Planetary Emergencies — 25th Session: Water — Pollution, Biotechnology — Transgenic Plant Vaccine, Energy, Black Sea Pollution, Aids — Mother–Infant HIV Transmission, Transmissible Spongiform Encephalopathy, Limits of Development — Megacities, Missile Proliferation and Defense, Information Security, Cosmic Objects, Desertification, Carbon Sequestration and Sustainability, Climatic Changes, Global Monitoring of Planet, Mathematics and Democracy, Science and Journalism, Permanent Monitoring Panel Reports, Water for Megacities Workshop, Black Sea Workshop, Transgenic Plants Workshop, Research Resources Workshop, Mother–Infant HIV Transmission Workshop, Sequestration and Desertification Workshop, Focus Africa Workshop

2001 — International Seminar on Nuclear War and Planetary Emergencies — 26th Session: AIDS and Infectious Diseases — Medication or Vaccination for Developing Countries; Missile Proliferation and Defense; Tchernobyl — Mathematics and Democracy; Transmissible Spongiform Encephalopathy; Floods and Extreme Weather Events — Coastal Zone Problems; Science and Technology for Developing Countries; Water — Transboundary Water Conflicts; Climatic Changes — Global Monitoring of the Planet; Information Security; Pollution in the Caspian Sea; Permanent Monitoring Panels Reports; Transmissible Spongiform Encephalopathy Workshop; AIDS and Infectious Diseases Workshop; Pollution Workshop

THE SCIENCE AND CULTURE SERIES
Nuclear Strategy and Peace Technology

INTERNATIONAL SEMINAR ON
NUCLEAR WAR AND PLANETARY EMERGENCIES
26th Session:

AIDS AND INFECTIOUS DISEASES — MEDICATION OR VACCINATION FOR DEVELOPING COUNTRIES; MISSILE PROLIFERATION AND DEFENSE; TCHERNOBYL — MATHEMATICS AND DEMOCRACY; TRANSMISSIBLE SPONGIFORM ENCEPHALOPATHY; FLOODS AND EXTREME WEATHER EVENTS — COASTAL ZONE PROBLEMS; SCIENCE AND TECHNOLOGY FOR DEVELOPING COUNTRIES; WATER — TRANSBOUNDARY WATER CONFLICTS; CLIMATIC CHANGES — GLOBAL MONITORING OF THE PLANET; INFORMATION SECURITY; POLLUTION IN THE CASPIAN SEA; PERMANENT MONITORING PANELS REPORTS; TRANSMISSIBLE SPONGIFORM ENCEPHALOPATHY WORKSHOP; AIDS AND INFECTIOUS DISEASES WORKSHOP; POLLUTION WORKSHOP

"E. Majorana" Centre for Scientific Culture
Erice, Italy, 19 – 24 August 2001

Series editor and Chairman: A. Zichichi

edited by R. Ragaini

World Scientific
New Jersey • London • Singapore • Hong Kong

Published by

World Scientific Publishing Co. Pte. Ltd.

P O Box 128, Farrer Road, Singapore 912805

USA office: Suite 1B, 1060 Main Street, River Edge, NJ 07661

UK office: 57 Shelton Street, Covent Garden, London WC2H 9HE

INTERNATIONAL SEMINAR ON NUCLEAR WAR AND PLANETARY EMERGENCIES 26TH SESSION: AIDS AND INFECTIOUS DISEASES — MEDICATION OR VACCINATION FOR DEVELOPING COUNTRIES; MISSILE PROLIFERATION AND DEFENSE; TCHERNOBYL — MATHEMATICS AND DEMOCRACY; TRANSMISSIBLE SPONGIFORM ENCEPHALOPATHY; FLOODS AND EXTREME WEATHER EVENTS — COASTAL ZONE PROBLEMS; SCIENCE AND TECHNOLOGY FOR DEVELOPING COUNTRIES; WATER — TRANSBOUNDARY WATER CONFLICTS; CLIMATIC CHANGES — GLOBAL MONITORING OF THE PLANET; INFORMATION SECURITY; POLLUTION IN THE CASPIAN SEA; PERMANENT MONITORING PANELS REPORTS; TRANSMISSIBLE SPONGIFORM ENCEPHALOPATHY WORKSHOP; AIDS AND INFECTIOUS DISEASES WORKSHOP; POLLUTION WORKSHOP

ISBN 981-238-092-2

Printed in Singapore by Uto-Print

CONTENTS

12. PERMANENT MONITORING PANEL REPORTS

1. OPENING SESSION

OPENING SPEECH

PROFESSOR ANTONINO ZICHICHI
CERN and University of Bologna, Geneva, Switzerland

Ladies and Gentlemen,

May I have the pleasure of welcoming you also on behalf of the Co-Chairmen, Professors T.D. Lee and Kai Siegbahn, to the 26th Session of the International Seminars on Planetary Emergencies.

This year we have a hundred and twenty participants from thirty-one countries and the topics will be restricted to the following areas. We cannot review the fifty-three Planetary Emergencies, but only a few of them. This year we will dedicate not only this Seminar, but all the activities of the Erice Centre to the Centennial Celebration of Enrico Fermi. I will say a few words later.

The topics of this year are:

a) AIDS – medication or vaccines for developing countries;
b) Missile proliferation and defense;
c) Chernobyl status, Mathematics and Democracy;
d) BSE status;
e) Coastal zone problems;
f) A very important topic: Science and Technology for developing countries. This is the only way we can ensure that the famous gap between G8 and the rest of the world can stop increasing every year;
g) The transboundary water conflicts;
h) The topic that attracts everybody's attention, namely the story about climatic changes;
i) Information security;
j) Pollution in the Caspian Sea which is a typical example of pollution in the world.

As I said before, these Seminars and the Centre's activities are dedicated to the greatest Galilean of the 20th Century, Enrico Fermi, the Italian Navigator, father of the weak forces.

Enrico Fermi entered in the history of the 20th Century with this coded sentence transmitted to President Roosevelt. While he was in Yalta with Churchill and Stalin, President Roosevelt received the news that the Italian Navigator has succeeded in reaching the New World. By this, it was meant that nuclear fire was finally achieved. This is how Enrico Fermi will be recalled in the history of the 20th Century. But he is also the father of the weak forces.

Why do we say that he is the greatest Galilean of the 20th Century? Because after a few hundred years of science since Galilei, this activity was showing clear signs of subdivision in three sectors: the sector of experimental Physics to be planned and implemented; the sector of technological inventions, needed for the advance of science; and the theoretical analysis. Enrico Fermi succeeded in being a star in the three sectors. This is why he is the greatest Galilean of the 20th Century.

In fact, Enrico Fermi invented and implemented many experiments, the last one was the discovery of what we call the $\Delta 33$ which is the first example of baryonic particle in addition to the two known baryons, the proton and the neutron. This particle was of great importance for the discovery of Quantum Chromium. So this is for experimental Physics.

In technology he invented many things and the most important one is how to slow down neutrons. This is how we succeeded with Wigner and other collaborators to start the nuclear fire. The first example of fire that does not depend on the sun. After ten thousand years of civilization, a new fire was implemented.

Enrico Fermi formulated the theory of weak forces, the control system which allows the sun and all stars to burn with extreme regularity. And this is how Bertrand Russell came to define Enrico Fermi, the Italian Navigator, father of weak forces. This is just an example.

Enrico Fermi discovered the statistical law named the Fermi Law. We are all made of Fermians. Protons, neutrons and electrons are all Fermians. He formulated the theory of cosmic rays and this is not all.

Enrico Fermi had two pupils, Ettore Majorana in Italy and Tsung Dao Lee in the United States. And now I give the floor to Professor Lee to recall his memories and give his testimony about Enrico Fermi.

ENRICO FERMI

PROFESSOR TSUNG DAO LEE
Department of Physics, Columbia University, New York, NY, USA

Enrico Fermi was a grand master of Physics, in both theory and experiment. In the History of Physics, perhaps we have to go back to Galileo to find his equal in both theory and experimental disciplines. In future generations, there may be none that rise to the same level of achievement.

Nino has already mentioned, and I will repeat, some of Fermi's great contributions. This is a partial list:

- Radiation from charged particles
- Fermi-Dirac statistics
- Thomas-Fermi model
- Theory of β-decay
- Scattering length (in atomic and nuclear collision)
- Slow neutron

Left Italy (December 6, 1938) arrived in the USA (January 2, 1939)

- Fission (through energy balance)
- Pile and nuclear reactor
- e-n interaction
- Origin of cosmic radiation
- Discovery of first Π-N resonance ($\Delta33$)
- Approach to equilibrium (Fermi-Pastor-Ulam)

The years in which Fermi made his transition from Italy to the U.S. were also a period of great unrest for the entire world. A few months after Fermi's arrival in the United States, Einstein wrote a letter to President Roosevelt (see annex).

Some time later, on December 2, 1942, a telephone conversation between A.H. Compton and F.D. Roosevelt ran thus: "The Italian Navigator has reached the New World," said Compton. "And how did he find the natives?" asked Roosevelt. "Very friendly" answered Compton. Like Columbus before him, Fermi came to the new continent and changed the world.

In the first part of this talk I would like to review the events related to Fermi during these particularly tumultuous years.

It is particularly moving to read of Fermi's concern and care for his colleagues in Italy at that time, as attested by a letter to Professor Pegram dated October 22, 1938, that Fermi went to Brussels to send (see annex). This is typical of Fermi as a person.

In September 1938 the Italian Anti-Semitic Law was passed which affected Mrs. Laura Fermi. Events started to move at a rapid rate:

December 6, 1938
Fermi left Rome for Stockholm

December 10, 1938
Fermi received the Nobel Prize.
After Stockholm, the Fermis traveled to Copenhagen and stayed with the Bohrs.

December 19, 1938
Otto Hahn wrote from Germany to Lisa Meitner in Sweden announcing the detection of "barium" from "n+u" collisions

December 24, 1938
Fermis sailed from Southampton by ship (the Franconia) and arived in New York on January 2, 1939.

Christmas 1938
Otto Frisch visited his aunt Lisa Meitner and, together, Meitner and Frisch formulated the fission theory to explain the Hahn-Strassmann discovery.

January 3, 1939
Frisch returned to Copenhagen and told N. Bohr of the Hahn-Strassmann discovery and the Meitner-Frisch theory of fission.

January 7, 1939
Niels Bohr left Copenhagen (by the Swedish ship Drotningholm).

January 16, 1939
Bohr arrived in New York, met by the Fermis. They had long discussions but...

January 21, 1939
Bohr lectured on the Hahn-Strassmann discovery and the Meitner-Frisch fission theory in Princeton. W. Lamb was in the audience and returned to New York that evening.

January 22, 1939
> Lamb told Fermi of the Bohr lecture. Fermi began his fission experiment which was completed three days later.

Fermi was Professor of Physics at Columbia from 1939 to 1946. In 1946 Fermi left Columbia and joined the University of Chicago. The same year, I received a Chinese Government fellowship which enabled me to go to the United States to further my study of Physics. At that time, I was 19 years old and had only two years of undergraduate training in China. Although I knew very little English, I had learned enough Physics to want to enter the graduate school directly. Fortunately, the University of Chicago would admit me.

Reminiscences of Chicago Days (1946 – 1949)

Faculty:
E. Fermi, E. Teller, M. Mayer, J. Mayer, H. Urey, G. Wentzel, S. Chandrasekhar, Mullikan, Zacherwein, Simpson...

Students:
O. Chamberlain, J. Steinberger, C.N. Yang, M. Rosenbluth, L. Wolgerstein, M. Goldberger, S. Treiner, R. Garcia, J. Orear, H. Agnew, G. Chew...

Right after the war the Chicago Physics Department was the best in the world.

Fermi and Urey had already received their Nobel Prizes

Mayer, Chandrasekhar and Mulliken would receive them later.

A few weeks later, I received a note from Fermi asking me to attend his special evening class (which was by invitation only). It was there that I had my first glimpse of Fermi in action. The subject matter ranged over all topics in Physics. Sometimes he would randomly pull a card out of his file, on which a subject title was usually written with a key formula. It was wonderful to see how, in one session, Fermi could start from scratch, give the incisive estimate and arrive at the relevant formula and the physics that could be derived from it. The freedom with which he moved from field to field was an inspiration to watch.

Another thing that comes frequently to my mind about student life in those times are the square-dancing parties held frequently at the Fermis' home. They were my first introduction to occidental culture. Enrico's dancing, Laura's punch and Harold Agnew's energetic calling of "do-si-do" all made indelible marks on my memory.

Soon afterwards, I became Fermi's Ph.D. student in theory. At that time, most of the Chicago Ph.D. students in theory were supervised by E. Teller; these included M. Rosenbluth, L. Wolfenstein and C.N. Yang.

The relationship between Fermi and his students was quite personal. I would see him regularly, about once a week. Usually we had lunch together in Commons, often with his other students. After that, Fermi and I would spend the whole afternoon talking. At that time, Fermi was interested in the origin of cosmic radiation and nuclear synthesis. He directed me first towards Nuclear Physics and then into Astrophysics. Quite often he would mention a topic and ask me if I could think and read about it, and then "give him a lecture" the following week. Of course, I obliged and usually felt very good afterwards. Only much later did I realize that this was an excellent way of guiding students towards independence.

Fermi fostered a spirit of self-reliance and intellectual independence in his students. One had to verify or derive all the formulas that one used. At one point I was discussing the internal structure of the sun with him; the coupled differential equations of radiative transfer were quite complicated. Since that was not my research topic, I did not want to devote too much time to tedious checking. Instead, I simply quoted the results of well-known references. However, Fermi though one should never accept other people's calculations without some independent confirmation. He then had the ingenious idea of making a specialized slide rule designed to deal with these radiative transfer equations :

$$\frac{dL}{dr} \propto T18 \quad \text{and} \quad \frac{dT}{dr} \propto L/T6.5$$

(where L is the luminosity and T the temperature). Over a week's time, he helped me to produce the magnificent six-foot-seven-inch slide rule in the photograph (in annex) with 18 log x on one side and 6.5 log x on the other. With that, even integration became fun and I was able to complete the checking quickly and then move on to a different topic for discussion. This unique experience made a deep impression on me. Even now, sometimes, when I encounter difficulties, I try to imagine how Fermi might react under similar circumstances.

All of us, especially at the very beginning, require personal guidance. I was fortunate to have Fermi as a teacher. In that sense, I was luckier than Fermi.

The legacy of Enrico Fermi will stay with the History of Humankind.

ANNEX

Letter to President Franklin D. Roosevelt, from Albert Einstein,
dated August 2nd 1939

Letter to Professor Pegram at Columbia University,
dated October 22, 1938.

Photograph of T.D. Lee holding the slide rule to calculate luminosity and
temperature of the sun, that Enrico Fermi helped him to build.

August 2nd, 1939

F.D. Roosevelt,
President of the United States,
White House
Washington, D.C.

Sir:

Some recent work by E.Fermi and L. Szilard, which has been communicated to me in manuscript, leads me to expect that the element uranium may be turned into a new and important source of energy in the immediate future. Certain aspects of the situation which has arisen seem
to call for watchfulness and, if necessary, quick action on the part
of the Administration. I believe therefore that it is my duty to bring
to your attention the following facts and recommendations:

In the course of the last four months it has been made probable -
through the work of Joliot in France as well as Fermi and Szilard in
America - that it may become possible to set up a nuclear chain reaction
in a large mass of uranium,by which vast amounts of power and large quantities of new radium-like elements would be generated. Now it appears
almost certain that this could be achieved in the immediate future.

This new phenomenon would also lead to the construction of bombs,
and it is conceivable - though much less certain - that extremely powerful bombs of a new type may thus be constructed. A single bomb of this
type, carried by boat and exploded in a port, might very well destroy
the whole port together with some of the surrounding territory. However,

I understand that Germany has actually stopped the sale of uranium
from the Czechoslovakian mines which she has taken over. That she should
have taken such early action might perhaps be understood on the ground
that the son of the German Under-Secretary of State, von Weizsäcker, is
attached to the Kaiser-Wilhelm-Institut in Berlin where some of the
American work on uranium is now being repeated.

Yours very truly,

(Albert Einstein)

CLUB DER UNIVERSITAIRE STICHTING

TELEFONE {CLUB (SALONS EN HOTEL) 11.97.89 / 11.97.80} {SECRETARIAAT 12.24.22 / 12.30.28}
POSTREKENING NR. 1038.48
TELEGRAMADRES : "FORUM-BRUSSEL"

BRÜSSEL, Oktober 22, 1938
11, LONDENSTRAAT

Dear Professor Pegram,

I cabled to you yesterday as follows:

L.C. Pegram Columeum Neuy York accept Professorship writing Fermi

I should like to express to you again my really very sincere thanks for your generous offer; and please extend my thanks also to Professor Butler.

I should like to come to New York, if possible, for the beginning of the spring term that starts, as far as I remember, at the end of February.

For reasons that you can easily understand however, I should like to leave Italy, without giving the feeling that this is due to political reasons. I could manage this much more easily if you could invite me officially

to teach at Columbia through the main Embassy in the U.S. Of course you need no mention, or stress, in this request, that it would be a permanent appointment.

In order to get a non quota visa for myself and my family, I should need besides an official letter from Columbia stating that I am appointed as professor and mentioning the salary. In case that you cannot invite me through the Embassy please send me only this second letter.

And in any case please do not give any publicity to this matter, untill the situation in Italy is finally settled.

I shall take the opportunity that I am writing to you from Belgium, in order to give to you some information about the situation of the Italian physicists, that have lost their positions on account of racial reasons. They are Emilio Segrè, whom you already know. He is now at Berkeley and has, so far as I know, a num usual research fellowship for one year from the University of California. I don't think, that I used to inform you about his scientific work.

Bruno Rossi, formerly professor at the University of Padova (married with no children; age about 32). He is one of our best young physicists; his work on the cosmic radiation is probably known to you. He has lately acquired some experience in high tension work, since he had built in Padova a one million volts Cockroft Walton outfit, that was just now being tested.

Giulio Racah, (formerly professor at Pisa (Not married; age about 30). He has a very extensive knowledge of theoretical physics. Has published many papers in atomic physics and quantum theory; in particular he has obtained independently and published only a few days after Heitler and Bethe equivalent results on the theory of the emission of high energy γ-rays from cosmic ray electrons colliding against nuclei.

Ugo Fano (age about 26; not married)

was my assistant for theoretical physics. Good knowledge of theory; very great enthusiasm for research. Has been lately very much interested for theoretical problems in connection with biology. Had several discussions on this topics with Heisenberg, Kronowski of Berlin and with P. Jordan of Rostock.

Leo Pincherle, formerly lecturer of theoretical physics at Padova (age about 30; married with 1 or 2 children). Has published rather interesting papers on intensity problems of X-Ray lines.

I might finally mention that Rasetti (although not for racial reasons), is trying to find a situation abroad. He would also like to be invited for some course next summer.

Please write to me to my home address

Via L. Magalotti 15 - Roma, Italy

looking forward to seeing you next winter, I am, with best greetings

Enrico Fermi

LETTER SENT FROM BRUSSELS, OCTOBER 22, 1938.

The slide rule that Fermi helped to build

$$\frac{dL}{dr} \propto T^{18}$$

$$\frac{dT}{dr} \propto L/T^{6.5}$$

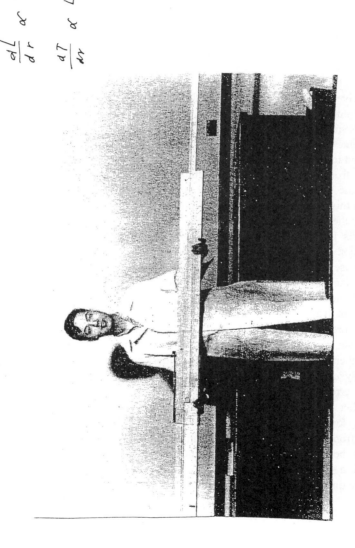

A dedicated slide rule for the internal temperature distribution of a main-sequence star, circa 1948.

ENVIRONMENTAL APPLICATIONS OF SPACE TECHNOLOGY

DAVID R. SCOTT

Scott Science and Technology, Inc., Los Angeles, California, USA

ABSTRACT

"What was most significant about the lunar voyage was not that men set foot on the Moon, but that they set eye on the Earth." This statement, by Mr. Norman Cousins, Editor of the Saturday Review, summarizes the most significant aspect of the first departure of humans from the environment in which they were born, and in which they must survive. Looking back at the Earth from the Moon, the view is both splendid and overwhelming. A small blue ball in the vastness of black space, dotted with millions of marvelous stars; an oasis that we must understand and protect. For, if one searches the heavens, one will find no other island for life as we understand it. If we humans do not protect and nurture this environment, it will disappear—just as quickly as the Earth can disappear from behind the outstretched thumb of a man on the Moon.

As the world continues to grow economically, more and more "power" is required to sustain this growth. These increased power requirements are currently being supplied mainly by the increased use of fossil fuels. Scientific consensus concludes that the use of fossil fuels for power is a primary cause of global warming and the consequential deterioration of the environment. Nuclear power is the only viable alternative to the increased use of fossil fuels for power. However, the increased use of nuclear power produces increased nuclear waste, the disposal of which is a serious environmental hazard. A method of *permanent* disposal of nuclear waste, especially high-level waste, is currently not available.

The present approach to "permanent" disposal of nuclear waste is to deposit it underground, or "geologic disposal." This approach is neither permanent nor safe. Among other problems, such disposal sites must be maintained for tens of thousands of years; they are subject to changing natural conditions; and they must be protected from theft and terrorism.

Rather than attempting to store high-level nuclear waste in Earth facilities for extremely long periods, this hazardous high-level nuclear waste can be permanently removed from the Earth by launching it into space. Studies by NASA have confirmed the feasibility, both technical and economic, of launching "safe canisters" of high-level nuclear waste into a space trajectory that departs the solar system or is sent directly to the Sun. This is the only known method of permanent disposal of nuclear waste.

The purpose of this paper is to summarize the concept of nuclear waste disposal in space and to describe the activities necessary to develop a comprehensive "case" for the disposal of nuclear waste in space. The activities necessary to define the case for nuclear waste disposal in space will be defined in the form of a series of feasibility analyses, the objectives of which will be presented in the form of "tasks" associated with a general statement of work. The statement of work will then present the objectives and supporting rationale for a subsequent major program to actually dispose nuclear waste in space. The subsequent "Nuclear Waste Disposal Program" (NWDP) will be planned and implemented independently.

The tasks (or feasibility analyses) will address each of the major issues that would or could be involved in implementing a program of nuclear waste disposal in space. Such issues include scientific (e.g., nuclear power and storage); technical (e.g., launching into space); economic (e.g., cost); societal (e.g., public acceptance); political (e.g., participation and contributions by nations); etc. These studies and analyses will be conducted by universities and other academic institutions insofar as is practical.

INTRODUCTION

Background
A View of the Earth

It is a pleasure to have the opportunity to participate in the 26th Session of the Erice International Seminars on Planetary Emergencies sponsored by the World Federation of Scientists. The activities of these Seminars exemplify the values of Earth and its inhabitants, which are so important to understand as we journey through space and time.

"What was most significant about the lunar voyage was not that men set foot on the Moon, but that they set eye on the Earth." This statement, by Mr. Norman Cousins, Editor of the Saturday Review, summarizes the most significant aspect of the first departure of humans from the environment in which they were born, and in which they must survive. Looking back at the Earth from the Moon, the view is both splendid and overwhelming. This small blue ball in the vastness of black space, dotted with millions of marvelous stars, is an oasis that we must understand and protect. For, if one searches the heavens, one will find no other island for life, as we understand it. If we humans do not protect and nurture this environment, it will disappear—just as readily as the Earth will disappear from behind an outstretched thumb of a man on the Moon. Everything that has meaning would disappear: science, history, music, poetry, art, literature, all existing on this small, fragile, and precious little spot out there in the vastness of the universe.

Almost 30 years ago I had the opportunity to view our Earth from a distant outpost on the Moon; the exploration site between Hadley Rille and the Apennine Mountains on the edge of the Imbrium Basin.

As we traveled through space and explored the Moon, we saw many new and fascinating physical sights and relationships among the Earth, the planets, the stars, and the black void of space. These views change not only our perspective of the Earth but

16

our value system as well. When we are on the Earth, we see the Moon track across the sky from horizon to horizon, always the same face, and always the same features. But from the Moon we see the Earth at the same point in the sky, day after day, but always turning, showing us new faces and changing features as the hours pass. Perhaps most significantly, from the Moon we see how the Earth is related to the rest of the universe, how it is not a separate "world" somehow isolated from the universe. We also see how the physical features of the Earth are interrelated. And in a sense, from the Moon or deep space we can "see the future" on Earth—we can see that tomorrow will be a sunny day in Erice, but that a giant storm is brewing in the Atlantic, and soon rain will come to Erice.

The Earth-Moon "System"

As we view the Earth from space, we see the earth as a dynamic and evolving system—a three-dimensional system that is ever changing in space and in time. The still "snapshots" that have been returned from space missions do not truly illustrate the changing nature of our dynamic and living planet. These snapshots rarely show the depth of the atmosphere or depth of the oceans—both of which compose a "shell" that contains the biosphere, the biota, and we humans. As viewed from space, we would see a whole, integrated planet on which these living things, as well as the air, the oceans, the mountains, the deserts, and even the rocks, all combine into a total and complete "system."

And this Earth system clearly has the capacity to regulate its temperature as well as the composition of its shell which keep it sustainable for life above, on, and below its surface. The self-regulation of this system is an active process in which living organisms and their material environment are tightly coupled. As this coupled system evolves, feedback mechanisms self-regulate the climate and the chemistry that comprise the basic nature of the system. This system is sometimes called a self-regulating super-organism.

Perhaps this living, dynamic, self-regulating system is best described by the "Gaia theory" developed by Dr. James Lovelock during the past 40 years. In his very informative book, The Ages of Gaia, he too takes us on a journey through space:

"As we move in towards the Earth from space, first we see the atmospheric boundary that encloses Gaia; then the borders of an ecosystem such as the forests; then the skin or bark of living animals and plants; further in are the cell membranes; and finally the nucleus of the cell and its DNA.....to recognize that the evolution of the species is strongly coupled with the evolution of their environment....

But suppose that the Earth is a super organism. Then the evolution of the organisms and the evolution of the rocks need no longer be regarded as separate sciences to be studied in separate buildings of the university. Instead, a single evolutionary science describes the history of the whole planet. The evolution of the species and the evolution of their environment are tightly coupled together as a single and inseparable process...."

Therefore, the basic Gaia approach views the Earth as a self-regulating super organism. However, we humans are the stewards of this superorganism for we are the only intelligent beings that can consciously affect a change in the nature of our planet, and not always benignly. And if we do not change our behavior and effect certain changes, the results could very likely be catastrophic. Let us discuss a specific example.

Power Growth and Environmental Deterioration

Picture yourself approaching the North Pole—what do you expect to see? Ice and more ice, of course! Wrong! Today you would see an expanse of ordinary looking open ocean with a few patches of ice and a distant view of ice floes.

One realistic conclusion from this observation is that environmental change in the form of global warming is causing the warmest period on Earth in several millennia. Many believe that if this trend continues, the Earth may very soon experience a catastrophic transformation of its environment: a return to high temperatures not seen since dinosaurs ruled the Earth.

Is human activity causing global warming? Answers are not conclusive, but a "yes" answer would appear to be convincing. As an example, we humans have consciously created artificial pollution in the atmosphere. According to consensus science this artificial pollution has prompted global warming and the potentially severe consequences thereof. And the activity causing the most damage would clearly be the use of fossil fuels for power.

As the world continues to grow economically, more and more power (or rate of energy utilization) will be required. Where will we obtain this additional power? Fossil, solar, wind, nuclear? Solar and wind power are new, relatively unproved sources of power, and will very likely be inadequate on the large scales required; they are therefore minor factors. And, furthermore, we are beginning to terminate the use of nuclear power—the replacement of which, even at current levels, will require more power from other sources, primarily fossil fuels.

If we continue this cycle—increased power; increased use of fossil fuels; increased pollution; increased global warming; increased consequences thereof—we may very well have begun what might be called the "Gaian Spiral"—an inward converging spiral which culminates in the destruction of the environment. As a result, human progress will come to an abrupt end; certainly the technological civilization that we have created will collapse; and eventually there will be no need for power at all.

Power Growth and Environmental Protection

Because a large-scale increase in power from solar and wind sources is unrealistic, the only practical alternative to the increase in the use of fossil fuels is the increase in the use of nuclear energy. However, in such a scenario, and due to adverse public perception, one must question the safety of nuclear energy.

During recent years, the concern over safety of nuclear energy has been growing. This lack of confidence comes at a time when the availability of energy is a worldwide problem for all nations. Two problems are of major concern: the operation of nuclear power stations, and the disposal of nuclear waste. Due to the conflicting attitudes toward the use of nuclear energy and the requirements for increased power, it is important to examine all viable approaches to increasing the safety of nuclear energy.

Nuclear waste is one of the most serious environmental hazards existing on Earth—today, and in the future. Disposal of nuclear waste is particularly troublesome, due to the extreme toxicity and long lifetimes of some waste components. Significant quantities of nuclear waste currently exist, and increased quantities are expected in the near future. At present there is no acceptable method of disposal of nuclear waste—the waste material will remain on the Earth, in a hazardous condition, for thousands of years regardless of the method used for "disposal." Guaranteeing the isolation of these high-level wastes from the biosphere for time periods ranging from hundreds to hundreds of thousands of years is one of the major problems facing our generation—because we must not, and cannot leave this to future generations.

However, the use of nuclear power can be enhanced by two specific initiatives: (1) improved management and operations of nuclear power stations; and (2) the permanent disposal of nuclear waste. The management and operations techniques of Apollo demonstrated that we have the capability to successfully manage highly complex and large-scale programs; nuclear power stations and the disposal of waste in space should be no exception. The permanent disposal of nuclear waste is the subject of the following sections.

HIGH LEVEL NUCLEAR WASTE

Definition
"Primary" (extremely dangerous) nuclear waste, or High Level Waste (HLW) is generated by two processes: (1) production of electricity in nuclear powerplants ("spent fuel"); and (2) production of nuclear materials for "defense" ("high-level radioactive waste"). A serious barrier to the use of nuclear powerplants is the problem of spent fuel disposal. Should such disposal be "clean" and permanent, nuclear electrical power could be vastly enhanced. All nations are attempting to reduce, or terminate the production of nuclear materials for defense. Even if such a goal can be achieved, the problem of high-level radioactive waste disposal remains uncertain, at best.

The primary nuclear waste consists of two general groups: fission products and actinides (i.e., radioactive elements above actinium, such as neptunium and plutonium). Actinides are the most hazardous. Compared to most fission products, they have exceedingly long half lives (hundreds of thousands of years) and are most dangerous to humans (if ingested will continually radiate the person from within). For practical purposes, and because actinides are most hazardous and comprise the lesser amount of primary waste, this paper will consider disposal of the actinides only, often termed "high level waste" (HLW).

Societal and Technical Challenges
Nuclear waste is one of the most serious environmental hazards existing on Earth—today,

and in the future. At present there is no acceptable method of disposal of nuclear waste—the waste material will remain on the Earth, in a hazardous condition, for thousands of years regardless of the method used for "disposal."

The U.S. National Academy of Sciences recently completed a detailed study on the "Disposition of High-Level Waste and Spent Nuclear Fuel"[1]. This study concludes that "Geological disposal remains the only long-term solution available." It further concludes that "The societal and technical challenges of geologicical disposal and, more broadly, of ongoing management of HLW (High-Level Waste) have turned out to be greater than anticipated when the United States and other nations established programs of HLW disposal starting some decades ago."

Among other findings of the study, the following are especially significant.

- Today's growing inventory of HLW requires attention by national decision makers.
- The feasible options are monitored storage on or near the earth's surface and geological disposition.... The major uncertainty is in the confidence that future societies will continue to monitor and maintain such facilities.
- Geological disposal remains the only long-term solution available.... As in all scientific work, progress in achieving geological disposal has been marked by surprises, new insights, and the recognition that for even the best-characterized sites, there always will be uncertainties about the long-term performance of the repository system.
- Today the biggest challenges to waste disposition are societal.
- Whether, when, and how to move toward geological disposal are societal decisions for each country.

The results of the activities to be conducted to present the case for the disposal of nuclear waste in space will show that most, if not all of these concerns can be circumvented. As an example, the disposal of nuclear waste in space is *permanent*, therefore there are no long-term problems associated with the disposal of waste in space; e.g., there need not be concern "in the confidence that future societies will continue to monitor and maintain such facilities." Also, there need not be concern about "the long-term performance of the repository system."

THE PERMANENT DISPOSAL OF NUCLEAR WASTE IN SPACE

Concept

A more permanent and reliable approach would be to dispose of nuclear waste in space. Space is, as we know, a radiation environment itself; for example, the Sun which generates its light and heat from nuclear energy of its own. Although in principle all nuclear waste could be transported to space, a more reasonable approach would be to use space only for the disposal of long-lived, hazardous isotopes. For this approach, space and geologic disposal would be complementary parts of a total waste management program using each disposal option where appropriate.

The primary requirement for any nuclear waste disposal system is that the system has acceptable risk in both the short and long term. Terrestrial and space disposal systems have much different risk characteristics. The space option has a finite short-term risk associated primarily with launch and injection into the disposal region; it has negligible risk thereafter. In contrast, the terrestrial options have a lower short-term risk, but a long-term risk, which is difficult to estimate with high confidence.

The minimization of short-term risk for space disposal demands some comprehensive and strict design requirements. However, we must remember that many years ago we designed launch vehicles for humans that had a very high degree of overall safety. As an example, the Saturn launch vehicles that were designed, tested, and used for Apollo were 100% successful—31 out of 31. Further, we can now design and construct capsules that can survive any type of launch catastrophe—this safety approach was also successfully developed during Apollo (although never used).

The burden of safety assurance for space disposal will be to demonstrate: 1) the ability of the payload to survive accidents intact; 2) the ability of rescue systems to locate and return the intact payloads from earth locations, and 3) from space locations. Thus, the nuclear waste payload will have to be designed to withstand the worst-case accident condition it may encounter, such as a launch vehicle explosion. Similarly, the requirement for assured recovery from any unplanned event will require the ability to locate and recover the payload from land, shallow water, or deep ocean with complete assurance following an aborted launch. And the assured recovery will also require the ability to locate and rescue payloads left in unplanned orbits following a failure or degraded performance of the launch vehicle.

During the past 20 years, each of these three areas has experienced significant technological improvement and corresponding reduction in risk. Nuclear waste can now be encapsulated in a package that could survive a major explosion. Deep submersibles can recover objects from great ocean depths (requiring launch at known sea-depth locations). And recovery of specific objects from unplanned launch orbits is currently being developed with no unforeseen uncertainties. Therefore, the risk of nuclear waste disposal in space has not only been reduced to acceptable levels technologically; but also the proven capability to manage large complex space projects (e.g. Apollo) will further minimize risk by incorporating the reliability of such demonstrated project management.

Disposal Program

The primary method of waste disposal currently being considered is isolation in stable geologic formation. Other means of isolation within the biosphere, such as sub-seabed or in very deep holes, also are being considered. However, any method which leaves all radionuclides on the Earth must rely upon the long-term behavior of constraint, containment, and surveillance systems and must consider release mechanisms into the biosphere that are difficult to quantify and analyze with high confidence. The long-term risks associated with the terrestrial options are probably undeniable, since they are dominated by future geological processes and by unpredictable social processes such as the decay or disappearance of control agencies or intrusive acts of sabotage.

NASA studies during the early 1980s verified the feasibility of nuclear waste disposal in space. Technology improvements since that time have enhanced this concept significantly. A general summary of the NASA studies was presented in a 1981 Position Paper by The American Institute of Aeronautics and Astronautics[2].

One design concept for disposing of nuclear waste in space begins with the processing and packaging of selected portions of high-level defense and commercial waste into a suitable storage form and containment vessel. The package would then be transported to the launch site and prepared for launch by the placement of radiation shields and accident protection systems. The waste package payload then would be placed in the cargo area of a launch vehicle. Once in orbit, a payload transfer vehicle would be used to deliver the waste package to its final destination. Among the several space destinations available, the complete transfer out of the solar system is both feasible and of the most permanent character. However, several other acceptable destinations were defined during the early NASA feasibility studies, including sending the waste payload directly into the Sun.

As an example, the waste is packaged in a modified rail/barge shipping "cask" that were developed for the U.S. Nuclear Regulatory Commission. They are certified safe (protection against potential exposure to radiation) for both normal transportation activities as well as accidents. The modified casks, or canisters, would weigh approximately 5,000 kg fully loaded and with shielding.

Certain commercial launch vehicles can deliver approximately 5,500 kg to a high earth orbit. From that orbit a space transfer vehicle would dock with the nuclear cask and inject it into a solar system escape orbit. The mass of each nuclear cask will be 5,000 kg, and the mass of each Space Transfer Vehicle (launched by a second launch vehicle) will also be approximately 5,000 kg. Therefore, for each "Disposal Mission", two launch vehicles will be required for each cask disposal to a solar escape trajectory. The nuclear waste payloads would be launched from Ocean platforms located at the equator in the Pacific Ocean. At each Ocean Launch Site, six platforms will be located around the circumference of a Launch Control Ship. In this manner, approximately one launch every three days can be conducted from a single Ocean Launch Site, or a total of 50 missions per year (2 launches per mission).

BENEFITS

Primary

- Reduction in the use of fossil fuels resulting in the reduction of global warming.
- Permanent disposal with no future concerns or implications.

Secondary

- Political—relieves political stress on the Middle East due to decreased reliance on oil.
- Economic—long term costs significantly less due to no need for maintenance and security of disposal sites.
- Economic—launch vehicles produced in Russia, thus stimulating economy.
- Psychological—no future concerns over the existence and disposition of nuclear waste.
- Legacy—for those who actively and substantially support this initiative, a legacy for the protection of the environment will remain for longer than the 100,000 years half life of the hazardous nuclear waste that would have been left on Earth.

ISSUES

General

To begin the development of the case for nuclear power, a number of issues must be addressed and resolved. This paper will consider some of these issues briefly; the proposed statement of work and tasks will cover each of these issues in detail.

Public Perception

The use of the word "nuclear" or "nuclear power" conveys very negative images in the minds of the public. This is associated with many problems related to the use of nuclear materials over the course of history. These include "the bomb," Chernobyl, Three Mile Island, and the temporary storage and transport of nuclear waste. Each of these problems must be shown to have been corrected and clarified in the minds of the public. The response to each of these issues as presented in the Statement of Work Plan must go beyond presenting a correction, it must provide an unqualified and positive description on the manner in which each of these issues has been corrected or should be perceived by the public.

Renewable Energy

"Renewable energy" is a favored approach by many to the need for increasing power. Such energy includes wind, solar, water and other non-fossil, non-nuclear concepts. Although these sources are currently being promoted heavily by their various supporters, none have really been developed to the extent that a comfortable "product" could be developed to support the increasing demand for power. In many cases, external factors negate the potential of such sources. Wind power would be of little use if there were no wind. Solar power would be greatly diminished during periods of heavy clouds.

Geologic Disposal

The favored approach to nuclear waste disposal is currently long-term storage

23

underground, or "geologic disposal. This method is currently being endorsed by the United States and many other countries, including Russia, Sweden, and Finland each of which is offering to provide disposal facilities to other countries.

Among the problems facing geologic disposal, the following will be discussed in detail within the tasks to be completed by the Statement of Work.

- Long term maintenance and storage stability requirements
- Protection against theft
- Protection against terrorist attacks
- Catastrophic events, natural and manmade
- Changing nature of natural conditions (geologic)
- Unknown and unforeseen events related to the presence of the material
- Numerous events and activities must be taken to complete storage
- Always there, causing psychological concern and debate
- Renewed controversy due to related or unrelated events; e.g., RTG or USSR Rorsat satellites reentry
- Anthropological change in human species
- Political protests in local area as well as international (as long as present on Earth)

Green Groups

Many environmental groups worldwide will oppose the use of nuclear power for a variety of reasons. Often it will be due to their own agendas of support for other measures. Often it will be due to their perception of the issues involved in the use of nuclear power. Often it will be just a matter of protest against the deterioration of the environment in general. The actual concerns of each of these groups must be addressed and an explanation must be prepared to either enlist their support or negate their protest.

FEASIBILITY ANALYSES

Objectives

The purpose of this document is to define the activities necessary to present a comprehensive "case" for the disposal of high-level nuclear waste in space. These activities are presented in the form of specific tasks to be conducted under a general statement of work. Each task will be assigned to a qualified educational institution for completion. The results of each task will be presented in the form of a document that can be used to support the overall objective of the statement of work. The results of these tasks will then be compiled into a single document that presents the overall case for the disposal of high-level nuclear waste in space.

The manner in which the subsequent nuclear waste disposal program will be implemented will depend on the results and conclusions of statement of work (feasibility analyses).

24

Statement of Work & Tasks

Statement of Work

To develop the concept and prepare the case for the disposal of nuclear waste in space, the following tasks will be used to prepare a "program definition" document that presents the overall program for nuclear waste disposal in space (NWDS). Each of these tasks will be implemented in the form of a feasibility analysis. Each task will be conducted by a qualified institution or organization. Universities and other academic institutions will be favored. It is presently contemplated that this work will be funded by a charitable foundation (Foundation).

Tasks (Feasibility Analyses).

1. Nuclear Power—The fundamentals of nuclear power shall be described, to include:
   a) Nuclear power stations, including power output capabilities, waste management, configuration, operations, maintenance, management, environmental impact, and especially risks and safety.
   b) Nuclear waste, all categories, including especially high-level waste, and the amount of existing waste and projected future waste.
   c) Current and planned methods of nuclear waste disposal, including environmental effects.
   d) Worldwide power requirements—define and describe the tradeoffs between the use of fossil fuels and the use of nuclear power for existing and future worldwide power requirements.
   e) Compare nuclear power with other sources of power such as wind, solar, water, and other "renewable" energy concepts.

2. Technical Feasibility—Investigate and describe the technical concept of waste disposal in space.
   a) Launch vehicles—production volume and cost
   b) Launch Site—Converted off shore launch platforms located at the equator
   c) Payload—encapsulated waste material
   d) Launch operations, including launch frequency.
   e) Logistics including establishment of the launch site and delivery of the payloads.
   f) Management—define an overall management plan
   g) Patents—existing patents and patent coverage of project concepts
   h) Environmental Impact Statement for anticipated launch site(s).

3. Economic Feasibility—investigate and define the economics of nuclear waste disposal in space.
    a) Estimate the cost of disposal in space (high-level waste only), including preparation of the payload, logistics of delivery and launch, and all launch facilities.
    b) Define other economic factors and include in cost estimates.
    c) Compare with the cost of geological disposal under current plans.

4. Environmental Feasibility—investigate and describe the environmental implications of NWDS.
    a) Describe the effects on the environment resulting from NWDS.
    b) Compare with the environmental effects of using fossil fuels.
    b) Prepare an Environmental Impact Statement for NWDS, including transport of waste and the launch site.

5. Develop a "Societal Plan"
    a) Assure public trust and public involvement.
    b) Eliminate uncertainty in the minds of the public.
    c) Obtain acceptance and consensus by scientists, critics, and members of the public.
    d) Define a public/private decision process on the implementation of the NWDS program.
    e) Describe the various approaches to public acceptance used by different countries.

6. Risk Management
    a) Describe and evaluate the risks of space disposal of nuclear waste.
    b) Compare with the risks of current methods of nuclear waste disposal.
    c) Define the manner in which the risks of space disposal can be minimized, including safety and security.

7. Strategic Partners, Trustees, and Advisors
    a) Identify and contact potential strategic partners (e.g., World Nuclear Association; Universal Studios; launch vehicle manufacturer; supportive governments, etc.).
    b) Identify and contact potential Trustees of the Foundation.
    c) Identify and contact potential sources of funding for the Foundation.
    d) Identify and contact sources of visible support for various phases of the project.

8. Government Involvement
    a) Define the potential roles of governments and other political organizations; e.g., UN.

b) Identify the type of support that could be provided by governmental organizations, e.g., financial, regulatory, resources.
c) Identify contact points and specific organizations that could provide such support.
d) Define the manner in which international cooperation can support NWDS.
e) Evaluate the effects and impact of national and international treaties, law, and regulations.

9. Prepare a Nuclear Waste Disposal in Space <u>Program</u> Plan
   a) Technical
   b) Economic
   c) Societal
   d) Environmental
   e) Management (including Risk)
   f) Political
   g) Implementation Plan

10. Funding
    a) Develop an approach to raise sufficient funds to initiate and operate the NWDS program.
    b) Prepare funding plans for both government and private sector sources as well as a combination of both.
    c) Describe the tradeoffs between and among public and private funding sources.

CONCLUSION

This document summarizes an approach to the permanent disposal of high-level nuclear waste by launching into space, preferably to deliver such waste to the sun. The Nuclear Waste Disposal in Space (NWDS) concept is far superior to the current alternative of geological disposal. Such permanent disposal in space will enhance the operations and prospects of nuclear power. This in turn will enable the reduction in the use of fossil fuels to support the growing needs of worldwide power. According to scientific consensus the reduction in the use of fossil fuels will reduce, or limit the increase of global warming. Thus the NWDS program will make a major contribution to the protection of the environment.

The first phase in the NWDS program will consist of a series of feasibility analyses to form the basis for the preparation of formal "case" for NWDS to be presented to responsible authorities and funding sources worldwide. The feasibility analyses will be conducted as tasks under an overall statement of work to be funded by a charitable foundation.

From the Moon we see the Earth as a "whole"—we see no borders, we see no boundaries, we see all humankind together and interrelated on this single small sphere.

This perspective from the Moon makes us realize that the Earth is dynamic and alive and evolving for the human presence—and we realize that if we care not for the life of the Earth itself, we care not for the life of its inhabitants. With this new perspective of the Earth and its place in human life, we must think of bold and visionary ideas to preserve our so limited and fragile environment. Temporary solutions to the problems of our times must be replaced by permanent solutions for future generations. For our generation did not inherit this marvelous environment in which we live, we borrowed it from our children, and our children's children. We owe them the best we can achieve; we owe them a conscious and substantial return on their investment in us.

REFERENCES

1.   "Disposition of High Level Waste and Spent Nuclear Fuel," Executive Summary, U.S. National Academy of Sciences, 2000.

2.   "Nuclear Waste Disposal in Space," AIAA Position Paper, American Institute of Aeronautics and Astronautics, January 14, 1981.

UPDATE ON BSE AND VARIANT CJD

PROF. ROBERT G. WILL
National CJD Surveillance Unit, Edinburgh, UK

Good morning everyone and I am very grateful to Professor Zichichi and the organisers for inviting me to come to the meeting although I am not so grateful that I have been asked to speak after Dr. Scott and after those fascinating presentations about Enrico Fermi.

However I will be talking about a subject that is of great political interest as well as scientific interest and has had many economic implications in Europe and now I think more generally.

I apologise, I have probably put in too many slides for the length of the talk, so I may whiz through them rather quickly.

There has been enormous scientific interest in the prion disease, or Transmissible Spongiform Encephalopathy, for many years, and indeed Dr. Gajdusek was awarded the Nobel Prize in the 1970s, as was Dr. Prusiner very recently for the Prion Theory. However these diseases were not of very great interest to the general public until there became a concern of a public health risk because of the cattle disease, BSE.

A paper was published in 1996 by the group in Edinburgh and in particular by my colleague, James Ironside, and of course a number of other individuals from other European countries because this disease was identified to a certain extent through a collaborative European surveillance system. The issue raised in this paper was: could it be this new disease was perhaps linked causally to the BSE epidemic in UK cattle. The hypothesis was that this was a new disease. That the UK population had been exposed to a new potential risk factor and that this had resulted in this new variant of CJD.

One of the ways of distinguishing the disease initially was the age of patients. Creutzfeld-Jacob Disease of the sporadic type is known world-wide, occurs with standard incidence of one case per million per year, and largely affects people of late middle age. Here is data from the UK for the last five or six years showing sporadic CJD and what has happened is that this new form of Creutzfeld-Jacob has been identified with a very young, or relatively young age at death. The mean age is only 29 years old. I am showing this just to show that the original hypothesis, which was based on ten cases, is now supported and it looks as if something has happened in CJD in the UK.

At that time my colleague in Edinburgh, James Ironside, thought that the newer pathology was unique. This is the newer pathological appearance of variant CJD. The brown staining that is shown on this slide is immunological staining of the disease associated prion protein, thought to be the causative agent. This is an area of the brain, and this is the thalamus with a vast amount of disease-associated protein. And Dr. Ironside had never seen anything like this before and thought it was new. And I think I can summarise the current situation by saying that a lot of work has been done to try and find similar conditions in other countries. Historically nothing has been found. This is a new disease.

The evidence of a link with BSE, and I will not have time to go through all this, is based on a range of laboratory studies, mainly transmission studies in mice carried out by Professor Collinge, Moira Bruce and more recently by Professor Prusiner's group. Essentially, what these studies show is that the transmission characteristics of variant CJD are the same as those of BSE and different from other forms of CJD. So there is now a consensus that BSE is the cause of variant CJD. This raises a lot of very difficult issues for public health.

Before discussing the number of cases, which of course is a very important issue telling what is happening in time, I will show you a brain scan showing a section of the brain using a MRI scanner. It is what is known as a T-2 weighted image and, in variant CJD, we have found that these white areas, which in a normal scan are a dark colour, are present in approximately 80 to 90% of all the patients with variant CJD. This allows us to diagnose cases in life. It has allowed us to formulate diagnostic criteria, This is not only important for the accurate reporting of cases, it is also important for families.

Regarding the question of what is happening over time, here is the recent data. This is UK data; I will come back to the other countries a bit later. 106 cases of variant CJD in the UK; many of these patients have died and post-mortems have been made, some have not and are classified as probable. But the mean age of death is only 29, the mean age of clinical onset is only 27, and you will see that there has been one individual aged 12 who developed variant CJD. We now have five individuals aged less than 16 who have developed this condition. What has changed in the last year or so is this case aged 74 which has significantly extended the age range. This is important scientifically, it is also important in relation to surveillance protocols.

The condition is rather different from classical CJD, it is an illness of much longer duration and 93 of the cases tested belong to a particular genotype. I will come back to that in a minute.

What is happening with time? This is an old slide but it shows the pattern, number of deaths, the clinical onsets per year. We have had a hundred cases and this shows the pattern of numbers increasing per year but with incomplete data for the year 2000. We have tried to analyse this data formally with the help of mathematicians, taking into account changes in delay in referral and diagnosis in order to compensate for these variables which might influence how these cases are identified. There have been changes in the delay between clinical onset and referral and diagnosis. This is because the system for identifying cases is becoming more efficient. In fact, there has been about an 8% reduction in the diagnostic delay each year.

Taking that into account, mathematicians can plot this. These are the actual data points, varying up and down, and these are the confluence intervals here, and this is the estimated rise in the quarterly incidence of vCJD deaths. These analyses are done quarterly; this is the most recent one and it shows a statistically significant increase with time.

When looking at clinical onsets, both the observational data and the estimates from the mathematical model, there is cause for great concern because it suggests that the number of cases of vCJD are increasing with time. This shows the mathematical estimate of the incidence of onsets and deaths in the UK. The doubling is estimated to occur every three years or so. The analysis of deaths is actually very similar. This means is that in future we might see double the number of deaths per annum. The question is how long is that going to go on for? We simply do not know. It depends on the incubation period of TSE in humans, which is an unknown.

Here is a terrible table from a paper in "Nature" estimating what could happen in the future. There are so many unknowns that it is very difficult to know what to make of this type of assessment. What Professor Andersons's group has done is vary the incubation period that may occur with BSE in humans and look at the number of cases we have observed then, using a number of models, they can predict what could happen in the future.

The future total number of cases in the UK ranges from around 70 to 136'000. The trouble with these very broad estimates is that they are not very helpful and, when they are published, they tend to be reported with the upper estimate in the popular press. However there is a great deal of uncertainly about what may happen in the future.

What is the cause of variant CJD? We think it is BSE. How do people get BSE? The favoured hypothesis is that this is through the food chain, probably through the consumption of high titer tissue, that is tissue containing very large amounts of infectivity from cattle, mainly brain, spinal cord and, I believe, in a product called "mechanically recovered meat". This is the pattern of disease onset in 87 of the cases of variant CJD and this analysis has been repeated on a regular basis to look for a number of things. One question is, can we find a geographic clustering of cases that might suggest an environmental source of infection? Secondly, what is the overall geographic distribution of cases and does it parallel population density?

There is a place called Queniborough, near Leicester in the middle of the UK, where a link was discovered between a cluster of five variant CJD deaths. This is a statistically significant cluster. An investigation has been carried out that suggests that at least four of these cases may be linked to particular butchering practices that probably occurred in the early 1980s. If that is correct, it is a matter of great concern because in the 1980s the incidence of BSE was exceedingly low in the UK. I have major doubts about how accurate this analysis is and how reliable it is, but it is certainly a matter for concern.

What if we look at the regional distribution of variant CJD in the UK? Briefly, if you split the UK into north and south geographically, you will find that the incidence of variant CJD is approximately double in the north of the UK compared to the south. This finding was not made on an a priori hypothesis, it has been sustained by each analysis in recent years. We do not know for sure what the explanation is. One possibility is that it may be our ascertainment, better identification in the north. However we don't believe that to be true because if you look at other forms of CJD, the incidence is equal north to south, the number of referrals is equal north to south. One analysis is being done looking at dietary intakes in the 1980s from a household survey, which showed that there was a relationship, not a very strong relationship, but a relationship, between the amount of non-beef meat products consumed per region and the incidence of variant CJD. It was almost linear. Some support for the dietary hypothesis.

So these are the risk factors we have identified:

- young age, we do not know why,
- methyonine-homozygocity, and
- residence in the UK.

But I will come back to that in a minute to say that it is not an absolute risk factor.

Genetic polymorphisms: There is a particular region of the prion protein gene that varies in the general population of the UK, about 50% are heterozygotes and 37% methyonine-homozygotes. Indeed, this is true of the population of Western Europe. In sporadic CJD there is an greater risk in having the methyonine-homozygote genotype but in variant CJD this is still true: 100% of the cases have this genotype. Therefore it appears to be a susceptibility factor. One important issue, not only for modelling, but also for public health: is it an absolute susceptibility factor?

One difficulty we have is that there are analogous situations in kuru for example and growth hormone-related CJD; where cases with this genotype occurred earlier and other cases later. Therefore this may be an incubation period factor and not only a susceptibility factor. So we wait to see whether the other portions of the UK population who were exposed develop variant CJD.

A further difficulty is that sometimes variations of this type affect the clinical and pathological appearance of the disease. We cannot be sure that BSE infection in this background would look the same clinically and pathologically so we spend a lot of effort trying to identify every case of prion disease in the UK and then trying to classify them.

One of the difficulties in the United Kingdom is that, if we are correct and this was a dietary exposure, it is quite likely that a significant proportion of the population in the UK was exposed to the BSE agent and probably to high-titer tissue. Currently a further effort is being made to find out about previous exposures to bovine high-titer tissue, but basically we do not know who is incubating the disease and we do not know how to identify such individuals because there is no test.

This is work from James Ironside showing a tonsil taken at post-mortem from a patient with variant CJD and again it shows this staining from prion protein. This means that in this individual there is probably infectivity in the tonsils. In fact, every variant CJD case that has been examined has shown this staining in the tonsils, lymph nodes and spleen and some other areas of the body, whereas in sporadic CJD there is no such staining.

The implication is: that would it be possible that if you had a surgical instrument that cut through a lymph node, and was then used on another patient, you might transmit the disease? What about someone who donates blood who has got this appearance, would they pose a risk to other individuals? These subjects are going to be discussed during our seminar.

There are a lot of major problems here for public health. What should be done about this issue? This is a theoretical risk, it could have very serious implications, but, if action is taken, that in itself can damage medicine and health care.

CJD has, in fact, been transmitted in the past. This is just a recent summary from Paul Brown, my friend and colleague, showing the numbers of cases in which CJD has been transmitted from person to person. It has never been transmitted by blood, only by neuro-surgical instruments. Never by ordinary instruments that are used in day to day medicine. However, as I said, sporadic CJD and iatrogenic CJD do not have this staining in the peripheral organs. Probably the risk is much lower.

What have we done about this? A range of action has been taken in the UK. We now leucophorese (take all the white cells out of all blood donations), and all plasma products like Factor-8 and immunoglobulins are produced from plasma imported from the United States.

The United States has banned anyone who has been resident in the UK cumulatively for six months between 1980 and 1996 from donating blood in the USA and there was a discussion earlier this year about extending that to the whole of Europe. This has major implications for the supply of blood products and implies that there may be a risk for blood products in Europe. This could be a major problem.

A lot of interest has been shown in the subject in the United Kingdom and a range of issues has been raised by the BSE Enquiry. The BSE Enquiry fears that MRM vaccinations could be a risk. And this leads to a further problem in that many medications were produced using bovine materials and some still are. Is there a possibility that medications might also pose a risk of onward transmission of these diseases? Again, a lot of action has been taken in relation to that.

"Health alert to 600 patients after CJD patient gives blood." This is a recent headline from the UK. What happened was that we had a donor who subsequently got variant CJD, the blood was traced back and it had been used for fractionation to produce various blood products such as Fraction-8 for haemophilia. All the individuals who received this had to be told in order for that product to be recalled. This is a problem for direct patient care.

The current status of BSE in Europe will be discussed in detail tomorrow by Dagmar Heim amongst others. So just briefly, this is the situation in the UK: a massive epidemic declining (hopefully), perhaps to disappear, although that is not yet known.

The problem in relation to European countries is that the epidemic in the UK was reduced effectively by a ban on feeding animal proteins to animals. It was an incomplete ban, otherwise it should have stopped about here, and it has continued. Public health measures were introduced in the UK around 1990. Again, they were not fully implemented and one of the main lessons of the BSE enquiry was that it was not simply good enough to have measures, you have to implement them.

In Europe the picture is slightly different because, although BSE had occurred in a small number of countries, particularly France and Portugal, there were some countries in which it was felt, certainly by the politicians, that it was exceeding unlikely to occur. Unfortunately, what has now happened is that BSE has been identified in Germany, in Italy and in Denmark and this has caused major concern to the general public about the potential for a risk in those countries because of BSE exposure. In many of these countries appropriate measures to protect the food chain were not taken fully until very recently.

Another problem that will be discussed by Dr. Ricketts next Friday at our seminar is why should this localised problem in the United Kingdom be causing so much concern? One of the reasons is that the BSE risk was inadvertently exported to other countries. That is why there is BSE in European countries. I think Dr. Ricketts will discuss this in greater detail.

Here is one example, these are the exports of cow feed, basically this is not all bovine material, but it is a surrogate marker for bovine material, and these are the quantities exported from 1979 to 1995 per thousand of tons of feed. Here we have exports to the EU, which suddenly decline around 1989, 1990, but these exports were fed to cattle. Then we have exports here to non-EU countries. This shows exports to Central and Eastern Europe and to some parts of Southeast Asia.

There was a meeting about a month ago, organised by the World Health Organisation, the Office International des Epizooties and the Food & Agriculture Organisation to determine what should be done about this. Because if potentially contaminated and infected cattle feed was exported, this could have been recycled in the bovine population and we may yet see epidemics of BSE in other countries.

If that is the case, you need to know and you also need to know whether there is a human disease developing as well. So the WHO is actively arranging and, in fact, funding a surveillance programme for BSE and variant CJD in a number of countries in Central and Eastern Europe and also currently in China.

This slide is slightly out of date, because this shows vCJD in Europe. It should now read 106 in the UK. The reason that I wanted to show this is because there has been one case of variant CJD in the Republic of Ireland. This was an individual who had lived in the United Kingdom for a period of four and a half years during the relevant period and may well have been exposed in the UK. There have been three cases in France, now I believe all confirmed cases of variant CJD, none of whom had even visited the UK, who must have been exposed to BSE in France. Probably, or possibly, to indigenous BSE but also possibly due to the import of contaminated human food in the 1980s which also was widely exported. Most recently there has been a case identified in Hong Kong of an individual who had also lived in the UK.

One of the findings of the BSE Enquiry was that they felt that the problem was that Science said one thing and action was taken on the basis of scientific advice, but was not fully implemented. Part of the reason for that was that there was not enough honesty and openness about the potential risks in the early days. There is very little written in the BSE Enquiry about media coverage of risk, and it is extremely difficult to discuss risks with all these uncertainties–as you may have gathered, I hope, from my talk. We do not know whether there is going to be a risk of secondary transmission, but should action be taken?

I would just like show you a few newspaper headlines, to give you perhaps a perspective of what I think are the difficulties:

Here we have a very balanced comment from Newsweek, "BSE—the creepiest of the Prion diseases—makes every other virus look like chickenpox".

Next: "Millions at risk from CJD, say EU scientists" which may be true but I think the statement also said that there were qualifications to that statement.

Next: This is something I would like to finish with because I think it shows you some of the difficulties that can arise: "Experts say that sheep may be at risk of getting BSE". Indeed, that is true, there is a theoretical concern that BSE material was fed to sheep populations. We know sheep are now susceptible orally experimentally, scrapie can be transmitted vertically, so you could have BSE in sheep, never identify it and it is maintained in the population. It has been a very difficult problem. Once at the CEAC meeting in the UK that I used to be on, I think we took seven meetings discussing this issue. But eventually a professor said that the public should be concerned and it was important to get this out in the public domain. Quite right, just what the BSE enquiry wanted everyone to do.

2. AIDS AND INFECTIOUS DISEASES — MEDICATION OR VACCINATION FOR DEVELOPING COUNTRIES

PREVENTIVE HIV VACCINES. CURRENT STATUS

GUNNEL BIBERFELD
Swedish Institute for Infectious Disease Control and Microbiology and Tumorbiology Center, Karolinska Institute, Stockholm, Sweden

AIDS was first described 20 years ago in 1981[1]. HIV, the virus that causes AIDS, was identified in 1983/1984[2,3]. WHO/UNAIDS estimated that at the end of year 2000 there were 36 million people living with HIV infection, 95% of them in developing countries[4]. Globally 22 million were estimated to have died of AIDS. During the year 2000 approximately 5,3 million people became infected with AIDS.

Highly active anti-retroviral therapy (HAART) has changed the course of HIV disease in developed countries by retarding progression of HIV-associated disease but HAART is not curative. However, HAART is very expensive and is usually not available in developing countries. An effective and safe prophylactic vaccine is urgently needed in order to decrease the spread of HIV infection, especially in developing countries where the incidence of HIV infection is high.

A major problem in the development of an HIV vaccine is the high variability of HIV[5]. Besides the two types of HIV, HIV-1 and HIV-2, there are several subtypes of both HIV-1 and HIV-2. HIV-1 occurs worldwide whereas HIV-2 has its epicentre in West Africa. The distribution of the various subtypes differs in different parts of the world. In Europe and the U.S. HIV-1 subtype B is the most common whereas C is the predominant subtype globally (50% of all HIV-1 strains) followed by subtype A. Multiple HIV-1 subtypes occur in African countries.

It is still not certain if it will be possible to develop an effective and safe preventive HIV vaccine for use in humans since no phase III efficacy trial of an HIV vaccine has yet been completed. However, the demonstration of HIV specific cellular immune responses in a proportion of highly exposed non-infected individuals, as for example prostitutes, and of correlations between strong cellular immune responses and long term survival in HIV-1 infected individuals make it likely that protective immunity can be achieved by immunization. Furthermore, studies in non-human primate models, including HIV-1 infection of chimpanzees, simian immunodeficiency virus (SIV) or HIV-2 infection of macaques or chimeric HIV/SIV (SHIV) infection of macaques have shown that it is possible to induce protective immunity by immunization with HIV-1, HIV-2 or SIV vaccine candidates[6,7,8]. Non-human primate models are useful but not ideal for HIV-1 vaccine studies since HIV-1 does not cause disease in these primates. In contrast, SIV and some strains of SHIV induce an AIDS-like disease in macaques. There

38

are different levels of vaccine-induced protection; a) complete protection of infection (so called sterilizing immunity) or b) reduction of virus production and prevention of disease.

An important aspect of the evaluation of HIV/SIV vaccine candidates in non-human primate models is to determine possible immunological correlates of protection, including neutralizing antibodies and cellular immune responses, especially cytotoxic T lymphocyte (CTL) activity. There is evidence that both antibodies and cellular responses play a role in protective immunity although a number of vaccine studies in monkey models have failed to identify a clear correlate of protection[8]. Passive immunization with neutralizing antibodies has been shown to protect against infection. Protective immunity has also been demonstrated in the absence of neutralizing antibodies but in the presence of CTL.

There are several types of possible HIV/SIV vaccine candidates, including live attenuated virus, whole inactivated virus, recombinant produced subunits, synthetic peptide, live recombinant vaccines and viral DNA. Live attenuated virus vaccines have been the most efficient in eliciting protective immunity in the SIV/macaque model[7,8]. However, these vaccines can induce disease in macaques and this approach will probably not be applicable to HIV vaccinations in humans for safety reasons. HIV subunit vaccines stimulate good antibody responses but poor cytotoxic T lymphocyte (CTL) response whereas live vector vaccines and DNA vaccines often generate CTL responses but low antibody responses. The so-called prime-boost vaccine regimen uses a combination of two candidate vaccines based on different principles. Prime-boost vaccine regimens using for example live recombinant virus vaccine followed by subunits or viral DNA vaccine followed by live recombinant vaccine have usually induced stronger immune responses and better protective immunity in macaque models than each of the vaccine candidates alone.

Many phase I/II trials have been done in humans to study the safety and immunogenicity of various HIV-1 vaccine candidates[9,10]. The first trial was done in 1987 in the U.S. Altogether more than 30 candidate vaccines have been tested in 70 phase I/II trials involving 10000 healthy volunteers. Most trials have been conducted in the U.S. and Europe but several trials have also been done in developing countries including China, Brazil, Haiti, Thailand, Trinidad and Uganda. Two phase III efficacy trials using HIV-1 envelope glycoprotein are ongoing in USA and Thailand.

Envelope subunits have been the most frequently tested HIV-1 vaccine candidates in humans, especially in the early trials. Subsequently, various live recombinant vaccines have been tested. So far canarypox virus has been the most commonly used vector in live recombinant vaccines but more recently vaccine candidates based on other vectors such as modified vaccinia Ankara (MVA), fowl pox, adenovirus, Semliki Forest virus, Venezuelan Equine Encephalitis virus, Salmonella and BCG, have been developed[11]. Prime-boost regimens with the use of canarypox based vaccine followed by envelope subunit have also been tested in humans[9].

Phase I/II trials have shown that the HIV-1 vaccine candidates tested so far are safe[9]. Envelope vaccines have induced antibodies in more than 90% of volunteers but these antibodies usually neutralize only homologous virus strains and fail to neutralize clinical virus isolates. Canarypox vaccines have induced CTL responses in approximately 40-60% of volunteers.

The first generation canarypox live recombinant vaccine expressed only the envelope protein. New generation live recombinant and DNA vaccine candidates are usually multiprotein vaccines, some of them including structural as well as regulatory proteins.

Most of the HIV-1 vaccine candidates tested so far have been based on subtype B strains. However, the vaccine used in the ongoing phase III trial in Thailand includes envelope glycoproteins based on both subtype B and E, the subtypes which are prevalent in Thailand. A phase I/II trial planned to begin in Kenya in year 2002 will use a prime boost protocol including DNA and MVA vaccine candidates based on subtype A.

An ideal vaccine should include long-lasting broad immunity against HIV-1 strains of various subtypes and protect against both systemic and mucosal infections since HIV is mainly transmitted sexually. It is not likely that the first HIV-1 vaccine candidates being tested in efficacy trials in humans will fulfil these criteria. However, even vaccines with limited efficacy may have an effect on the spread of HIV infection.

There will be a need for future testing of various HIV-1 vaccine approaches based on different subtypes in efficacy trials in humans. These trials will have to be performed mainly in developing countries where the incidence of HIV infection of the corresponding subtype is high. Thus several years of work remain before we can have HIV-1 vaccines of proved efficacy for use in different parts of the world.

REFERENCES

1. Gottlieb, M.S., Schroff, R., Schanker, H.M., et al. Pneumocystis carinii pneumonia and mucosal candidiasis in previous healthy homosexual men: evidence of a new acquired cellular immunodeficiency. N Engl J Med 1981; 305:1425.
2. Barré-Sinoussi, F., Chermann, J.C., Rey, F., et al. Isolation of a T-lymphotropic retroviruses. Nature 1985; 317-395.
3. Gallo, R., Shearer, G.M., Kaplan, N., et al. Frequent detection and isolation of cythopathic retroviruses (HTLV-III) from patients with AIDS and at risk for AIDS. Science 1984; 224:500-3.
4. UNAIDS/WHO. AIDS epidemic update. December 1999.
5. Peeters, M., Sharp, P.M. Genetic diversity of HIV-1: the moving target. AIDS 2000, 14 (suppl 3): S129-S140.
6. Heeney, J.L. Primate models for AIDS vaccine development. AIDS 1996, 10 (suppl A): S115-S122.
7. Almond, N,M,, Heeney, J.L. AIDS vaccine development in primate models. AIDS 1998, 12 (suppl A): S133-S140.
8. Bogers, W.M.J.M., Cheng-Mayer, C., Montelaro, R.C. Developments in preclinical AIDS vaccine efficacy models. AIDS 2000, 14 (suppl 3): S141-S151.

9.   Mulligan, M.J., Weber, J. Human trials of HIV-1 vaccines. AIDS 1999, 13 (suppl.A): S105-S112.

10. Esparza, J., Bhamarapravati, N. Accelerating the development and future availability of HIV-1 vaccines: why, when, where, and how?. The Lancet 2000, Vol 355, 2061-2066, 2000.

11. Voss, G., Villinger, F. Adjuvant vaccine strategies and live vector approaches for the prevention of AIDS. AIDS 2000, 14 (suppl 3): S153-S165.

THERAPEUTIC HIV VACCINES

ALF A. LINDBERG, M.D., PH.D.
Executive Vice President R&D, Aventis Pasteur, Marcy l'Etoile, France

MICHEL KLEIN
Vice President Science and Technology, Aventis Pasteur, Marcy l'Etoile, France

INTRODUCTION

The HIV/AIDS epidemic has exceeded all predictions regarding its severity and scale of its impact: in 2000, there were an estimated 36 million individuals living with HIV/AIDS and more than 20 million had already died[1]. The epidemic is still spreading and in 2000, and there is an estimated 5.3 million new infections globally. The highest prevalence is seen in the sub-Saharan region of Africa where there are 7 countries with a prevalence of 20% or more.

The administration of antiretroviral therapy causes a rapid reduction in HIV RNA levels, from 10^3 to 10^6 copies of HIV RNA per milliliter of plasma down to undetectable levels (< 50 RNA cpies/mL). However, there is no clearance of HIV from the body and hence therapy should be life-long. In the infected patient, not yet treated, at least 10^{10} virions are generated daily and approximately 140 generational cycles occur annually[2]. Consequently, even during asymptomatic infection the magnitude and dynamics of virviral infection is considerable

The introduction of triple/quadruple therapy (highly-active antiretroviral therapy or HAART) in 1996 had a profound impact on AIDS events, in particular opportunistic infections. Death rates dropped dramatically and the clinical burden decreased. On the immunological side, CD4$^+$ lymphocyte counts rise, and a partial restoration of the immune system is observed. Although, HAART works very well to control the viral load it is: (i) expensive, and therefore has gained wide spread use only in Western countries which account for only 5% of the epidemic, (ii) not well tolerated by the patients, and actually more than three quarters of those treated experience long-term side effects, especially metabolic; (iii) demanding on patients, and require a high level of compliance, and (iv) selecting for drug-resistant viral mutants. The fact that there is no proof-reading mechanism to correct for replication mistakes increases the risk for selection of drug-resistant mutants, which can hide in latently infected cells.

42

HIV VACCINE DEVELOPMENT

Over the last decade, HIV vaccine concepts have evolved and consequently, the composition of an HIV vaccine, whether prophylactic or therapeutic, has been a moving target[5]. Lessons learned from the natural history of HIV infection and from monkey studies have revealed that neutralizing antibodies protect against live virus challenge. Such antibodies are dependent on memory B cells which can divide and differentiate into antibody-producing plasmacytes upon re-exposure (boosting) to antigen. This confers long-term protection by neutralization (=inactivation) of the virus before it has a chance to infect host cells. The antibodies can also use effector functions, like complement activation and Fc receptor binding to neutrophils and antigen-presenting cells to mobilize the inflammatory system. The problem with antibodies is that they must be broadly cross-reactive to neutralize genetically divergent HIV isolates. The extensive genetic diversity among different strains and clades of HIV remains a major obstacle to vaccine development.

Monkey studies with chimeric simian/human immunodeficiendyvirus (SHIV) have convincingly shown that high concentrations of passively transferred antibodies of the correct specificity confer protection. Antibodies also confer protection at mucosal sites. However, natural HIV infection is mediated by HIV-infected cells (90% of cases) and not by free virus. Thus, a vaccine should elicit cytotoxic T-cell (CTL) responses to destroy incoming HIV-infected cells. Similarly, in chronically infected patients, cytotoxic T lymphocytes (CTL) recognize, bind to and kill cells that display foreign HIV epitopes.

It has also been shown that strong HIV-specific cellular immunity (both CD4[+] and CD8[+] T-cells) correlates with long-term non-progression in humans and protection in monkeys. More specifically, CD4[+] T cell responses inversely correlate with viral load[3]. Virus-specific CTL possess a range of antiviral activities which include the ability to kill infected cells and to produce cytokines and chemokines[4]. However, it is not yet clear which functions of CD8[+] T-cells are most important in controlling HIV. It is presently surmised that induction of CD8[+] T-cells cannot prevent infection by HIV, but could abort infection before it becomes established, or control the level of viremia. Potent CTL responses have been seen in highly-exposed, seronegative Nairobi sexworkers which shows the importance of CD8[+] T-cell responses[3]. A requisite for protection and non-progression is that CTL should be maintained in an active state for long periods of time, and rapidly expanded to respond to the challenge of an acute viral infection. This is likely to require frequent booster vaccinations.

CANDIDATE IMMUNOGEN(S) FOR AN HIV VACCINE

HIV encodes several gene products which can serve as targets for immune recognition. Several of the viral proteins contribute to the structure of the virus and are synthesized in large quantities such as the envelope (Env) and the matrix proteins (Gag) (Fig. 1). The regulatory proteins that modulate viral gene expression, are synthesized in lower quantities.

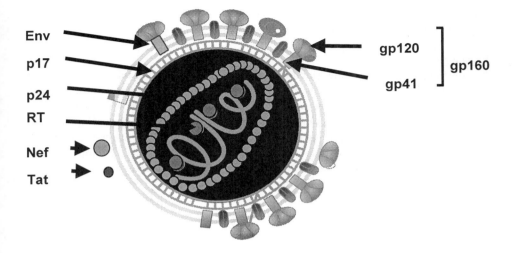

Fig. 1. HIV virion.

Env is expressed on the surface of HIV particles, and is the only target for neutralizing antibodies. It is probable that CTL responses against Env will be beneficial as well. Env exists as a gp160 glycoprotein composed of gp120 (surface exposed) and gp41 (membrane anchored).

Gag is a major internal viral protein and a target for CTL responses. It is cleaved by the protease into p17 and p24 that are the matrix proteins.

Nef is an early viral product of importance in viral replication. It is a major target for CTL. However, it is uncertain how important it will be as a vaccine target. Nef down-regulates the expression of CD4 and HLA-class I antigens which as cell-surface proteins are essential for the recognition of virus-infected cells by CTL.

Tat is an HIV virulence factor which facilitates the spread of the virus and contributes to HIV-induced immunosuppression. The goal is to produce a detoxified Tat vaccine which elicits neutralizing antibodies to inhibit deleterious Tat activities.

Pol is the polymerase and a major CTL target in HIV infection. It is a conserved protein and may add value to a vaccine.

Pro is the protease which cleaves the Gag protein into its functional matrix proteins p17 and p24.

(iii) Genetic diversity

There are 12 genetic HIV-1 subtypes characterized by their Env glycoprotein sequences, where 11 (labeled A, B, C, D, E, F, G, H, J, K, L) clades are in one family and the remaining O is an outlier. Within HIV-1, there are also sub-subtypes: F and F2. In addition, another virus HIV-2 circulates in West Africa. The increasing emergence of inter-subtype recombinants like AB (Russia), AC (Africa), AD (Uganda), AG (West Africa), BC (China), AJG (Africa) AJGE (Africa) must also be taken into account. In spite of these disturbing facts from a vaccine development viewpoint, there is a high degree of CTL epitope conservation between clades.

Escape mutants
The challenges involved in developing an HIV vaccine go beyond the issue of finding "the ideal vaccine". HIV is a virus that evolves in the presence of antibodies, and has been over time subjected to selective pressure to evade immune detection. Mechanisms responsible for escape from immune surveillance include the heavy glycosylation of the envelope proteins and the masking of critical parts of proteins such as the CD4 and co-receptor binding sites. It is also possible that the envelope has developed decoy mechanisms. HIV can also escape detection by cellular immune responses through multiple mechanisms like Nef-mediated reduction in HLA-class I antigen expression. To counter the escape abilities of HIV and be efficacious, a vaccine must elicit strong immune responses, both humoral and cellular, against several antigens such as Env, Gag, Nef, Pol and Tat.

THERAPEUTIC HIV VACCINES

The introduction of highly-active antiretroviral therapy in 1996 made therapeutic HIV vaccination possible. Before HAART, the marked depression in HIV-infected individuals of $CD4^+$ T-cell functions which play a central role in immune defenses, made vaccine attempts futile. The availability of potent anti-viral therapy changed the scene since it was found that under treatment, the $CD4^+$ T-cell count rises and responses to recall antigens (PPD, tetanus toxin, EBV, CMV) are restored. However, even under HAART, immune restoration is incomplete in particular with respect to HIV-specific $CD8^+$ T-cell responses which are lost. The rationale for therapeutic vaccination (Table 1) is to restore, amplify and broaden CD4 and CD8 T-cell responses in aviremic patients with a CD4 count > 350 cells/mm^3, then stop therapy to allow for therapeutic windows. The end points for vaccine efficacy are still debated: absence of or delay in viral rebound which usually occurs within one month after HAART discontinuation; viral set point below the original set point and/or maintenance of $CD4^+$ T-cell count compatible with prolonged drug interruption. (Table 1).

Table 1. Rationale for therapeutic vaccination of patients receiving triple therapy.

• To restore pre-exisiting HIV-specific immunity
• To induce new $CD4^+$ and $CD8^+$ T-cell specificities
• In order to:
− Further reduce the viral reservoir
− Control HIV replication

For an HIV vaccine to succeed, it most likely must induce both humoral (neutralizing antibodies) and cellular ($CD4^+$ and $CD8^+$ T-cells) immune responses that are broadly cross-reactive, polyepitopic, long-lasting and boostable. Presently, it appears that an HIV vaccine should stimulate multiple components of the immune system. It should induce sustained cross-neutralizing antibody responses and durable memory T-cells (of both CD8$^+$ CTL and CD4$^+$ helper T-cells)[5,6]. In order to have the desired broad coverage and the ability to handle escape mutants, the vaccine should contain multiple antigens such as Env, Pol, Gag, Nef and Tat (Fig. 1).

To reach this goal, the current strategy is to prime the immune system with one immunogen (viral or DNA vector expressing multiple HIV genes) and boost with a different immunogen (another viral vector, lipopeptides or protein subunits formulated in a strong Th1 adjuvant). This approach is flexible and referred to as prime-boost or mixed modality vaccine strategy. Experiments in monkeys have shown that mixed-modality vaccination regimens were more efficacious in controlling viremia and protecting animals against disease after lethal viral challenge than either immunogen alone[7-10].

Evaluation of the first therapeutic vaccine with HAART discontinuation has been recently concluded by Aventis Pasteur and the Aaron Diamond Center in the U.S. in an open phase I study[11]. Twenty individuals treated with HAART within 120 days of acute HIV-1 infection were included. Fifteen of them were given the Canary poxvirus vector vCP1452 (expressing *env, gag, pro* plus DNAsequences coding for immunodominant Pol- Nef CTL epitopes) in four doses at months 0, 1, 3 and 6) and two doses of rp160 recombinant protein (months 3 and 6).

Thirteen of the fourteen patients who completed the vaccination exhibited an increase in antibody response to gp120 and p24 and in 9 (60%) of them, and T-cell proliferative responses to HIV antigens were augmented. Eleven patients (79%) developed HIV-specific CD8$^+$ T-cell responses as measured by production of IFN-__(ELISPOT) After cessation of HAART therapy, 6 of 11 vaccinated subjects maintained more than 1 log reduction in viremia levels relative to the peaks of viremia at the time of treatment discontinuation whereas reduction in virus load was observed in only one out of the five patients in the control group.

Recently, the transactivator Tat protein, which is also a toxin capable of suppressing the activity of non-infected T-cells, has gained interest as a vaccine component. Since 1997, four clinical trials have been carried out using Tat toxoid[12]. The experience accumulated has proven the Tat toxoid vaccine to be safe and well tolerated. Vaccinees responded with anti-Tat neutralizing antibodies, even when treated with HAART. Tat-vaccinated patients have benefited from a stabilization of their clinical, virological and immunological parameters whereas open comparison patients showed progressive deterioration of their immune system status and evolution to AIDS.

Based on these preliminary findings, Aventis Pasteur has initiated seven therapeutic phase I/II trials with several collaborators in 2000. The studies encompass more than 400 patients in the US, Europe, Canada and Australia. Interim analyses indicate that the vaccines elicit strong cellular immune responses. The results of the trials are expected in mid to late 2002. A significant advantage of therapeutic vaccine trials is that efficacy can be addressed six months after cessation of HAART. Information gained from these trials should also help accelerate the development of an efficacious prophylactic HIV vaccine.

CONCLUSIONS

Advances in the understanding of the immunopathology of parasite-host interactions in acute, latent and chronic HIV infection have increased the probability of vaccine development. It is likely that therapeutic HIV vaccines will be the first to be developed. This is so because the arrival of HAART in 1996 that not only decreases the patient's virus load but also causes an increase in CD4$^+$ T-cell count and a partial if not full, restoration of the immune system, makes vaccination possible. The goal of therapeutic vaccination is to restore and broaden pre-existing HIV-specific immunity (in particular CD4 Th1 and CD8 responses) in HAART-treated aviremic patients, then discontinue treatment to allow for long therapeutic windows. By using mixed-modality (or prime-boost) regimens where the immune system is primed with one immunogen (a viral vector or DNA with several HIV genes) and boosted with another immunogen (another viral vector, lipopeptides or several recombinant proteins) robust humoral and cellular responses have been obtained in macaques and shown to be protective against lethal doses of pathogenic SIV or SHIV.

The first phase I study conducted in aviremic patients treated early after acute infection revealed that the vaccine could restore immunity in immunocompromised individuals and had a protective although short-lived effect in controlling HIV rebound in a couple of patients who stopped therapy. Several phase I/II trials designed to evaluate different immunogens in mixed-modality schedules are currently in progress and results are expected in the second half of 2002.

REFERENCES

1.	P. Piot, M. Bartos, P.D. Ghy-s, N. Walker and B. Schwartländer. 2001, "The Global impact of HIV/AIDS." *Nature.* **240**:968-973.
2.	A.T. Haase. 1999, "Population biology of HIV-I infection: viral and CD4+ T-cell demographics and dynamics in Lymphatic tissues." *Annu. Rev. Immunol.* **17**:625-656
3.	C. Brander C, and B.D. Walker. 1999. "T lymphocyte responses in HIV-1 infection: implications for vaccine development." *Curr Opin Immunol.* **11**:451-459.
4.	R.M. Zinkernagel. 1995, "Are HIV-specific CTL responses salutory or pathogenic?" *Curr. Opin. Immunol.* **7**:462-470.
5.	A.J. McMichael and S.SL Rowland-Jones. 2001, "Cellular immune responses to HIV." *Nature.* **410**:980-987.
6.	M. Klein. 2001, "Current progress in the development of human immunodeficiency virus vaccines: research and clinical trials." *Vaccine.* **19**:2210-2215

7.   Robinson, H.L. *et al.* 1999. "Neutralizing antibody-independent containment of immunodeficiency virus challenges by DNA priming and recombinant poxvirus booster immunizations." *Nat. Med.* **5**:526-534.

8.   Amara, R.R. *et al.* 2001. "Control of a mucosal challenge and prevention of AIDS by a multiprotein DNA/MVA vaccine." *Science.* 292:69-74.

9.   Osterhaus, A.D. *et al.* 1999. "Vaccination with Rev and Tat against AIDS." *Vaccine.* **17**:2713-2714.

10. Hanke, T., *et al.* 1998, "Enhancement of MHC class I-restricted peptide-specific T-cell induction by a DNA prime/MVA boost vaccination regime." *Vaccine.* **16**:439-445.

11. X. Jin, L. Zhang, A. Hurley et al., "Prolonged HAART and therapeutic vaccination with canarypox vCP1452 and rgp 160 followed by drug cessation in individuals newly infected with HIV-I." Abstract 336, *AIDS Vaccine 2001*, September 5-8, 2001, Philadelphia, PA. USA.

12. A. Gringeri, H. LeBuanec, A.B. Oschini et al. 2001, "Four year follow-up of Tat vaccination : correlation between anti-Tat neutralizing activity and clinical outcome." J. Human Virol. **4**:136, Abstract 58.

SCIENTIFIC NETWORKING IN AIDS CONTROL

PR. GUY DE THÉ
Institut Pasteur, 28, rue du Dr Roux, 75728 Paris Cedex 15, France

This session on AIDS presented different facets of the human tragedy that the HIV-AIDS epidemics imposes on Africa, the hardest hit continent, but also on India and Asia. The burden paid by infected mothers and their newborns is particularly tragic with 600 to 700.000 new-born babies being infected yearly by the HIV, the great majority of them dying within a year of life.

Facing this tragedy, many western countries both individually and collectively, are responding, mostly by providing funds, but also in trying to adapt the patented rights to the very low and low income countries, for critical drugs involved in AIDS control.

Hope is thus emerging that highly active Anti Retroviral Therapy (HAART) will progressively become available for populations of the developing countries. In parallel, preventive vaccines preparations are being tested in Phase I, and Phase II and even Phase III clinical trials. As we know little about the best correlates of protection to HIV infection, involving neutralizing antibodies and specific T cell responses, we cannot design with precision the ideal vaccine preparations.

Furthermore, the presently available HIV vaccine preparations are mostly directed against HIV-1,B subtype prevalent in the industrialized world, while the prevalent subtypes in the most affected countries are the A, C and E subtypes. Fortunately, the most recent vaccine preparations are being directed against subtypes prevalent in Africa and Asia, as well as against the "occidental" B subtype.

In view of the many unsolved questions, we, at the World Federation of Scientists, feel that in parallel to the governmental and intergovernmental financial support, it is critical that scientists from the high-income countries come together with scientists of the developing world to collaborate and answer critical questions over the epidemic and its control, which can only be resolved by working in the field, with the population concerned.

We all agree that it is the duty of concerned scientists to participate in the sustainable development of third world countries, particularly those severely affected by HIV-AIDS. As Nehru once said: "The poorest countries need up-most to engage in research."

The World Federation of Scientists at the Erice Center has taken important steps in the last three years to get together selected scientists and clinicians from the developing countries and from the industrialized world, to assess the available data, to define most critical questions and to make specific recommendations to governments and non-governmental organizations.

In 1999, the World Federation of Scientists organized the first "think-tank" *on the urgent need for a third world adapted HIV vaccine*. This World Federation of Scientists' recommendation has been an item of evaluation at UN AIDS meetings, and of a media "event" on December 1st: AIDS's day at the Eiffel Tower in Paris.

In August 2000, the World Federation of Scientists and its program on AIDS and Infectious Diseases, together with the PMP on Mother-Child Health, organized a second "think-tank" on 'Mother to child transmission of HIV', focusing on the new hope raised by the drug named 'Nevirapine', which significantly decreased the HIV viral load in the mother at the time of delivery, thus preventing mother to child transmission of HIV. This has no significant beneficial impact on the evolution of HIV infection in the mother, and the virus can be transmitted by subsequent breast feeding which cannot be avoided in large parts of the developing world (by lack of sterile water to prepare bottles). The report August 2000 think tank was published in Acta Pediatrica 89:1385-6,2000.

In 2001, a third Erice Workshop was organized jointly by both the PMPs on Aid and infectious diseases on mother and child health, focusing the discussions on the new developments created by the substantial decrease of the anti-retroviral drugs and the foreseen possibility of having in not a too distant future, an HIV vaccine with hopefully both preventive and therapeutic properties.

The assessment and recommendation of the 2001 workshop focussed therefore on two aspects: the first being the impact that anti-retroviral therapy will have on the structure of the health system in these countries (in most low income countries), the present system is poorly adapted to the delivery of drugs, as these need sophisticated blood surveillance. The report of this year's workshop is given in Annex 2. A second aspect of critical importance for setting research priorities concerns the preventive and therapeutic vaccine strategies, aimed at preventing mother to child transmission of HIV. This might imply the vaccination by therapeutic vaccine of pregnant women which raise critical ethical issues as well as the eventuality of preventive therapeutic vaccination of neonates, hoping that even a low efficiency vaccine preparation could be instrumental to prevent primary infection with small infectious dose.

In complement to the activities of the World Federation of Scientists, we have been promoting, under the aegis of the InterAcademy Panel, of the French and Swedish Academies of Sciences, a "scientist to scientist network" to boost research on mother and child health, with the aim of establishing better collaboration between scientists of high and low income countries.

In conclusion, the two PMPs on AIDS and Infectious Diseases and Mother to Child Health, of the World Federation of Scientists have been very active since three years, promoting useful think tanks on this tragedy of AIDS, that hits the developing world so badly. We must pursue our activities in Erice. The "scientist to scientist network" and its web site is very complementary to the FS activities and jointly we will continue to work for the sustainable development of the low income countries and against the HIV-AIDS epidemics.

3. MISSILE PROLIFERATION AND DEFENSE

ERICE RECOMMENDATIONS UNHEEDED BY WASHINGTON AND MOSCOW

PROFESSOR ANDREI PIONTKOVSKY
Strategic Studies Center, Moscow, Russia

This morning we heard a fascinating historical account of the Manhattan Project and the personalities involved in it. The task of today's session is to deal with some long-term repercussions of this project—80 years of repercussions. The repercussion today is a nuclear world and the question is how to deal with today's nuclear world in a changing strategic environment.

The task of the Erice Planetary Emergency Seminars is, as I understand it, not to follow events and crises but to anticipate these crises and present the scientific community and decision-makers with ways out of them. I am convinced, in as far as the global defense issue is concerned, that our Erice community was up to this task.

In the 1992, 1993 and 1994 Seminars our joint group of American, Russian, and European strategic analysts elaborated a new concept of transition from adversarial principles of strategic stability, the famous mutually assured destruction paradigm of strategic stability, to a cooperative one of mutually assured protection. As we defined it: from much stability to net stability.

The mathematical apparatus for this was presented in many papers by my American friend, Greg Canavan, and myself. We wrote the same equation, we drew the same diagrams and we understand the task of transition from one strategic paradigm to another in the same way. But it was not enough because, as Professor Zichichi has repeatedly emphasized to us, it is not enough to write a good scientific paper. It is necessary to implement these ideas, to make decision-makers grab these ideas and put them into practice. I also think that our group was very constructive and successful in this part of the job. Take for example the Seminar in Erice in the spring of 1998 when we assembled not only experts from the scientific community but the top representatives of the political establishments of Russia and the United States, from the Foreign Office, the Senate and the Douma. After very heated discussions, we signed our joint declaration which, in my view, is still the blueprint for this new strategic framework about which there is so much talk now.

I would just like to remind you of some of the key principles of our declaration. Point No. 2 reads: "The relationship between the United States and Russia should not be based on adversarial principles as the current strategic paradigm based on mutually assured destruction is not adequate to address emerging security concerns of Russia and the West." It sounds like a quotation from President Bush's speech, but it is quite the reverse, President Bush's speech is a quotation from this Erice Statement.

However at the same time, as experts, we realize that this transition from one paradigm to another cannot be implemented at the moment. A transition period is necessary. In declaring the elaboration of new paradigms to be our medium term goal, we also addressed the immediate issue of global defense.

In our Point No. 4 we stated that, in the near term, the United States and Russia share the same strategic objective: protection of their interests and populations from limited ballistic missile threats. Also neither the United States nor Russia intend to jeopardize each other's deterrent capability. This was the solution to the debate on national missile defense because a simple question arose during these debates: are the two goals: the American goal to protect itself from "rogue states" and the Russian goal to preserve its deterrent, its retaliatory capability during this transitional period really compatible? It is neither a political nor an ideological issue; it is an operational resource issue. Our group's answer was, "Yes, of course they are compatible", so modification of the ABM treaty is possible. Firstly to allow the Americans to satisfy their security needs and to deploy the system protecting them from the so-called "rogue states", and secondly, not to damage or undermine the Russian deterrent.

Another important point of the debate is the readiness of both sides to use Russian technology especially C300 and C400 systems in these projects of both the American national missile defense project and regional, for example European, missile defense projects.

The only task remaining was to present this blueprint, this strategic package, to our governments. This was my part of the group and for three years my friends in the Russian strategic community and I pushed the case for modification of the ABM Treaty as being an action which would benefit the common interest. This was not easy at first because of the strong inertia of Cold War thinking. For many years the Russian top official repeated the same mantra maintaining that the 1972 ABM Treaty was a corner-stone of strategic stability and did not move an inch from this position.

However, through our persistent efforts the idea of modification became more and more prevalent and popular in the Russian strategic community. Until about a year ago, during the last year of the Clinton Administration, I was able to say that a compromise could be reached, that the ABM Treaty is dead, and its modification is in the Russian interest.

It seemed that all the tasks of the Erice group were fulfilled. The strategic paradigms are defined, the political path of realization suggested and politicians are more or less driven to accepting this idea. But new obstacles have arisen with the advent of a new American administration.

Now, when Moscow is ready to say, "OK, we understand your strategic concerns and we are ready to talk seriously about modifications to the ABM Treaty", the message from Washington is quite clear: "No, wait a minute, we are not interested in modifying the ABM treaty any more. Moreover, we are not interested in any treaty at all. The reason being that the Cold War is over, we are friends, there are no treaties between friends and we do not want to restrict ourselves by any international obligations, we want to define our security needs for ourselves. We haven't made final decisions. Yes, we need to make some tests and we need to hold some debates." After they take a final decision they will certainly want you to know and redesign a new strategic framework for the world. Meanwhile our ex-enemy enjoys life.

I think that the unilateral stance contained in this message from Washington is unconstructive, unhelpful and counter-productive even from an American perspective, because I think that such a philosophy will be very difficult to sell to even the closest American allies.

So what we are witnessing is a chain reaction of inadequate political posturing in both capitals, first in Moscow then in Washington, and now I am concerned that there will be another cycle to this chain reaction. The Americans say they are not interested in any treaty and do not want to make any arms control agreements. The prevailing mood in Moscow (which I also view as very unconstructive and which I will do everything to counter) is that if the Americans abandon the 1972 treaty and do not want to engage themselves, then we will also abandon all existing arms control treaties like START 1, like START 2, intermediary nuclear force and conventional forces use. It is no less an unreasonable action in my view than this American philosophy of unitary stance. Because this treaty is in Russian interests.

What can be done in this situation from the Russian perspective? First of all, I think that unfortunately Moscow now has no negotiating position. One day Mr. Putin and his Foreign Defense Minister send a clear signal that they are flexible and ready to consider the American plan of a national defense system. On the next day some of them are repeating the same old formula of 1972 of constant strategic stability which practically all of them have already abandoned. I think that the Russians would have had a very strong position had we said, "As it is formulated in the famous Erice Protocol, we understand your strategic concerns, we understand that you should protect yourselves from "rogue states," and we note that your point is objective and does not jeopardize the Russian deterrent. Let's sit down, let's take paper and pencil and fix the parameters of your plausible system so that it answers both these objectives. What number of interceptors shall we allow, what limitation on the capacity of radars and so on? It is a very technical, practical issue. Let's fix it and let's discuss, according to the Reykjavik Protocol, our further steps on transition from adversarial to cooperative principles of strategic stability." The next speaker, Dr. Canavan, gives a modified, more sophisticated model of such strategy.

If the World Treaty, and by the way "World Treaty" has somehow become a politically incorrect term in English in Washington (however as the English language is the richest in the world, they can call it whatever they wish: provisions, agreements, acts, principles, anything), but I think that we should clearly formulate and put our common objectives down on the paper to avoid ambiguity and mutual suspicion. And I think that this active Russian position is very convincing and would obtain the support of the majority even in the American political establishment. If the old mantra is simply repeated and finally the Americans declare that they have abandoned the 1972 Treaty and do not want to sign any other treaty (we also live in a regime of many important treaties) it will be the road to strategic instability in the world.

I want to make my point clear: I am not concerned with American withdrawal from the 1972 ABM Treaty, I am not concerned with American problems of national missile defense, because I know that all these plans that are now on the design tables in institutes do not threaten Russian deterrents and do not change strategic commitments. My concern is with this philosophy of abandoning this cooperative approach to strategic stability, with this tendency to unilaterally declare a new strategic principle differing from the Russian one and just informing other people about it. I would have said that it is counter-productive and the spirit of Erice demands another approach to this problem of us.

TRANSITION FROM ADVERSARIAL TO COOPERATIVE INTERACTION

GREGORY H. CANAVAN

Los Alamos National Laboratory, Los Alamos, New Mexico, USA

ABSTRACT

Game theoretic analysis of strategic conflicts is extended to interactions between two or more adversaries with realistic target sets. Current offensive configurations are shown to be stable, insensitive to reductions in offensive forces, deployment of limited defenses, and the exchange of offenses for defenses. The transition from adversarial to cooperative interaction improves stability monotonically. The shift of targeting to high value targets stabilizes trilateral configurations, in which defenses lead to a balance between large and small sides resembling a small scale version of that between large sides.

INTRODUCTION

This note extends the game theoretic analysis of strategic conflicts begun in earlier Seminars on Planetary Emergencies to interactions with and without defenses between two or more adversaries with more realistic target structures.[1] It reviews the essentials of game theory as applied to the analysis of strategic decisions, the application of first and second strike costs as payoffs, and solution optimization, which resolves several inconsistencies seen with earlier metrics. The stability of the current bilateral offensive configuration is shown to be high and insensitive to deep reductions in offensive forces, the deployment of limited defenses, and the exchange of significant offensive forces for defenses.

The transition from adversarial to cooperative interaction is represented by the progressive reduction of the parameters representing each side's preference for damaging or deterring the other, which monotonically improves stability. Estimates of strike incentives in bilateral and trilateral configurations are reduced by the inclusion of high value targets in both sides' force allocations, which dominates the details of offensive and defensive forces. The shift to high value targets stabilizes trilateral offensive configurations, a result that differs with that from analyses based on military costs only. When defenses are included, they lead to a balance between a large defended side and small undefended side that resembles the balance between two large sides. Including the large side's preference for defense of high value targets in the analyses reduces its strike

58

incentives and thus the small side's incentive to preempt. However, it also removes the large sides' ability to deter, so the stability of multi-polar configurations continues to be controlled by the least stable dyad, which places constraints on the size of defenses that can be deployed stably that could be more stringent than those from the bilateral balance.

GAME THEORETIC FRAMEWORK

Damage and cost models[2] are derived and discussed in earlier papers.[3] Strikes are estimated with conventional exchange models. Costs are taken to be exponential approximations to the aggregate damage to self and incomplete damage to other, so first and second strike costs are of the form Cost = damage to self + L(1 − damage to other), where L is a parameter that measures the attacker's relative preference for inflicting damage on the other and preventing damage to self.[4] If neither side strikes, the cost of inaction to each is L, which is also a measure of the damage each wants to be able to inflict on the other, i.e., the amount of value it wishes to hold at risk to deter the other from untoward action or to force it to comply with some demand. During the cold war, L was large because deterrence was a dominant role of strategic forces. As conflict is replaced by cooperation, L falls as the deterrent role diminishes and the role of strategic forces is in response to damage to self.

The side that considers striking first minimizes its first strike cost, C_1, and then decides whether to strike depending on whether C_1 is smaller than the cost of inaction, L. This minimization also determines the optimal allocation of the first striker's weapons between missiles and value targets and the cost of the side that strikes second, C_2.[5]

The essentials of game theory for crisis stability are summarized in Figure 1, which defines a graph of play, the decision nodes, which side decides at each node, and a set of payoffs to both sides for traversing each path.[6] The nodes represent decisions whether to strike first and strike back, so the first and second strike costs described above are appropriate payoffs. The two sides are identified only as U (unprimed) and P (prime), corresponding to the un-primed and primed symbols used for their forces, strikes, and costs. Identification with specific countries cannot be established without specifying their damage objectives, which are imperfectly known.

In the three nodes at the upper right corner, U can strike first, but P plays a key role in U's decision. U determines how P would optimally respond to each of its choices and then chooses the action that minimizes U costs, which also determines P's actions and costs. The top branch is a first strike by U followed by a second strike by P, which has costs (C_1, C_2'). The second is inaction by U followed by a strike by P and restrike by U. On it, P would not strike for $C_1' > L'$, but would for $C_1' < L'$. Thus, for $C_1' > L'$, U chooses between C_1 from the top branch and L from the second. If $C_1 > L$, U would also choose inaction, neither side would strike, and the costs for the crisis would be (L, L').

If C_1' 'fell to L', P would see an incentive to strike. However, U, anticipating that transition, would preempt P before C_1' reached L'. In this way a rational decision by P to reduce its costs slightly when L' reaches C_1' would induce a rational decision by U to preempt that imposes larger costs on both sides than inaction, the preferred path if either side controlled all decisions. Moreover, U's decision to preempt depends on its evaluation of P's cost C_1' and damage objective L', neither of which U knows with precision. Thus, imperfect knowledge of the other's decision parameters could cause U to strike by accident. The difference $C_1' - L'$ represents the margin of safety against such accidental exchanges.

The lower half of the decision tree contains the symmetric branch on which P decides first. The two halves are combined by a decision as to which of the two sides could strike first in a crisis that is conventionally modeled as a random decision by Nature (N),[7,8] which was explored in earlier notes and is represented here by the probability u that U can strike first in a crisis.

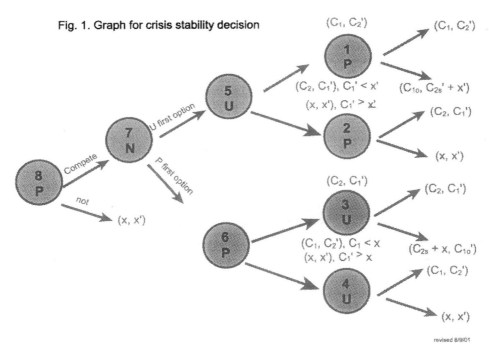

Fig. 1. Graph for crisis stability decision

revised 8/9/01

BILATERAL ENGAGEMENTS WITH DEFENSES

Figure 2 shows the cost to U of reaching node 7 as a function of U's defenses and the probability that it can strike first in a crisis, u.[9] Small numbers of interceptors have no impact; even 600 interceptors do not change strike incentives, because the decision variable $C_1 - x$ remains positive. The decision to act is made when the difference becoming negative, which produces a strike incentive. Larger numbers of interceptors produce large costs at small u because P has and incentive to preempt and does so. At large u, U can strike first and use its defenses to negate P's suppressed second strike. At very large defenses, U's costs are reduced below those of inaction for all u, but those reductions are gained through strong interactions and at large cost to P.

Fig. 2. Cost to U of reaching node 7 (7U)

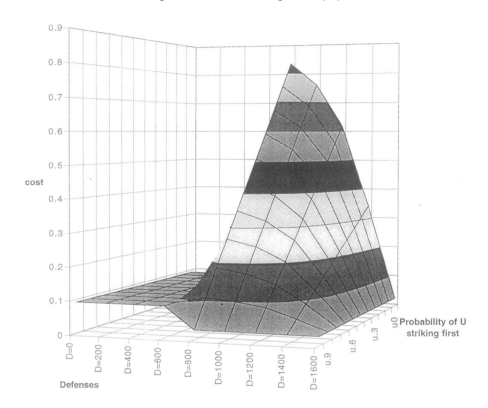

Figure 3 shows the ratio of first to second strike cost, which is often used as a stability index.[10] The composite index falls rapidly with the number of interceptors. It is reduced about 40% by 400. That is largely because U's index falls; P's is largely unchanged. In the game theoretic analysis, small defenses have no impact on stability, but large ones produce an incentive for P to preempt. In the cost ratio metric, even small defenses degrade indices, and large ones only treat the increase in U's incentive to strike first, which is a less important and unstable mechanism.

Fig. 3. Cost ratio Indices

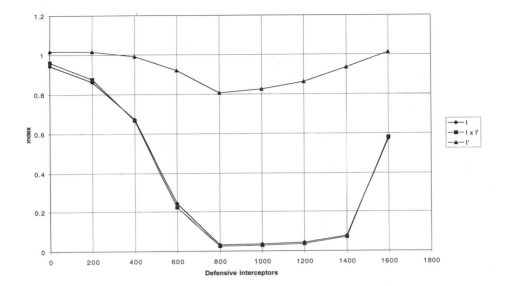

TRADING OFFENSES FOR DEFENSES

Freedom to trade offenses for defenses makes it possible to deploy large defenses without inviting preemption or raising costs. Figure 4 shows the impact of U trading offenses for defenses, starting with START III level U and P offenses, L = 0.5, and L' = 1, i.e., fairly aggressive opponents.[11] U and P's first strike costs for those forces and damage preferences are the two central curves at large numbers of weapons. If U unilaterally reduced its offensive forces from that level while P maintained its forces at 2,000, by U = 500, that would cause U's first strike cost to increase to ≈ 1.25 while P's fell to 0.4. Such a discrepancy would be undesirable, as it could incentivize strikes, given current estimates of damage preferences.

Parenthetically, this result could be inverted to suggest that Russia reducing its forces for economic or other reasons to 500-1,000 weapons while U.S. remained at current levels would produce a situation that Russia could interpret as giving the U.S. an incentive to strike first, depending on the damage preference it attributed to the U.S.

Side U unilaterally adding 1,500 interceptors would reduce U's first strike cost to ≈ 0.1 and increase P's to ≈ 1.5, which would be unacceptable to P. But U deploying those defenses and reducing its offensive weapons to 250 weapons would produce the intersection at $C_1 \approx C_1{}' \approx 0.8$. Thus, for these conditions it is possible to trade large offenses for defenses without adversely impacting the first strike costs of either side to a significant extent. As first strike costs are their primary decision variables, this means it is possible to trade offenses for defenses without impacting strike incentives, margins, or stability. In this example it is possible to trade 1,500 defensive interceptors for about 1,500 offensive weapons, which implies a tradeoff ratio of roughly 1:1.

Fig. 4. First strike cost vs weapons with and without defenses

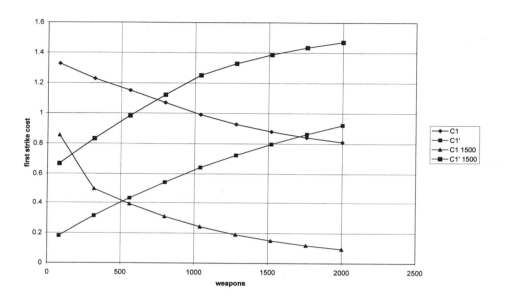

Reduced damage preferences make it possible to increase stability margins while trading offenses for defenses. Figure 5 has the same weapons levels as Figure 4 but less aggressive opponents with $L = L' = 0.25$. The curves without defenses are little changed, so the intersection at large numbers of weapons is at about the same level of cost. However, the margin there is now $C_1 - x \approx 0.85 - 0.25 \approx 0.6$, which is much larger than that the $\approx 0.85 - 0.5 \approx 0.35$ of Figure 4. Shifting from adversarial to cooperative interaction, as modeled by the reduction of damage preferences in the game theoretic framework, strongly reduces incentives, increasing stability margins.

Comparing Figures 4 and 5 shows that with defenses, reducing damage preferences significantly increases U's and reduces P's first strike costs. The result is that for 1,000 interceptors, the intersection shifts to about 1,000 weapons, where the costs are $C_1 \approx C_1' \approx 0.5$. In contrast to Figure 4, first strike costs do fall by $\approx 0.45 - 0.75 \approx -0.3$, but the resulting value of 0.45 still has a margin of $\approx 0.45 - 0.25 \approx 0.2$, which is almost 100% with respect to their reduced damage objective. The exchange ratio is again roughly 1:1. It appears possible to deploy large defenses stably by trading offenses for defenses at constant first strike costs, reducing damage objectives, or some combination of the two.

Such trades should also make it possible to accommodate changes in the opponent's offenses and defenses with exchange ratios on the order of unity.

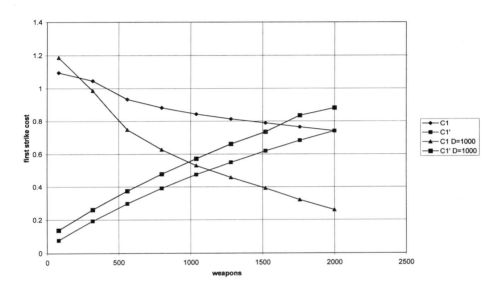

Fig. 5. First strike cost vs U offensive weapons w/ith & without defenses
(L = L' = 0.25)

BILATERAL OFFENSIVE ENGAGEMENTS AT LOW FORCES

First and second strike costs at historical and projected force levels were discussed in earlier Workshop reports using cost ratio metrics.[12] Those analyses imply significant strike incentives at present force levels, explain the modest stability improvements from arms reductions to date, and suggest that strike incentives could grow in the deep reductions to 1,000 to 300 weapons postulated by the U.S. National Academy of Sciences.[13] As weapons become scarce, second strike forces survive, so C_2 falls more rapidly than C_1 increases. That causes the u-weighted average of C_1 and C_2 to drop rapidly towards L, which raises the possibility that low force levels could be less stable. The assessment is complicated by the shift of targeting to higher value targets at low forces, which was ignored in earlier analyses. Both effects can be treated by the extension of the previous game theoretic formalism for military value to address target sets that contain both military and high value targets. The formal extension is straightforward. The resulting analysis indicates that the shift to high value targets and fall in second strike costs at low forces do not adversely impact stability.[14]

Figure 6 breaks first and second strikes from symmetric forces at progressively lower total numbers of weapons into strikes on military and high value targets.[15] For large numbers of weapons, first and second strikes are primarily on military targets, as assumed in previous analyses. At about 1,000 weapons the strikes on military and high value targets are comparable. At lower weapons the strikes shift to high value for obvious reasons: as weapons become scarce, they are reserved for targets with the highest value. The previous conversion of costs into strike incentives remains valid when some or all of the costs are from high value targets. It predicts the low weapon configurations of Figure 6 remain stable to very low numbers of weapons; however, it also shows that for a as forces are reduced, targeting shifts from military to high value targets for a wide range of damage preferences, which has implications that are not captured by numerical indices.

Fig. 6. First and second strikes on military and high value targets as functions of total weapons

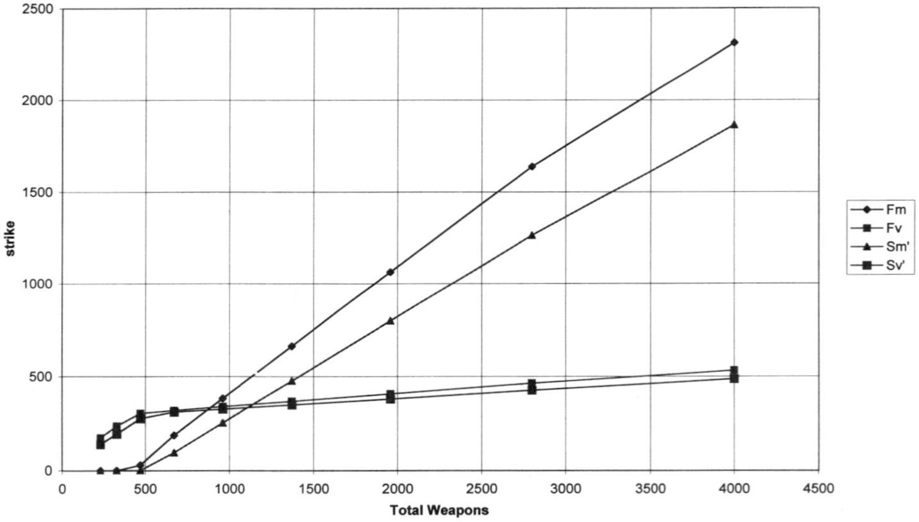

The extended analysis also makes it possible to study the variation of costs and strike incentives with damage preferences. Figure 7 shows the variation of costs at nodes 1 and 2 with L (= L'), assuming that the two sides have equal preferences for defending (K = K' = 2) and attacking (V = V' = 1) high value targets.[16] The cost to U for inaction, i.e., not damaging either military or high value targets, is L + V, which is the bottom curve on Figure 7. The top curves are the cost to U and P for action. As they are far above the costs of inaction for all L, there is no interaction. Moreover, the stability margin grows from $\approx 3 - 2 \approx 1$ or 50% at large L to $2.5 - 1$ or 150% at small L, so sensitivity to uncertainties is also reduced.

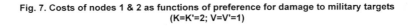

Fig. 7. Costs of nodes 1 & 2 as functions of preference for damage to military targets
(K=K'=2; V=V'=1)

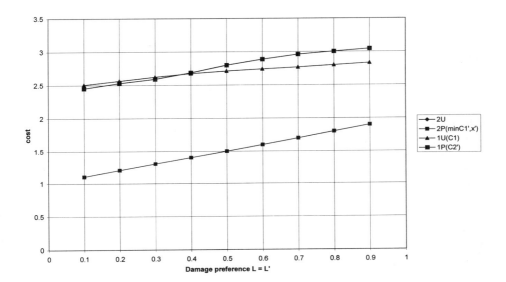

The parameter K represents U's preference for survival of its high value targets relative to that for the survival of military targets. When K is small, the analysis defaults to that based on military costs. When it is large, it produces a cost to U that grows with K to levels that make inaction preferable. That typically occurs for K > 1, as seen above. A strong preference for protecting one's high value increases first strike cost, which undercuts the attacker's motivation for action. V represents U's preference for destruction of P's high value targets, so V/L is U's relative preference for the destruction of high and military value targets. When V is small, U has little incentive to strike high value targets, and the analysis defaults to the military targets treated earlier. When V is large, U's ability to do significant damage to P's few high value targets gives U an incentive to strike, which typically occurs for V > 1.[17]

At small forces, attacks shift for obvious economic reasons to higher value targets. As both sides are generally seen as being more interested in the survival of its high value targets than in the destruction of the other's, the values of K and V in Figure 7 should be appropriate through the transition to a cooperative arrangement, in which the primary emphasis is the reduction of L, their damage objectives for military targets.

TRILATERAL OFFENSIVE ENGAGEMENTS

Previous studies treated the impact on START I level U and P forces of third country (T) forces varying from NAS to START levels under the assumption that all sides target military value.[18] The principal result is that the interaction between U and P remains essentially bilateral, and hence stable, but that both see a strong incentive to strike T at force levels below START II, which in turn gives T a strong incentive to preempt them. For the reasons discussed above, those incentives are eliminated if T is allowed to target U and P's high value targets, given U and P's apparently strong preferences for survival of their high value targets. Thus, if T can hold a significant number of them at risk with its relatively few weapons, U and P's strike incentive is eliminated, which in turn eliminates T's incentive to preempt.

Figure 8 shows these trends quantitatively.[19] As U's preference to reduce damage to its high value targets, K, increases, the cost to U for striking first increases, and the advantage to T for preemptively striking U's high value targets decreases. For K < 4, U has a strike incentive like that for military targets discussed earlier and T has a strong incentive to preempt, so there is a strong and costly preemption and restrike. By K ≈ 4, U has little incentive to strike first, but T still has an incentive to preempt, although doing so increases costs to both of them far above those for inaction. For K > 5, U no longer has an incentive to strike, which removes T's incentive to preempt.

Fig. 8. Costs of reaching node 7 as functrions of preference to preserve high value targets

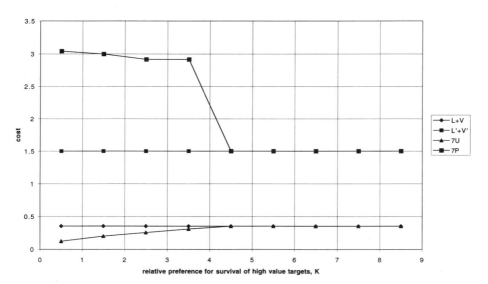

The elimination of U's strike incentive at $K \approx 5$ is due to the increase with K in the cost associated with a given level of damage to U's high value targets. Increasing K ultimately produces a level of cost that removes U's incentive to strike first. Stability is achieved by T being able to deter U with its relatively few weapons by directing them to U's high value targets; thus, the balance between U and T resembles that between U and P, although on a much smaller scale. The stability of the triad is still determined by that of the least stable dyad, although it is improved by U's concern for high value targets.[20]

TRILATERAL INTERACTIONS WITH DEFENSES

Previous reports studied the impact of defenses on trilateral force structures in which U and P have START I and T has NAS 1 (1,000 weapon) level forces.[21] The interaction between U and P resembles their bilateral interaction, i.e., modest defenses had no impact, but large defenses produce preemption by P, in accord with Figure 2. The interaction between U and T gives U an incentive to strike, but T no incentive to preempt, so U would use its offenses in conjunction with its defenses to suppress T, if it could. However, for nominal damage preferences, the barrier from P's potential preemption at large defenses prevented U from deploying the defenses needed to suppress T. The result is to limit U's defenses, produce a largely offensive balance between U and P, and allow T to increase its forces from current low levels without constraints.

However, P's preemption barrier to U's deployment of large defenses can be removed by the freedom to trade offenses for defenses discussed above and illustrated in Figs. 4 and 5. Thus, it is useful to examine the resulting relationship between U and T. Figure 9 shows the costs to U and T at node 7.[22] For up to ≈ 4 interceptors, the interaction between U and T is stable; neither sees an incentive to strike. At 6 interceptors, U has an incentive to strike, which decreases its cost slightly but increases T's greatly. At 8 to 10 interceptors, T has an incentive to preempt, which strongly raises costs to both. By 12 interceptors, T's strike costs increases enough that its strike incentive is lost, but U still sees an incentive to strike, which it does, reducing its cost but maintaining T's at a high level thereafter.

The basic shape of the curves for U and T in Figure 9 is the same as those for U and P in Figure 2 for intermediate u, as are the underlying reasons. Allowing T to target U's high value targets gives it an incentive to preempt at levels of U defenses that would negate T's second strike forces, which produces a barrier in the U-T relationship similar to that in the U-P relationship, which restricts the levels of defenses that U can stably allocate to the U-T portion of the P-U-T triad. These calculations use preferences $K = 20$ and $V' = 1$, which tend to favor the number of interceptors that can be allocated stably. Values that emphasize targeting military value predict instability, as before. Increasing U's preference for the protection of high value targets removes its incentives to strike by increasing the cost of damage to U's high value. That increases stability, but at the price of U's ability to deter untoward action by T.[23]

68

the shift of targeting from military to high value targets at low force levels, which is more important than the details of offenses and defenses in determining strike incentives. The analysis indicates little risk of reduction in bilateral stability in deep offensive force reductions and appropriate defensive deployments.

Inclusion of the shift to high value targets stabilizes trilateral offensive configurations, even when one side's forces are very weak, which differs with the results of analyses based on military costs only. It produces a similar shift when defenses are included, leading to a balance between a strong, defended side and a and weak, undefended side that resembles, on a smaller scale, the balance between the two strong sides. Including the strong, defended side's preference for survival of its high value targets reduces its incentive to strike, which removes the weak side's incentive to preempt. However, that also reduces the strong sides' ability to deter the weaker. The stability of multi-polar configurations are still controlled by that of the least stable dyad, which places constraints on the defenses that can be deployed stably which could be more confining than those from the bilateral balance.

REFERENCES

1.   G. Canavan, "Missile Defense and Proliferation," A. Zichichi, ed, *25th Session of the International Seminars on Planetary Emergencies, 19-24 August 2000* (London, World Scientific, 2000).
2.   G. Canavan, "Crisis Stability and Strategic Defense" *Proceedings of the Military Modeling and Management Session of the ORSA/TIMS National Meeting, November 12–14*, S. Erickson, Ed. (Operations Research Society of America: Washington, 1991).
3.   G. Canavan, "Analysis of Decisions in Bi- and Tri-Lateral Engagements," U.S. *State Department Stability Workshop* (Institute for Defense Analysis, November 2000); Los Alamos National Laboratory Report LA-UR-00-5737, November 2000.
4.   G. Kent and R. DeValk, "Strategic Defenses and the Transition to Assured Survival," RAND Report R-3369-AF, October, 1986
5.   A. Piontkovsky, "Global Defense and Strategic Stability," A. Zichichi, ed, *16th Session of the International Seminars on Planetary Emergencies, 19-24 August 1992* (London, World Scientific, 1993).
6.   R. Powell, *Nuclear Deterrence Theory* (Cambridge, University Press, 1990).
7.   T. Schelling, *The Strategy of Conflict* (Cambridge, Mass, Harvard University Press, 1960).
8.   T. Schelling, *Arms and Influence* (New Haven, Yale University Press, 1966).
9.   G. Canavan, "Analysis of Decisions in Bi- and Tri-Lateral Engagements," op. cit;
10.   G. Kent and R. DeValk, "Strategic Defenses and the Transition to Assured Survival,"
11.   G. Canavan, "Freedom to Mix Defenses in Modest Forces," Los Alamos National Laboratory Report LA-UR-01-4563, June 2001.

Fig. 9. Cost of node raching node 7 as function of interceptors allocated in tralateral interaction.

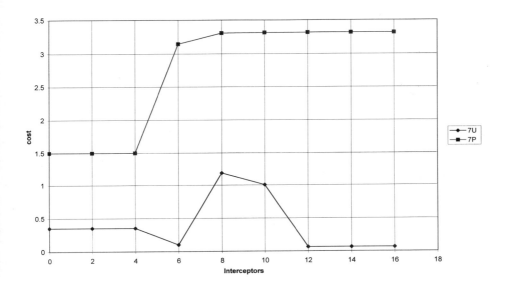

SUMMARY AND CONCLUSIONS

This note extends the game theoretic analysis of strategic conflicts begun in earlier Seminars to interactions with and without defenses between two or more adversaries with more realistic target structures. Game theory is sufficiently flexible to treat all current issues in strategic force reduction and defensive deployments under discussion with sufficient fidelity through the variation of a few readily identifiable preference parameters. The analysis sheds light onto a number of contradictions in previous cost ratio analyses such as the U.S. appearing to be the source of instability in the current bilateral offensive configuration, small defenses seeming to have a large impact on stability, and the current trilateral balance being unstable. The stability of the current bilateral offensive configuration is shown to be high and insensitive to deep reductions in offensive forces, deployment of limited defenses, and the exchange of offensive forces for defenses.

The transition from adversarial to cooperative interaction, as represented by the progressive reduction of each side's damage preference parameters, monotonically improves stability and margins. Inclusion of high value targets reduces estimates of strike incentives in bilateral and trilateral configurations with and without defenses by capturing

12.   G. Canavan, "Analysis of Decisions in Multilateral Engagements," Los Alamos National Laboratory Report LA-UR-01-1219, February 2001.

13.   G. Canavan, "Considerations in Missile Reductions and De-Alerting," L. Kruger, ed., *Missile Proliferation and Defence Seminar 4-9 April 1998* (Erice, World Federation of Scientists, 1998).

14.   G. Canavan, "Cost of Addressing Targets of Unequal Value," Los Alamos National Laboratory Report LA-UR-01-4561, June 2001.

15.   G. Canavan, "Variation of Strike Incentives in Deep Reductions," Los Alamos National Laboratory Report LA-UR-01-4596, June 2001.

16.   G. Canavan, "Variation of Strike Incentives with Damage Preferences," Los Alamos National Laboratory Report LA-UR-01-4597, June 2001.

17.   G. Canavan, "Cost of Addressing Targets of Unequal Value," op. cit.

18.   G. Canavan, "Analysis of Decisions in Bi- and Tri-Lateral Engagements," op. cit.

19.   G. Canavan, "Variation of Strike Incentives with Preference for Survival of High Value Targets," Los Alamos National Laboratory Report LA-UR-01-4590, June 2001.

20.   G. Canavan, "Cost of Addressing Targets of Unequal Value," op. cit.

21.   G. Canavan, "Analysis of Decisions in Bi- and Tri-Lateral Engagements," op. cit.

22.   G. Canavan, "Limits on Defenses in Interactions Between Disparate Forces," Los Alamos National Laboratory Report LA-UR-01-4598, June 2001.

23.   G. Canavan, "Analysis of Decisions in Multilateral Engagements," *69th Military Operations Research Society Annual Meeting, Annapolis Maryland, 12-14 June 2001* (Alexandria Va, MORSS, 2001); Los Alamos National Laboratory Report LA-UR-01-3115, June 2001.

BALLISTIC MISSILE DEFENSE: A RISK FOR STABILITY AND INCENTIVE FOR NEW ARMS RACES?

REINER K. HUBER
Institut für Angewandte Systemforschung und Operations Research
Fakultät für Informatik, Universität der Bundeswehr München
D-85577 Neubiberg, Germany, huber@informatik.unibw-muenchen.de

BACKGROUND

During the Cold War, strategic stability was the result of a reliable deterrence relationship between the United States and the Soviet Union based on the paradigm of *Mutual Assured Destruction* (MAD). That paradigm implied the survival of a retaliatory strike capability sufficient to inflict unacceptable damage to the party that launched the first strike. The Anti-Ballistic Missile (ABM) Treaty of 1972 was the attempt to preserve this bilateral US-Soviet strategic relationship since the build-up of anti-ballistic missile defenses (ABM) would undermine strategic stability because it eventually might provide one or both sides with the capability to intercept and neutralize the opponent's retaliatory strike and, therefore, an incentive for both sides to strike first in a crisis. Only if both sides' ABM systems were to get to the point where they could block the other side's first strike—considered as infeasible at that time for technological and financial reasons—would the reason to preempt be removed for either side.

While the ABM Treaty was successful in preventing a defensive arms race, it did not, however, stop the offensive arms race between the U.S. and the USSR. Nevertheless, being the first arms control agreement successfully negotiated and signed by both governments, the ABM Treaty is still widely regarded as a cornerstone of strategic arms control and, therefore, essential for keeping the strategic arms control framework alive. However, it is questionable whether the terms of the treaty are still adequate considering the dramatic changes in the post-Cold War security environment. For one thing, the ABM Treaty's underlying principle of mutual vulnerability seems inappropriate as the U.S. and Russia seek to forge a new strategic relationship of cooperation. For the other, new strategic threats outside the bilateral US-Russian relationship are emerging because of the widespread proliferation of weapons of mass destruction (WMD) and ballistic missile technology since the end of the Cold War.

MOTIVES BEHIND PROLIFERATION AND ALTERNATIVE OPTIONS FOR RESPONSE

Other than the five declared nuclear powers, nearly 25 states have acquired, or are about to acquire, ballistic missiles and/or the know-how and technology required for their domestic production (Rumsfeld[1], Dean Wilkening[2] and Schilling[3]). Most notable among them are India, Pakistan, and the rogue states North Korea, Iran, Iraq, Syria and Libya. Even though today's delivery accuracy of ballistic missiles produced in emerging missile states may be rather low, in conjunction with nuclear, bacteriological, or chemical warheads, they represent a formidable threat to urban targets and military facilities alike. Most of the current arsenals of rogue states consist of SRBM (Russian Scud and Scud improvements) with ranges up to about 600 km. However, there are several indigenous IRBM development programs such as the North Korean Taepo-dong and the Iranian Sahab. If launched from North Africa, the 2,000 km range (presumably operational) of Taepo-dong-1 and the Sahab-4 would be sufficient to cover most of Southern Europe, the Balkans, and Turkey. The IRBM Taepo-dong-2 and Sahab-5 expected to become operational in about 2005 will have a range of 5,000-6,000 km capable of reaching, from the Middle East, targets anywhere in Europe, Western Siberia, Central and South Asia and North and Central Africa. North Korea, Iran, and Iraq are suspected of developing ICBMs with ranges of up to 12,000 km that could become operational within 10-15 years.

Keith Payne[4] lists five main motives underlying the acquisition of ballistic missiles and WMD by proliferating governments:

- obtaining coercive leverage against regional rivals;
- deterring intervention by the international community, especially by the U.S. and its allies, to aid victims of aggression by the respective government;
- taking advantage of the military and cost effectiveness of ballistic missiles (limited logistical and manpower requirements; difficult to locate and target; intercept may be possible only by BMD deployed on the territory of third parties);
- earning hard currency from exporting or re-exporting;
- enhancing power and prestige.

Other than active defense, there are essentially four options to respond to the proliferation of ballistic missiles and WMD:

- arms control in form of agreements among governments, of industrialized countries, on export limitations of crucial technologies;
- preemptive/preventive strikes against missiles and WMD storage sites and manufacturing facilities;
- reducing vulnerability to BM/WMD;

- deterrence by threatening punishment in case of aggression and/or use of WMD.

Arms control suffers from the difficulties of achieving agreement among potential supplier states on technology limitations. These difficulties are related to: economic competition and differences in political goals and interests; problem of enforcing compliance with agreed-upon limitations if rewards of non-compliance exceed possible sanctions by some significant margin; and the declining efficacy of controls as the technologies spread beyond the group of signatories to the arms control agreement.

Preemptive/preventive strikes in the form of counter-force attacks by air or special forces, if not outright infeasible for constitutional reasons, are politically difficult to justify and militarily risky because their operational effectiveness may be limited and trigger immediate counter-attacks. The likelihood of preemptive non-nuclear air attacks succeeding is low given that most SRBM and IRBM are mobile and fixed ICBM-sites usually hardened. The first and only successful attack of this kind in history was carried out in 1981 by Israel on a highly visible and relatively soft stationary target, the Osiraq reactor nearing completion at the Tuwaitta nuclear research center near Baghdad. Carried out with devastating precision by 14 attack aircraft, the raid had been carefully planned and rehearsed for some time (Armitage[5]). Nevertheless, even if militarily successful, preventive raids of this kind against BM/WMD production and storage facilities of proliferating states could be quite counterproductive. While their effects are merely temporary, they might serve proliferating governments to justify their actions vis-à-vis their populations, not to mention the immediate adverse reactions that must be expected from the international community and possibly even from allied governments.

Reducing vulnerability involves measures and strategies to protect populations and military forces from WMD-effects. However, even though population centers, especially in Europe and the Middle East, seem like prime targets for BM/WMD attacks by rogue states in order to avert formation of coalitions against them or defeat in a coalition war with them, passive defenses against WMD are typically not available for civilian populations for reasons of practicality and cost. In contrast, most military forces are equipped to deal with chemical and biological weapons. In addition, the U.S. forces are exploring operational concepts in which a projected force under BM/WMD threat would minimize its footprint by employing dispersed forces of high mobility in theater that are more difficult to target, and reaching back to forces and resources in sanctuary out-of-range of BM/WMD. This concept is at the heart of the Joint Vision 2020 as a means of countering so-called asymmetric threats under the rubric of the Revolution in Military Affairs (RMA). An operational concept that would essentially employ current technology and could be implemented within five years has been proposed by Gritton et al[6].

Deterrence is considered to be inherently unreliable vis-à-vis proliferating states in the post-Cold War era. It is difficult to establish reliable rules for deterrence with governments of proliferating states whose cultural and social environment are not well understood in most cases, and whose behavior is evidence of their unfamiliarity with the essential prerequisites of effective deterrence which emerged during the Cold War (rational behavior of antagonists; mutual understanding of motives; effective communication between antagonists; credibility of threats). Moreover, a strategy of deterrence by retaliation vis-à-vis rogue states implies the retaliator's willingness to victimize some part of his own population and, therefore is hard to sell in democracies if an alternative is becoming available that would protect the population from the missile threat in the first place[7].

The forgoing observations illustrate that the alternative response options to BMD have more or less serious operational limitations when considered in isolation. But together with at least some active defense systems, preemptive strikes may become more attractive and deterrence of proliferating governments more reliable. Paul Davis[8] asserts that deterrence by threat of punishment would be much enhanced by missile defenses in conjunction with well exercised and advertised military options for severe punishment.

However, testing and deployment of BMD is severely limited under the terms of the ABM Treaty. Designed to preserve mutual deterrence between the U.S. and USSR, the treaty prohibits the U.S. and Russia to defend themselves against strategic missile threats by third parties even though historical records leave ample room for doubt about some of them being at all susceptible to deterrence by retaliation. Recognizing the lack of viable alternatives for dealing with proliferating states in turn raises the question as to whether the ABM Treaty can still be regarded a reasonable basis for defining the bilateral strategic relationship between the U.S. and Russia.

CHANGING THE BILATERAL US-RUSSIAN STRATEGIC RELATIONSHIP

As regards the US-Russian strategic relationship, Melvin Best and Andrei Piontkovsky[9] argue that preserving the MAD-stability paradigm involves the risk of codifying hostile attitudes between Russia and the US. It is for this reason that they recommend to adopt a proposal that Piontkovsky and Skorokhodov[10] had presented first in 1992, during an international workshop in Munich[11], to replace MAD by the paradigm of *Mutual Assured Protection* (MAP) by implementing radical nuclear arms reductions and constructing a global BMD system. In other words, they proposed to integrate offensive and defensive strategic arms control in a manner that eventually would make the territories of the U.S. and Russia, and of all the world for that matter, invulnerable to ballistic missile attacks. In order to maintain strategic stability throughout the transition from MAD to MAP, the build-down of offensive strategic forces and the build-up of defense systems need to be closely coordinated among both sides so that mutual vulnerability is maintained, albeit at ever decreasing levels[12,13].

Aware of the fact that US-Russian relations in 1992 were still characterized by some degree of mutual distrust that made public acceptance of a sweeping change in the strategic paradigm from MAD to MAP rather unlikely, Piontkovsky and Skorokhodov suggested an intermediate option that implied both concepts in form of a "*development of an ABM defense system on a level that would be sufficient to protect both the U.S. and Russia from terrorist threats and unauthorized launches, but at the same time to preserve mutual deterrence, i.e., not to infringe on both, the first and second strike capabilities of both states."* However, they regarded this option, of building what amounts to national missile defense (NMD) systems to protect the U.S. and Russia from missile threats outside their bilateral relationship, as a transitory one on the road to MAP because the implied balance between MAD and MAP would be rather shaky. Moreover, it might threaten the deterrence potentials of other nuclear powers while preserving the deterrence potentials of the U.S. and Russia.

The Russian proposal of February 2001 for a common NATO-Russia BMD system appears to fall considerably short of the intermediate option proposed by Piontkovsky and Skorokhodov because its architecture seems to have little or no growth potential for an eventual transition to a global defense system and, thus, for the "normalization" of the bilateral US-Russian relationship. Meant as an alternative to the American BMD plans, the proposal acknowledges that Russia and Europe need to be protected from terrorist missile threats while implying that MAD will continue to define the strategic relationship between the U.S. and Russia.

COUNTER-ARGUMENTS

There are a number of objections to BMD arguing, among others, that:

- the missile threat is greatly exaggerated in order to provide a rationale for embarking on expensive new programs for the benefit of arms industries after the end of the Cold War;
- the world would experience a new division between countries being under the protective umbrella of BDM and those being left outside;
- BMD systems are infeasible on technical grounds because the engineering problems that need to be solved in order to make BMD systems reliable are insurmountable, especially since BMD countermeasure development may be always one step ahead;
- the cost of BMD are enormous and may turn out to be too much of a burden that taxpayers even in well-to-do nations would be willing to shoulder, not to speak of the opportunity cost of BMD for societies and mankind;
- the benefits for international security and strategic stability are doubtful, if not outright negative;
- the implementation of BMD would stimulate new arms races;
- there are ways and means other than ballistic missiles for delivering WMD.

There may be some truth to the first objection insofar as BMD development programs would certainly help the respective industries to maintain technological leadership regardless of whether or not the threat materializes. For a somewhat skeptical assessment of the BM threat and the rationale for BMD in general, and national missile defense (NMD) and theater missile defense (TMD) in particular, the reader is referred to Wilkening[2]. In contrast, the Rumsfeld Commission[1] concludes that BM are not a distant threat to the US. In the new strategic environment, each of the emerging BM powers is considered to have *"the capacity, through a combination of domestic development and foreign assistance, to acquire the means to strike the U.S. within about five years of a decision to acquire such a capability (10 years in the case of Iraq)"*.

Regarding the second objection one may argue that the world has always been divided into zones of unequal security. While not providing equal security for all, at least as long as a global protection system as envisaged by Piontkowsky and Skorokhodov has not materialized, BMD may yet improve the security even for those outside its protective umbrella because of more credible extended deterrence that may be provided by states under the umbrella.

The arguments that effective BMD systems are infeasible on technical grounds are usually based on the assumption that countermeasures are easy to implement but impossible to overcome. According to a report in the Sueddeutsche Zeitung, the assertion that the systems cannot differentiate between warheads and the numerous decoys that will be deployed from the BM during exo-atmospheric flight, and therefore will minimize the time available for intercept after reentry, figure prominently in a paper recently submitted to President Bush by a group of nuclear physicists around Noble laureate Hans Bethe (Martin Urban: Weniger Sicherheit durch ein Raketenabwehrsystem, *Sueddeutsche Zeitung* Nr. 150, 3 July 2001, p. V2/9). Engineering experts, however, point out that solving the still existing technical problems is not a matter of technical feasibility but of funding the necessary engineering development and test programs (Karl-Heinz Allgaier: Verteidigung gegen Gefechtskoepfe und Taeuschkoerper. *Frankfurter Allgemeine Zeitung* Nr.111, 14 May 2001, p.11).

The monetary cost of BMD will certainly not be insignificant. Nevertheless, at this time assertions about the cost being too high seem quite speculative and deserve some serious analysis on the benefits that may accrue out of BMD and the eventual cost that the U.S. and the international community might possibly have to face because of not having invested in BMD. However, this is an issue beyond the framework of this paper which is to address the last three objections that refer to processes important for international security, namely deterrence and defense, arms races, and alternative means of WMD delivery.

DETERRENCE AND DEFENSE

In the context of strategic nuclear deterrence, stability may be measured in terms of a so-called "first-strike stability" index as defined, for example, by Kent and Thaler[14,15] which implies that striking first becomes less attractive as the cost incurred by the first striking party increase and/or the cost of striking second decrease[16].

In contrast, models of conventional stability usually include both, the notion of deterrence and defense. For example, Huber[17] considers the conventional balance between two opposing forces to be stable if either:

- the attacker is risk-averse to the extent that he is deterred from attacking because the estimated probability of victory is lower than minimum success probability that needs to be assured of before attacking, or
- the defender can defeat the undeterred attacker at conditions that meet his defense success criteria.

By analogy with Huber's model, a situation characterized by missile threats would be considered as stable by the targeted country if its capability to intercept any missile launched by a proliferating government is high enough to defeat the very objectives of its BM/WMD threat meant, for example, as a deterrent in a desperate situation such as, for example, an imminent defeat in war with the targeted country.

The principal concerns about strategic nuclear stability regarding BMD are reflected by the observations of Piontkovsky and Skorokhodov about the fragile offense-defense-balance of limited BMD and its implications for the deterrent potentials of other nuclear powers. While their argument appears plausible in the light of Bracken's analysis results of 1986, numerical results from a model for calculating a multipolar first-strike stability index presented by Melvin Best and Jerome Bracken at the mentioned 1992 workshop in Munich indicate that overall strategic stability among nuclear powers might actually improve as the U.S. and Russia deploy small to medium size ABM defenses. Their calculations took into account all theoretically possible combinations of first-striking coalitions and second-striking coalitions involving the nuclear arsenals of the US, Russia, United Kingdom, France, and China. For each combination, overall strategic stability was defined in terms of the lowest value of the first-strike stability index that resulted among all pairs of opposing coalitions (Best and Bracken[18,19]). These findings were confirmed in a later investigation using a different stability index based on utility rather than cost ratios used by Kent and Thaler (Bracken[20]).

These results suggest that, contrary to conventional wisdom projected from analyses of the bipolar situation during the Cold War, limited ABM defenses might not adversely affect the stability of strategic relations among nuclear status-quo powers in a multipolar world as long as they share the rationality underlying the concept of mutual deterrence and have the capability for effective communication among each other in a crisis.

At the same time, however, a limited BMD capability sufficient to intercept the missiles that proliferating states can be expected to be launching for sake of coercion would remove pressures for politically and militarily risky preemptive strikes in a regional crisis and provide time for responding to the challenge of the respective situation in an appropriate manner. In particular, protection of population centers against missile-borne WMD may soon become a *conditio-sine-qua-non* for building and preserving coalitions in response to aggression by proliferating governments.

ARMS RACES

In a nutshell, arguments that the development and deployment of BMD systems would contribute to, or cause, arms races revolve around the assumption that the protective shield provided by them permits the states owning BMD to coerce other states, or deny other states strategic leverage, thus forcing them to expand their missile inventories and WMD, and develop BMD countermeasures, in order to overcome the protective shields. In other words, BMD would make a bad situation worse. Moreover, a renunciation of the ABM Treaty by the U.S. involves the risk that the entire strategic arms control framework established over the past 40 years might collapse. The principal plausibility of this reasoning notwithstanding, it may not necessarily be relevant, however, when considering the circumstances of real situations and the interests of the actors involved. Therefore, a brief look shall be taken at the two most formidable opponents of the BMD plans of the U.S.: Russia and China.

In case of a unilateral withdrawal of the U.S. from the ABM Treaty, Russia may threaten offensive countermeasures, including the development of ABM countermeasures, and WMD proliferation[21,22]. In fact, the Duma has already pronounced it will not ratify START II which requires both sides to reduce their deployed nuclear arsenals by almost half to a level of 3,000-3,500 warheads each. Talks on START III aimed at a limiting the number of warheads to at most 1,500 for both sides would almost certainly be suspended immediately. However, even if Russia were to follow through on these threats, the likelihood for an arms race between the two countries is small since Russia will be hard pressed to maintain, let alone modernize, its aging strategic nuclear arsenal on the level permitted by START II. Part of the required funds might be obtained from proliferation of BM/WMD and BMD-countermeasure technology and systems which, by the way, are seen as an appropriate and sufficient response to the U.S. challenge by many a Russian military expert. However, such a policy would result in Russia's self isolation and may endanger Russian security in the long term. For these reasons, and because it badly needs Western support in rebuilding its economy and infrastructure, Russia is likely to agree to a modification of, or a follow-on to, the ABM Treaty that permits the U.S. to deploy a limited national BMD system (NMD) that would satisfy U.S. security concerns vis-à-vis rogue governments while leaving Russia with a credible second strike capability vis-à-vis the U.S.[23].

A model describing, in mathematical terms, how comparable second strike capabilities, or a strategically stable offense-defense relationship between two nuclear powers, might be maintained when one or both of them were to build up BMD systems is presented in the annex to this paper. An analysis of the model's mathematical relations illustrates that the solution to the problem of comparable and sufficient retaliatory capabilities is rather straightforward, at least in principle, as long as the structures of the offensive strategic forces of both sides are similar. In that case, the offensive BM arsenal of the power unilaterally deploying BMD would have to be reduced to the same degree that its BMD system can neutralize BM launched by the other power. The problem is more complicated, however, if the structures of the offensive forces of both sides are dissimilar and, therefore, may involve significant differences in the first strike effectiveness of both sides and the vulnerability to the opponent's first strike. It is these differences that, together with the desired BMD capabilities, determine the necessary reductions of offensive forces.

However, neither the effectiveness of nor the vulnerability to first strikes can be measured directly. For that purpose, more or less complex dynamic nuclear exchange models are required that permit to run virtual experiments on the mutual strategic force structures and future development options. Thus, sharing information on system architectures and performance data as well as assessment models would seem like a logical step to take for the United States and Russia in order to overcome mistrust and arrive at an agreement for the modification or replacement of the ABM Treaty that would preserve strategic stability in the traditional sense of MAD while permitting them to provide protection from emerging rogue threats.

More so than Russia's reaction, however, it is China's response to the deployment of a U.S. national BMD that causes concerns because it could trigger a cascading offensive arms race that eventually might involve India, Pakistan, Iran, Iraq, Israel, and others. However, considering the cost to its economic development which depends on foreign trade and access to Western technology, it appears highly unlikely that China would respond by modernizing and expanding its ICBM arsenal in order to maintain whatever strategic leverage it may derive vis-à-vis the U.S. from its current arsenal of some 20 nuclear-armed ICBMs capable of reaching targets on the West Coast[24].

This view was confirmed at a recent symposium in Munich[25] by Dr.Yang Xiyu, the deputy director of the Institute of World Development in Bejing, when he explained China's attitudes toward the BMD plans of the US. He pointed out it is not NMD that worries China, but TMD cooperation with, or deployment to, countries on China's periphery, in particular Taiwan. He argued that under the protective shield of TMD the Taipei government would be encouraged in its moves to independence while China would be deprived of strategic leverage for dissuading the U.S. from military intervention in regional conflicts, in particular the defense of Taiwan. Therefore, one may conclude that China's dedicated position, that the ABM Treaty must not be terminated unless replaced by a "better" convention, appears to be motivated less by concerns for the strategic arms control framework, and strategic stability among the established nuclear powers, than by the desire to have TMD included in a follow-on treaty, presumably in form of deployment limitations and the like, in exchange for China agreeing to limited NMD deployment by the US.

China's response to a renunciation of the ABM Treaty by the U.S. and TMD cooperation in its neighborhood might include both, the development of ABM-countermeasures, perhaps in cooperation with Russia, and an expansion of its SRBM and IRBM arsenal of currently some 400 missiles a large part of which is deployed in Fujian province opposite Taiwan. While being less costly than increasing the ICBM arsenal and not implying a direct threat to the US, such a response would very likely intensify, and possibly spread to adjoining regions, the arms race that has been going on in the region for some time, especially between the nuclear powers China, India, and Pakistan. In addition, the proliferation problem is likely to become worse as nations involved in the cascading arms race will attempt to pay for their build-ups by selling BM/WMD technology and systems on the world's arms markets.

Interestingly enough, however, it is the availability of ABM-countermeasures that might reduce or even eliminate the risk of cascading arms races in China's neighborhood and beyond. Effective penetration aids would permit China to do without a sizeable expansion of its BM force and, therefore, remove the pressure from its neighbors to expand their BM arsenals in response, albeit at the risk of a qualitative offense-defense arms race between China and the U.S. and eventual proliferation of ABM-countermeasure technology. This, in turn, would require more or less expensive improvements for the BMD system of the U.S. and its allies depending on the countermeasure techniques that proliferating states are able to handle[26,27,28]. It goes without saying that the most effective manner for countering ABM-countermeasures would be to intercept incoming ballistic missiles before they can activate any of their penetration aids, i.e., in the boost phase. For this reason, it seems the adoption of a BMD architecture that has the potential to grow into a global protection system as postulated by Piontkvosky and Skorokhodov[9], capable of boost phase interception, would minimize the risk of a qualitative arms race between BMD and BMD countermeasure technology erupting in the long term.

ALTERNATIVE MEANS OF WMD DELIVERY

Finally, there is the frequently voiced argument that BMD systems are counterproductive and a waste of money because there are other means of WMD delivery that BMD can not address such as, for example, cruise missiles (CM), special operations forces (SOF), and terrorists.

CM equipped with WMD may indeed become a formidable threat in the near future. Mathew Ganz[29] estimates that about 30-35 nations have land attack cruise missile programs. CM have similar payloads to BM, ranges of currently up to 600 km and possibly 2000 km by 2005, and cost significantly less than BM. They may be launched from a variety of small and highly mobile platforms such as fast patrol boats, submarines, and trucks. Other than for attacking projected forces, however, CM may not be regarded as a viable alternative to BM. Considering the geographic locations of the states which are likely to employ BM/WMD threats for the purpose of coercion, CM would have to be transported to their launch areas by ship or submarine and thus run the risk of being detected, tracked, and sunk during transit. Given effective surveillance systems, detection of the launching platform would be almost certain after the CM were launched. Thus, because of the comparatively long flight times of CM there should be a good chance to intercept and kill CM before reaching land[30].

SOF and terrorist operations are dependent upon some infrastructure and logistics in the targeted countries and, therefore, are subject to detection, penetration and surveillance. This reduces their reliability of being available and able to respond as planned in order to accomplish the political objectives of coercion. Intimidation and coercion as mechanisms in negotiations between parties require that the coercing party is able to closely control the tools of coercion. It is highly unlikely that the feed-back control loops for SOF or terrorists will ever arrive at response times on the order of BM flight times from launcher to target, not to mention the risk of their being neutralized in the course of the negotiation process. Besides, once released for their mission, terrorists may decide to do their own thing. The events in the Middle-East illustrate that controlling terrorists is quite difficult, if not impossible.

That is not to say that alternatives means of WMD delivery can be dismissed. On the contrary, terrorism especially must increasingly be reckoned with. However, if they were efficient tools for accomplishing whatever purpose BM may serve in the hands of rogue governments, why then would these governments ever bother to spend money on the acquisition of comparatively expensive BM? Besides, by the very logic of the argument it would appear quite pointless to invest in any kind of protection system. It is well known that not very many people follow such a logic in their personal lives, nor do local administrations and governments responsible for the security of their citizens.

CONCLUSIONS

Summarizing the observations presented above, one arrives at the conclusion that it is highly unlikely that limited BMD will endanger strategic stability as defined by the deterrence concept of MAD that was at the heart of the US-Soviet strategic relationship in the rather predictable bipolar world of the Cold War period. On the contrary, in conjunction with reductions of nuclear arsenals, the deployment of limited BMD by the U.S. or Russia, or both, may reinforce multipolar first strike stability among the established nuclear powers. However, in today's rapidly changing security environment, the static stability of nuclear deterrence of the Cold War period, maintained through a set of commonly shared principles codified in the form of arms control agreements, appears to become increasingly irrelevant regardless of BMD.

The uncertainty and unpredictability of the new security environment requires that the established "rules of the road" are more or less continuously reviewed and, if necessary, rewritten in order to manage the highly dynamic processes of change in a manner that hopefully provides stability in the sense of keeping the security environment within manageable bounds and minimizing the probability of major war. Proliferation of WMD and missile technology are regarded as the most dangerous obstacle in that process. Therefore, BMD is likely to become an indispensable tool for preserving both strategic and crisis stability in the 21st century provided it is implemented in a constrained and cooperative manner. Short of BMD, there are only two options for responding to coercive BM threats, capitulation or highly problematical preemptive counter-missile operations.

Provided that it is implemented in a constrained and cooperative manner, the risk that BMD will stimulate proliferation of WMD/BM and new arms races, or intensify ongoing regional arms competitions such as in Asia, is regarded to be low. On the contrary, the prospects of BMD cooperation among potential victims of BM threats would greatly diminish the value of BM for rogue governments and, therefore, remove incentives for proliferation of WMD and missile technology.

True, there are alternatives to BM for the delivery of WMD that BMD cannot address. However, none of them compares to BM as a means of coercion by rogue governments in terms of reliability and controllability as well as cost-effectiveness.

ACKNOWLEGMENTS

The research underlying this paper had been motivated by an invitation of the Hanns-Seidel-Foundation (HSS) to discuss stability implications of BMD at the International Symposium "Aspects of NMD/BMD" held in Munich, 15 May 2001. The author is indebted to several friends and colleagues for discussions and advice during the genesis of this paper and review of the manuscript, in particular Col. Heinrich Buch, German Army (ret), Prof. Jerome Bracken of Yale University, Dr. Paul Davis of RAND, Prof. (em) Klaus Gottstein of the Max-Planck-Society, Prof. Klaus Lange of HSS, Dr. Daniel McDonald of the Potomac Foundation, Dr. James Scouras of DynCorp, Dr. Stuart Starr of MITRE, and Prof. Andrei Piontkovsky of the Russian Academy of Science.

REFERENCES

1. Rumsfeld, Donald H. et al.: Report of the Commission to Assess the Ballistic Missile Threat to the United States http://www.fas.org/irp/threat/missiles/rumsfeld
2. Dean A. Wilkening: Ballistic-Missiles Defence and Strategic Stability. *Adelphi Paper 334*, London 2000, The International Institute for Strategic Studies.
3. Walter Schilling: Die Proliferation von ballistischen Raketen und Massenvernichtungswaffen. *Europäische Sicherheit* Nr. 5, May 2001, pp. 49-51.
4. Keith Payne: Diplomatic and Dissuasive Options (Counter-Proliferation, Treaty based Constraints, Deterrence and Coercion). In: Ranger (Ed.): *Extended Air Defence & the Long-Range Missile Threat.* Bailrigg Memorandum 30 Lanchester University 1997, Centre for Defence and International Security Studies, pp. 38-43.
5. Michael Armitage: History of Airpower. In: Dupuy (Ed.): *International Military and Defense Encyclopedia.* Washington-New York 1993, Brassey's, pp.82-93.
6. Eugene C. Gritton, Paul K. Davis, Randall Steeb, and John Matsumura: *Ground Forces for a Rapidly Employable Joint Task Force.* Santa Monica 2000, National Defense Research Institute, RAND.
7. It should be remembered that it was the desire to avoid the moral dilemma associated with MAD, of taking the populations of both sides hostage, that motivated President Reagan in 1983 to put forward the Strategic Defense Initiative (SDI).
8. Paul K. Davis: Protecting Weak and Medium-Strength States: Issues of Deterrence, Stability, and Decision Making. In: Huber and Avenhaus (Eds.):

Models for Security Policy in the Post-Cold War Era. Baden-Baden1996, Nomos, pp. 43-81.

9.   Melvin R. Best and Andrei A. Piontkovsky: New Stability Paradigms in the Post-Cold War World. In: In: Huber and Avenhaus (Eds.): *Models for Security Policy in the Post-Cold War Era.* Baden-Baden 1996, Nomos, pp. 141-149.

10.   Andrei A. Piontkovsky and Arkadii Skorokhodov: Global Defense and Strategic Stability: From MAD-Stability to MAP-Stability. In: Huber and Avenhaus (Eds.): *International Stability in a Multipolar World: Issues and Models for Analysis.* Baden-Baden 1993, Nomos, pp. 217-222.

11.   Workshop on "Problems of International Stability in a Multipolar World" held 23-25 November 1992 at the Universität der Bundeswehr München, Germany, sponsored by the VW-Foundation.

12.   As a contribution to the SDI debate, Jerome Bracken has analyzed strategies of transition from the strategic force structures of the mid-eighties based on the MAD paradigm to alternative force structures characterized by BMD, air defense systems, and deep cuts in the offensive nuclear arsenals of both, the U.S. and USSR. He observed that the adoption of BMD would not make sense unless accompanied by deep cuts in the offensive arsenals of both sides in which case the bipolar strategic system would reach a state of what he called "Mutual Assured Survival" (MAS). He concludes that maintaining strategic stability during transition is not impossible, but difficult and dependent upon mutual perceptions about each others BMD capabilities.

13.   Jerome Bracken: Stability, SDI, Air Defense and Deep Cuts. In: Avenhaus, Huber, and Kettelle (Eds.): *Modelling and Analysis in Arms Control.* Berlin-Heidelberg-New York 1986, Springer-Verlag, pp.183-206.

14.   Glenn A. Kent and David Thaler): First-Strike Stability: Methodology for Evaluating Strategic Forces, R –3765-AF, Santa Monica 1989.

15.   Glenn A. Kent and David E. Thaler (1990): First-Strike Stability and Strategic Defenses, Part II of a Methodology for Evaluating Strategic Forces, R-3918-AF, Santa Monica 1990.

16.   The Kent-Thaler index assumes the value 1 if the cost of striking first and second are identical for both sides, cost being defined in terms of the difference in the damage levels suffered by both sides in each case, and the value 0 if there is no sizeable damage associated with striking first for one or both sides.

17.   Reiner K. Huber: Military Stability of Multipolar International Systems: Conclusions from an Analytical Model. In: Huber and Avenhaus (Eds.): *Models for Security Policy in the Post-Cold War Era.* Baden-Baden 1996, Nomos, pp. 71-80.

18.   Melvin L. Best and Jerome Bracken: First-Strike Stability in a Multipolar World. In: Huber and Avenhaus (Eds.): *International Stability in a Multipolar World: Issues and Models for Analysis.* Baden-Baden 1993, Nomos, pp. 223-254.

19.   Melvin L. Best and Jerome Bracken: First Strike Stability in a Multipolar World. Management Science, Vol. 41, 1995, pp. 298-231.

20. Jerome Bracken: Multipolar Nuclear Stability: Incentives to Strike and Incentives to Preempt. *Military Operations Research*, Vol 3, No 1, pp.5-21.

21. Less than 24 hrs after the Secretary of the Russian Security Council, Sergei Ivanov, had rejected U.S. plans for NMD at the Munich Conference for Security last February, the Chief of Russia's General Staff, Sergeiev, made an explicit statement to that extent and threatened that Russia would activate three "powerful" systems (Wolfgang Fechner, 2001).

22. Wolfgang Fechner: Ohne die USA geht nichts – Ohne die Europäer auch nicht. *Europäische Sicherheit*, Nr. 3, March 2001, pp. 10-13.

23. According to a Russian source, Russia would have been ready for such a deal with the U.S. some time ago if it hadn't been for European, in particular German, concerns about NMD that raised hopes in Russia to drive a wedge between the U.S. and its European allies. Disruption of Euro-Atlantic ties may have been one of the motives underlying Russia's counterproposal for a joint development of a Russian-NATO theater missile defense system (TMD) which was received, however, with considerable reservations by the European NATO allies.

24. However, there are estimates that China has the capability and plans to expand, within the next 10-15 years, its strategic nuclear force to about 200 BM warheads that could reach the US, and about 300 capable of reaching the Russian heartland (Wilkening[1], pp. 83-84), provided that it found a way to fund the expansion program. Therefore, the question has been asked whether China's future role will be that of a large rogue or a small Russia whose deterrent deserves to be preserved.

25. Expert Symposium on "Aspects of NMD/BMD" organized by the Hanns-Seidel-Foundation, 15 May 2001.

26. Expert opinions differ on that issue. It seems that not many of them share the opinion of Lewis and Postol that the technologies and skills required for developing and implementing these countermeasures are less demanding than those needed to design, build, and deploy rudimentary ballistic missiles (see also Huber[28]).

27. George Lewis and Theodore Postol: Future Challenges to Ballistic Missile Defense. *IEE-Spectrum*, September 1997, pp. 60-68.

28. Reiner K. Huber: Missile Technology Developments: Countermeasures against Ballistic Missile Defenses. In: Krueger (Ed.): Seminar Proceedings of the Working Group on Missile Proliferation and Defense Seminar. Lausanne 1998, The World Federation of Scientists, pp. 20-23.

29. Mathew Ganz: Cruise Missile Defense. In: Ranger (Ed.): *Extended Air Defence & the Long-Range Missile Threat*. Bailrigg Memorandum 30. Lanchester University 1997, Centre for Defence and International Security Studies, pp. 20-22.

30. Flight time would be 35-40 minutes if launched from a distance of 600 km.

REDUCING THE BALLISTIC MISSILE THREAT

LT. GEN. V.J. SUNDARAM
National Design and Research Forum, The Institution of Engineers, Bangalore, India

INTRODUCTION

The ballistic missile threat to planet Earth has continuously increased in the last fifty years due to the very large number of nuclear weapons in the world today, multiple nuclear powers, regional tensions and nuclear doctrines. While total elimination of ballistic missiles with nuclear warheads will not be feasible at present in the current political scenario, the policies of "No First Use," deterrence and the ABM treaty have reduced their threat so far and any changes must be viewed with caution. In addition, confidence building measures, robust design, fissile material control, monitoring, effective maintenance, and rigorous training are essential to achieve the aim of a safer world.

THE THREAT

The ballistic missile threat can be from ICBMs, IRBMs, or SRBMs and may be classified as follows:

a) From countries which may not be deterred by threat or retaliation:
- North Korea—Taepodong I. Nuke 2500km
                             Chem 4000km
           Taepodong II.   >5000km
- Iran
- Iraq
- Libya

b) Accidental/unauthorized launch from:
- Russia
- China
- USA
- NATO Country
- India
- Pakistan

86

- Israel
- North Korea
- Iran
- Iraq
- Middle East Country

c) Triggered by use of Tactical Nuclear Weapon by a commander in the field faced with annihilation by a force with superior conventional strength.

THE TARGETS

The targets which can be considered for IRBMs and SRBMs are:

a) U.S. Outposts, troops and assets
- Aleutian Islands
- Hawaii
- Alaska
- Japan
- NATO Countries

b) Western, Russian, Chinese troops and assets.

c) Indian, Pakistani, Israeli, Taiwanese assets, troops, and cities.

d) Iran, Iraq, Libya, North Korea.

ICBMs can reach targets in U.S. Mainland from North Korea, Russia and China and of course the reverse is true.

BALLISTIC MISSILE DEFENSE SYSTEMS

Nuclear Anti-Ballistic Missiles
These have now been dropped for defense systems.

Theatre Defenses (Land, Sea and Air)
These consist of layered defenses with space-based sensors and are considered effective against IRBMs and SRBMs.

- Point defense is covered by lower tier systems like Patriot and Hawk.
- Area defense is based on upper tier systems:
  - Army—THAAD
  - Navy—Theatre Wide Defense (on Aegis Cruisers)
  - Israel/U.S.—Arrow.

Defense support program (DSP) satellites have detected thousands of launches including Scuds in Iran-Iraq war and Gulf war. The DSP satellites would be replaced by Space Based Infra-Red Systems.

Boost-Phase Defense
DF lasers have been used to shoot down Katyushas. Air Borne Lasers using COIL will probably be available with U.S. Air Force by 2006 for ranges of 200–300 km. Missiles with 50 km range to be launched from UAVs are under development.

National Missile Defense (NMD)
NMD proposed to be developed and deployed by the U.S. has had recent successes and the program managers are very confident. The system has to overcome counter measures like sub munitions, decoys, stealth (radar and IR) and evasive maneuvers. Estimates of cost vary from $100 b to $200 b. There is also strong opposition from Russia and China and misgivings in some other countries as well that it will be a gross violation of the 1972 ABM Treaty and will open the doors for an ICBM race again. Thus, technological, economic, and political factors will all play an important role in implementing NMD.

TRIGGERING ACCIDENTAL NUCLEAR WAR

No First Use
Acceptance by all of "No First Use" of nuclear weapons would be a major factor in reducing the threat of ballistic missiles.

Deterrence
The holding of a minimum number of ballistic missiles for effective retaliation will continue to be the cornerstone of nuclear powers to achieve deterrence.

Fissile Material Control
Fissile material can be separated into:

a) Civil/Power—with full safeguards and international inspections
b) Military—where acceptance of safeguards may take longer time

CONFIDENCE BUILDING MEASURES (CBMs)

Exchange of documents on nuclear doctrine and advance notification of missile and satellite launches will go a long way in reducing fears of being caught unawares.

Strong command and control should be exercised over own arsenals and nuclear accidents/incidents should be thoroughly investigated and explained. Hotlines between countries at the highest decision-making level as well as at the principal executive level are essential.

Verification, mutually, by non-intrusive methods using advanced technology would be acceptable to start with. Intrusive bilateral verifications will be more difficult.

Tactical nuclear weapons should be banned totally, world-wide. Decisions on their employment may be taken by field commanders to avoid imminent defeat/capture and can very well trigger the launch of ballistic missiles in response. People should be educated on the dangers of such weapons to own troops and populations.

Air dropped bombs which provide a temptation for pre-emptive strikes should also be banned.

Command and control centers should be hardened with alternate centers and "Permissive Action Links" (PALs) established with reliable communication, interrogation and response systems without the fear of hidden bugs. Joint development and verification could be done to promote confidence.

"Alert States" should have common terminology and definitions which are clear and unambiguous. Reliability and certainty of a second strike should be established with smaller numbers. *Give confidence that response can be effective even with small numbers of "reliable" missiles.*

Joint seminars/exercises should be held to propagate above concepts and give stability in nuke related crisis.

MONITORING

Verification mutually by non-intrusive methods using advanced technology may be acceptable to a limited extent. An International Monitoring System applicable to all nations should be the aim. It must be kept in mind that tags are vulnerable to tampering. Utilization of RF MEMS can be explored.

UNAUTHORIZED/ACCIDENTAL LAUNCHES

Risk of unauthorized/accidental launches of ballistic missiles can be reduced by the following actions:

- Good robust design for safety
- Harden and protect
  - Safety locks
  - Communication and video links including Hot Lines (optical links with optical routers could be used)
- Don't mate warhead to delivery system
- Consider:
  - Fire
  - Transport accidents
  - Lightning
  - Mishandling

ALERTING SYSTEMS

An integrated alerting system utilizing radars, satellites and UAVs can be established to give warning through hardened links and data fusion with adequate training in interpretation. Action for alerting should follow a clear line of command.

TESTING

Testing is required because elements other than the nuclear core degrade with time and make the missile unsafe. Hence, periodical inspection, testing and maintenance is essential.

Avoid system modification for "improvements" unless adequate testing to prove reliability is possible at "system" level.

Testing to keep missiles safe and reliable can be done up to a point where *core nuclear explosive will not go critical.* The missile should then be formally recertified as safe after clearing all tests.

If there are unanticipated technical problems in an existing design/stock pile, it may be necessary to conduct small nuclear tests of a few kTs to ensure safety of the larger arsenal.

CTBT which is contemplated could be signed by all countries with provision for the above contingency suitably incorporated.

MAINTENANCE IN SERVICE

The following actions at gross roots level must be implemented:

- Training of technical officers and staff from services.
- Inspection.
- Inventory control.
- Accident/defect reporting.
- Periodical withdrawal of weapon for refurbishment. Resources and fabrication facilities must be available to undertake this task.
- Quality Assurance.
- Lifting studies on all devices/material.
- Reliability experiments on all components.
- Evolve maintenance/repair procedures.
- Independent verification of hardware and software.

OPERATIONAL TRAINING

Training must be imparted to reduce operational response time safely. Warfare Analysis Laboratory exercises should be conducted at different levels. A virtual reality environment can be provided so that operational staff are exposed to crisis situations more realistically and troop drills are honed towards perfection.

CONCLUSIONS

The various types of threats, defenses doctrines, confidence building measures as well as the importance of design, testing, maintenance and training have been discussed and it is felt that all nations should work towards the implementation by mutual cooperation of these suggestions, as well as others which will surely emerge in the conference, to reduce the ballistic missile threat to the world.

4. TCHERNOBYL — MATHEMATICS AND DEMOCRACY

DEMOCRACY AND MATHEMATICS IN LITHUANIA

Z.R. RUDZIKAS
Lithuanian Branch of the ICSC - World Laboratory, State Institute of Theoretical Physics and Astronomy, A.Gostauto 12, Vilnius, 2600, Lithuania, E-mail: tmkc.plls@wllb.lt

This paper is aimed at the discussion of the historical path of Lithuania, at the mathematical expression of its economical and political changes. The main attention is paid to the last decade, to the period of restoration of its independence and transformation of the Soviet type state regulated economy to the free market one as well as of the authoritarian political regime to the democracy. The paper is mainly based on books[1,2,3].

INTRODUCTION

Lithuania is not widely known in the world community, therefore at the beginning of this paper we shall present some geographical and historical data about it.

Lithuania is located on the eastern shore of the Baltic Sea. In the north, it is bordered by Latvia (a 576 km border); in the south - by Belarus (660 km); in the south-west - by Poland (103 km) and the Kaliningrad district (273 km) of the Russian Federation. 91 km of Baltic sea shoreline form most of the Lithuanian's western border. Vilnius is the capital of Lithuania. The territory of Lithuania covers an area of 65,300 km². It is larger than Latvia, Estonia, Belgium, Denmark, the Netherlands and Switzerland. In 2001, its population was 3,5 million, Lithuania extends 373 km from east to west and 276 km from north to south. According to the French National Geographical Institute, the centre of Europe lies at 54°51' N and 25°19' E, i.e. about 20 km north to Vilnius. This point is marked by a special monument.

The country's landscape is diverse, mainly consisting of plains and forests (30.3% of the area). The average height above sea level is 99 m. The highest point is 294 m. In Lithuania, there are about 2,830 lakes larger than 0.5 ha. Lithuania's climate is transitional between maritime and continental. The average temperature in January is -4.8°C and in July 17.2°C.

The Lithuanian language belongs to the group of the Baltic languages that, together with Hindi, Greek, Armenian, Albanian and German, forms the Indo-European family of languages. Of the Baltic languages, only Lithuanian and Latvian are still alive and spoken still today. The Lithuanian literary language began to form in the middle of the 16th century. The first Lithuanian book was printed in 1547.

As concerns religion, the Roman Catholic church prevails. However, there are also communities of Evangelic Lutherans, Orthodox, Judaic and many others.

The national currency is Litas, reintroduced on June 25, 1993. So far it is bound to USD (1 USD=4 Litas), but soon it will be reoriented into Euro.

HISTORICAL HIGHLIGHTS

Lithuania was first mentioned in 1009 in the ancient Annales of Quedlingburgenses (town in Germany close to Magdeburg). Thus, Lithuania is going to celebrate its millennium in 2009.

The Lithuanian people are a part of the Baltic group, which is a division of the larger Indo-European group. Indo-European tribes settled in this territory approximately 5,000 years ago. Eventually they mixed with the local inhabitants, who had been there since the end of the last Ice Age (some 10,000 years ago).

The Lithuanian State started to form in the second half of the first millennium. The domains of individual sovereigns were united into the Grand Duchy of Lithuania by Grand Duke Mindaugas, who became the first and the only King of Lithuania. He reigned from approximately 1236 to 1263. On July 6, 1253 envoys from the Pope placed a crown on his head. Thus, already in the 13th century Lithuania became equal to other Christian states of Europe.

Lithuania was considerably expanded by Grand Duke Gediminas, who ruled from 1316 to 1341, as well as by his sons Algirdas and Kestutis. Particular prosperity of Lithuania is linked with the reign of Kestutis' son Vytautas (1392-1430). Under his reign the Grand Duchy of Lithuania stretched from the Baltic to the Black Sea. He became especially famous in Europe after the Zalgiris (Grünwald or Tannenberg) battle, where on July 15, 1410, the joint forces of Poles and Lithuanians defeated the Teutonic Order that used to attack their lands for centuries. This victory opened the long term prospects of peaceful life for Lithuania.

Lithuania maintained tolerant religious and ethnic policies in the extensive Slavic lands, which it annexed. In 1387, the formerly pagan country adapted Christianity (last pagans in Europe?) and through Catholicism, drew closer to the culture of Western Europe.

In the 16th century, Renaissance and Reformation ideas began to spread in Lithuania. Vilnius University was founded in 1579. However, politically the state began to weaken, whereas the neighbouring states started to grow and strengthen. In 1569, Lithuania was united with the Polish Kingdom into a federal state Rzeczpospolita (People's Republic).

In 1795 Lithuania was annexed by Russia. Vilnius University was closed in 1832, massive Russification was the general policy of the Tsarist autocracy. The efforts to restore the independence of Lithuania ended by the relevant declaration on February 16, 1918.

However the further progress of Lithuania was soon stopped by the secret protocols of the Molotov-Ribbentrop Pact, signed on August 23, 1939. On October 10, 1939 the Soviet Union forced Lithuania to sign a mutual assistance pact. In accordance with it Lithuania was given back Vilnius, occupied by Poland since 1920, but it also had to accept 20,000 Red Army soldiers. Moreover, according to the ultimatum of June 14, 1940, the Soviet Union brought additional military forces, and Lithuania was completely occupied and integrated by force into the Soviet Union.

The Soviet period ended in Lithuania in 1990, when on March 11 the Supreme Council of Lithuania announced the Act for the Renewal of Independence of the Republic of Lithuania. In 1991 the independence of Lithuania was officially recognized by a vast majority of countries.

Soviet and nazi occupations caused mass deportations of the inhabitants of Lithuania as well as the numerous killings of resistance fighters (see Table 1).

Table 1. Human losses in Lithuania in the middle of the 20th century[1].

1941 (first six months)	repatriation of Lithuania's Germans (50,000 people); the first mass deportations to the Soviet Union (23,000 people);
1941-1944	Nazi genocide of the Jews (220,000 people);
1943-1944	10,000 people taken into forced labour in Germany; 60,000 people flee to the West;
1945	140,000 inhabitants of the Klaipeda district emigrate;
1945-1946	200,000 Lithuanian Poles deported ("repatriated") to Poland;
1945-1953	mass deportations to Siberia and other eastern parts of the Soviet Union (250,000 people);
1941-1951	about 25,000 resistance fighters and 10,000 Soviet activists and supporters killed.

MATHEMATICS AND DEMOCRACY IN THE CONSTITUTION OF THE REPUBLIC OF LITHUANIA[2]

The constitution was approved by the citizens of the Republic of Lithuania during the Referendum on 25 October 1992. Here we shall discuss its main ideas mainly following a book[2].

Chapter 1. The State of Lithuania.
The State of Lithuania shall be an independent and democratic Republic.

In Lithuania, the powers of the State shall be exercised by the Seimas (Parliament), the President of the Republic and the Government, and the Judiciary. The most significant issues concerning the life of the State and the People shall be decided by referendum.

Citizenship of the Republic of Lithuania shall be acquired by birth or on other bases established by law. With the exception of cases established by law, no person may be a citizen of the Republic of Lithuania and other state at the same time. The State of Lithuania shall protect its citizens abroad.

Lithuanian shall be the State language. The capital of the State shall be the city of Vilnius.

Chapter 2. The Individual and the State.

The rights and freedoms of individuals shall be inborn and protected by law. The person and his freedom, his private life, property and dwelling place, shall be inviolable. Human dignity shall be protected by law. Freedom of thought, conscience, and religion shall not be restricted.

All people shall be equal before the law, the court, and other State institutions and officers. A person may not have his rights restricted in any way, or be granted any privileges, on the basis of his or her sex, race, nationality, language, origin, social status, religion, convictions, or opinions.

Every person shall be presumed innocent until proven guilty according to the procedure established by law and until declared guilty by an effective court sentence.

Citizens may move and choose their place of residence in Lithuania freely, and may leave Lithuania at their own will.

Citizens who, on the day of election, are 18 years of age or over shall have the right to vote in the election. Citizens shall be guaranteed the right to freely form societies, political parties, and associations, provided that the aims and activities thereof do not contradict the Constitution and laws.

Citizens who belong to ethnic communities shall have the right to foster their language, culture, and customs.

Chapter 3. Society and the State.

The family shall be the basis of society and the State. Family, motherhood, fatherhood, and childhood shall be under the care and protection of the State. Marriage shall be entered into upon the free consent of man and woman. The State shall register marriages, births, and deaths. The State shall also recognize marriages registered in church. In the family, spouses have equal rights. The law shall provide for paid maternity leave before and after childbirth, as well as favourable working conditions and other privileges.

State and local government establishments of teaching and education shall be secular. At the request of parents, they shall offer classes in religions instruction. Education shall be compulsory for persons under the age of 16. Citizens who demonstrate suitable academic progress shall be guaranteed education at establishments of higher education free of charge.

There shall not be a State religion in Lithuania.

Censorship of mass media shall be prohibited.

The State shall support ethnic communities.

Chapter 4. National Economy and Labour.

Lithuania's economy shall be based on the right to private ownership, freedom of individual economic activity, and initiative. The law shall prohibit monopolization of production and the market, and shall protect freedom of fair competition.

Land, internal waters, forests, and parks may only belong to the Citizens and the State by the right of ownership.

Every person may freely choose an occupation or business, and shall have the right to adequate, safe and healthy working conditions, adequate compensation for work and social security in the event of unemployment. Trade unions shall be freely established and shall function independently. Employees shall have the right to strike in order to protect their economic and social interests.

The State shall guarantee the right of citizens to old age and a disability pension, as well as to social assistance in the event of unemployment, sickness, widowhood, loss of breadwinner, and other cases provided by law. The procedure for providing medical aid to citizens free of charge at State medical facilities shall be established by law.

Chapter 5. The Seimas (Parliament).

The Seimas shall consist of representatives of the People. The Seimas consists of 141 members, elected for a four-year term on the basis of universal, equal, and direct suffrage by secret ballot. Any citizen of the State who is not bound by an oath or pledge to a foreign state and who, on the election day, is 25 years of age or over and has permanently been residing in Lithuania, may be elected a Seimas member. Pre-term elections to the Seimas may be held on the decision of the Seimas adopted by three-fifths majority vote of all the Seimas members. The President of Lithuania may also under certain conditions announce pre-term elections to the Seimas.

The duties of Seimas members shall be incompatible with any other duties in State or private institutions and organisations. A Seimas member may be only appointed as Prime Minister or Minister.

Every year, the Seimas shall convene for two regular sessions - one in Spring and one in Autumn.

The Seimas shall consider and enact amendments to the Constitution, enact laws, announce presidential elections, form State institutions and appoint and dismiss their chief officers, approve or reject the candidature of the Prime Minister proposed by the President, consider the programme of the Government, supervise the activities of the Government, appoint judges to and Chairpersons of the Constitutional Court and the Supreme Court, announce local government Council election, approve the State budget and supervise the implementation thereof, ratify or denounce international treaties and consider other issues of foreign policy, etc.

The laws enacted by the Seimas shall be enforced after the signing and official promulgation thereof by the President. Within days of receiving a law passed by the Seimas, the President shall either sign and officially promulgate the said law, or shall refer it back to the Seimas together with relevant reasons for reconsideration.

Chapter 6. The President of the Republic.

The President of the Republic is the head of the State. Any person who is a citizen of Lithuania by birth, who has lived in Lithuania for at least the past three years, who has reached the age of 40 prior to the election day, and who is eligible for election to a Seimas member may be elected President. The same person may not be elected President for more than two consecutive terms.

A person elected President must suspend his or her activities in political parties and political organisations until a new presidential election campaign begins.

The President shall settle basic foreign policy issues and, together with the Government, implement foreign policy, sign international treaties and submit them in the Seimas for ratification, appoint or recall, upon the recommendation of the Government, diplomatic representatives of Lithuania, appoint or remove, upon approval of the Seimas, the Prime Minister, charge him or her to form the Government and approve its composition, accept resignations of the Government, appoint or dismiss individual Ministers upon the recommendation of the Prime Minister, declare states of emergency, make annual reports to the Seimas about the situation in Lithuania and the domestic and foreign policies of Lithuania, etc.

Chapter 7. The Government of the Republic of Lithuania.

The Government shall consist of the Prime Minister and Ministers. The Prime Minister, within 15 days of being appointed, shall present the Government, which he or she has formed and which has been approved by the President, as well as its programme to the Seimas for consideration. The Government shall return its powers to the President after the Seimas elections or upon electing the President. A new Government shall be empowered to act after the Seimas approves its programme.

The Government shall administer the affairs of the country, ensure State security and public order, implement laws and resolutions of the Seimas as well as the decrees of the President, coordinate the activities of the ministries and other governmental institutions, prepare the draft budget of the State and submit it to the Seimas, execute the State Budget, establish diplomatic relations with foreign countries and international organisations, etc.

Upon the request of the Seimas the Government or individual Ministers must give an account of their activities to the Seimas.

Chapter 8. The Constitutional Court.

By definition the Constitutional Court shall decide whether the laws and other legal acts adopted by the Seimas are in conformity with the Constitution and legal acts, adopted by the President and the Government, and do not violate the constitution or laws. It shall consist of 9 judges appointed for an unrenewable term of 9 years. Every three years, one-third of the Constitutional Court shall be reconstituted.

In fulfilling their duties, judges of the Constitutional Court shall act independently of any other State institution, person or organisation, and shall observe only the Constitution of the State. Its decisions shall be final and may not be appealed.

Chapter 9. The Court.

In the Republic of Lithuania, the courts shall have the exclusive right to administer justice. While administering justice, judges and courts shall be independent, they shall obey only the law.

The court system of Lithuania shall consist of the Supreme Court, the Court of Appeal, district courts, and local courts. For the investigation of administrative, labour, family and other litigations specialized courts may be established.

Judges may not hold any other elected or appointed posts, and may not be employed in any business, commercial, or other private institution or company. They may not participate in the activities of political parties and other political organisations.

In all courts, the investigation of cases shall be open to the public. Closed court sitting may be held in order to protect the secrecy of a citizen's or the citizen's family's private life, or to prevent the disclosure of state, professional, or commercial secrets. Court trials shall be conducted in the State language. Persons who do not speak Lithuanian shall be guaranteed the right to participate in investigation and court proceedings through an interpreter.

Chapter 10. Local Governments and Administration.

Administrative units provided by law on State territory shall be entitled to the right of self-government. This right shall be implemented through local government Councils. Members of local government Councils shall be elected for a two-year term on the basis of universal, equal and direct suffrage by secret ballot by the residents of their administrative unit who are citizens of Lithuania. Councils shall form executive bodies. The State shall support local governments. Local governments shall draft and approve then own budget.

Chapter 11. Finances, the State Budget.

In the Republic of Lithuania, the central bank shall be the Bank of Lithuania, which is owned by the State. It shall have the exclusive right to issue bank notes.

The budgetary system of Lithuania shall consist of the independent State Budget of the Republic of Lithuania and the independent local governments budgets. State Budget revenues shall be accrued from taxes, compulsory payments, dues, receipts from State property, and other income.

The Government shall prepare a draft budget of the State, and shall submit it to the Seimas for approval. During the budget year the Seimas may change the budget.

Chapter 12. Control of the State.

State control shall supervise the legality of the management and utilisation of State property and the realization of the State budget. The State Controller shall give an account to the Seimas on the annual execution of the State Budget.

100

Chapter 13. Foreign Policy and National Defence.

In conducting foreign policy, the Republic of Lithuania shall pursue the universally recognized principles and norms of international law, shall strive to safeguard national security and independence as well as the basic rights, freedoms and welfare of its citizens and shall take part in the creation of sound international order based on law and justice. War propaganda shall be prohibited.

Weapons of mass destruction and foreign military bases may not be stationed on the territory of Lithuania. The defence of the State from foreign armed attack shall be the right and duty of every citizen. Citizens are obliged to serve in the national defence service or to perform alternative service.

The Seimas shall impose martial law, shall announce mobilization or demobilization, and shall adopt decisions to use armed forces in defence of the homeland or for the fulfilment of the international obligations of Lithuania. States of emergency shall be regulated by law.

Chapter 14. Amending the Constitution.

In order to amend or append the Constitution of the Republic of Lithuania, a proposal must be submitted to the Seimas by either no less than one-fourth of the members of the Seimas, or by at least 300,000 voters. Amendments must be considered and voted upon in the Seimas twice. There must be a lapse of at least three months between each vote. Amendments must be approved, in each of the votes, at least by two-thirds of all members of the Seimas. The most important amendments must be approved even by Referendum, in which at least three-fourth of the electorate vote in favour thereof.

Table 2. The structure of state authority of the Republic of Lithuania[1].

	CONSTITUTIONAL	COURT	
	LEGISLATIVE Issue of laws	EXECUTIVE Implementation of laws	JUDIClARY Realization of justice
	SEIMAS	PRESIDENT	COURTS
State Authority		Government Prime Minister 13 Ministries, Institutions subordinated to the Government 10 Counties	Supreme Court Court of Appeal Country Courts District Courts Administrative Court
Local Authority		Elective Authority Executive Authority Controlling Institutions	

The Constitutional Act "On Non-Alignment of the Republic of Lithuania to Post-Soviet Eastern Alliances" of 8 June 1992 is considered as a part of the Constitution. It says: The Supreme Council of the Republic of Lithuania, seeing the attempts to preserve any semblance of the former Union of Soviet Socialist Republics with all of its conquests, as well as the intentions to include Lithuania into the defence, economic, financial, and other "domains" of the post-Soviet Eastern block, resolves:

To develop mutually advantageous relations with every state which was formerly a constituent part of the USSR, but to never and in no way join any new political, military, economic or any other state alliances or commonwealths formed on the basis of the former USSR.

PRACTICAL STEPS OF LITHUANIA IN TRANSITION (1990–...) TO FREE MARKET ECONOMY AND DEMOCRACY

Now **Lithuania** is a democratic Republic with its **President** elected by general voting for a period of five years and **Seimas** (Parliament) elected for a period of four years, and **Government** formed at the proposal of the Prime Minister nominated by the President.

Now the ultimate challenge of Lithuania is to become a member of NATO and European Union, as well as to maintain good relations with neighbouring countries, ensuring in such a way its independence, security and prosperity. This will require fundamental changes in all areas of life. The 50-year Soviet annexation is a vivid reminder that, in order to survive and secure its cultural and national identity, Lithuania needs to participate in a wider European Framework.

Foreign Affairs and Security Policy.
Lithuania applied for European Union (EU) membership on 8 December 1995. The Agreements signed with the European Community created conditions for the participation of Lithuania in the Pre-Accession Strategy for candidate countries. They also define the regulations according to which the movement of services, capital and persons is liberalized, as well as those on cooperation in law harmonization, finance, environmental protection, culture, and other areas. Lithuania is determined to complete the EU accession negotiations in 2002, and be ready to assume the obligations of EU membership as of 1 January 2004, together with the first group of new members.

The main conclusions of the "Regular Report 2000" of the European Commission specify Lithuania as a country with a functioning market economy able to cope with the competitive pressure and market forces within the EU in the medium term, provided that it continues the implementation of the current structural reform programme. The recommendations in the report point out the same direction: preserving fiscal discipline, continuing structural reforms (including privatisation), law harmonization and implementation.

Another strategic objective of Lithuania is integration into NATO. Lithuania applied for membership in January 1994. The country's aspirations to join NATO are based on solid foundations: a stable democracy, the rule of law, economic growth, good relations with neighbouring states and a solid commitment to the development of its defence structures. All the main political parties support the country's integration into NATO. Lithuania is already participating in the Partnership for the Peace Programme. Defence expenditure in the state budget for 2001 amounts to 1.95% of GDP. The government will seek to increase defence spending to 2% of GDP in 2002. Internal stability and respect for the rights of ethnic minorities enable Lithuania to maintain good relations with all neighbouring countries, Poland, Belarus, Russia with the Kaliningrad region included.

Lithuania is a member of the United Nations since 1991 and participates in its various activities. Lithuania is also a member of all major universal and regional conventions for the promotion and protection of human rights, and implements them in its law and practice.

Mathematics of the Lithuanian Parliament

Even a brief analysis of the Constitution of the Republic of Lithuania leads to the conclusion that all necessary prerequisites to built a prosperous democratic country are created there. However, the practical implementation of its ideas is a non-trivial challenge and requires much time and, sometimes, painful unpopular measures and decisions. The latter, in their turn, cause negative consequences for politicians and political parties in power, changing the mathematical distribution of political forces. Let us illustrate this statement by considering the evolution of the Parliament of the independent Lithuania.

Table 3 illustrates the political composition of the Lithuanian Parliament (Seimas) and its evolution during three terms (1992-1996, 1996-2000 and 2000-2004) as well as at the beginning and the end of each term. We will not describe the programmes of each party, because it is not important for our consideration, we will restrict ourselves by saying that the main ideas of their programmes resemble those of such parties in the Western countries. Also during certain periods of time some parties have been making coalitions with the smaller parties or a group of independent members of the Seimas. We will be most interested in the mathematical changes of the strength of the parties in the Seimas and, as a consequence, in the changes of the Governments formed practically by the Seimas.

It is interesting to mention that during the 2000 elections, two candidates collected exactly the same number of votes and, according to the rules, the eldest was declared elected to Seimas. By the way, the present President of the Republic of Lithuania was elected by a fraction of percent majority.

The Government of the 1992-1996 term was formed by one (Lithuanian Democratic Labour) party of the left wing. Two parties of the right wing, namely, the Conservative party and Christian-Democrats formally established the Government of the 1996-2000 term. Practically the first party has dominated. The Government of the 2000-2004 term was for the first time formed by a real coalition of two parties - Liberal and Social-Liberals - with a few members of small parties. However, the coalition was not stable and in eight months it was replaced by the coalition of the Lithuanian Democratic

Labour party, Social-Democratic party and Social-Liberals, shifting the power from the centre to the left. Thus, we see the oscillations of power—left, right, left again.

The positive aspect of this process illustrating the democratic nature of the State is a smooth and peaceful character of transferring the political power from position to opposition.

However, many parties are not stable enough. As it is seen from Table 3, even during a term, remarkable changes occur in a number of members of the Seimas, belonging to one or another party. It is not surprising to leave one party and to join the other. Some parties split into two parties (e.g., occurrence of the Moderate Conservative party, etc.).

Table 3.   Mathematics and democracy of the Lithuanian Parliament (Seimas). Distribution of its members with respect to parties.

Party or coalition	1992 -	1996	1996 -	2000	2000-2004
	Beginning	End	Beginning	End	Beginning
Lithuanian Democratic Labour party	73	66	12	13	51
Social-Democratic party	8	7	12	7	
Conservative party	30	23	70	53	9
Moderate Conservative party	–	–	–	14	1
Christian-democrats	18	15	16	12	2
Social-Liberals	–	–	–	–	29
Liberals	–	–	1	1	34
Central Union	2	2	13	17	3
Polish Union	4	3	1	2	2
Nationalistic party	4	2	2	1	1
Peasant's party	–	1	1	2	4
Others (independent)	2	18	13	15	5
Total number	141	137	141	137	141

The still turbulent character of political life in Lithuania is illustrated by the occurrence and jump to power of two political parties: Social-Liberals and Liberals. The Liberal party was founded long ago, however it was small and unpopular, whereas the Social-Liberal party was actually founded just before elections. Worsening economics, unpopular decisions of the Conservative party have paved the way to power for the above-mentioned two parties. However, their coalition was not stable enough.

Thus, the mathematics of the Lithuanian Seimas, the evolution of the parliamentary system in Lithuania indicate the growing role of democracy. However, there are many aspects to be improved., e.g., the women in the Seimas comprise only 10.6%. The ideological basis of each party must play a more important role. And, of course, the politicians must serve to the people, must devote themselves to the studies of the needs of the electors and to the improvement of economic, social and spiritual conditions for the population. Up to now it is rather far from the reality.

Education and Science.

For many centuries the Vilnius University, founded in 1579, was the intellectual centre of Lithuania. In 1773 an educational institution for the Polish-Lithuanian state—the Education Commission—was set up, actually the first Ministry of Education in Europe. Between the wars (1918-1940) the independent Lithuania had to build a new educational system. After Soviet occupation the educational system was reshaped according to the Soviet norms.

After restoration of independence of Lithuania, education became one of the top priorities. In 1991 a Law on Science and Education was passed and in 1992 the Government adopted the **Education Concept of Lithuania**, based on the values and experience of European culture and education. Training and qualification programmes for teachers were upgraded and expanded. The transition to the basic ten-year secondary education and to the school structure 4+(4+2)+2 was implemented. Pre-school education (up to the age of six) is available. General secondary education is obtained at three levels: primary (forms 1-4), lower secondary (forms 5-10) and secondary (forms 11-12).

In September 2000, there were 2,311 general education schools. Of these, 2,031 provide an education in Lithuanian, 74 in Polish, 68 in Russian, and 72 are mixed. There are also schools in Yiddish, Belorussian and German. There are also vocational schools providing general knowledge and a primary professional education.

Higher education institutions in Lithuania now are of two types: universities and colleges. Up to now there are 19 state (ten universities, five academies and four colleges) and seven non-state (four university-type institutions and three colleges) higher education institutions. The duration of university undergraduate studies (Bachelor's programme) is four years, the duration of non-university undergraduate studies is not less than three years. The duration of Master's studies is not more than two years. After completion of Master's studies, Doctoral studies last not more than three years. Thus, the education system has come closer to the Western standards. In 2000, there were 258 students per 10,000 inhabitants. According to the Constitution, higher education is free for "good" students. There is a system of adult education, as well.

Studies must go together with research. Scientific research is currently carried out by approximately 10,000 scientists. About 6,800 of them work at universities and other institutions of higher education. The others work at 29 national research institutes. Distribution of academics across fields is as follows: humanities 18.8%, social sciences

14.3%, technological sciences 21.5%, physical sciences 18.6%, agricultural sciences 6.7%, natural sciences 10%, medicine 10.1%.

The country's scientific potential has formed over a number of decades and depends very much on the competence and management skills of leading academics. Cooperation with foreign research institutions has increased dramatically. Many scientific institutions and individual scientists in Lithuania belong to international organisations and participate in leading international programmes. The number of research institutions and groups that participate in the European Union and NATO programmes is steadily growing.

Health Care and Social Security.
The state programme for 2000-2004 in the health care sector is patient-oriented, and seeks to provide timely, safe and efficient health care, enabling the cost-effective use of resources, and continuity of health reform. Special attention is paid to the health of mother and children. It helped to lower the infant mortality rate from 16.4 per 1,000 live births in 1992, to 9.2 in 1998. Average life expectancy in 2000 is 67 for males and 77 for females.

In 1999 4.6% of GDP was spent on health care. The high level of medicine may be illustrated by the fact that heart and kidney transplants are routinely conducted in Lithuania.

Since 1990 the social security system in Lithuania has consisted of two main parts: Social Insurance and Social Assistance. Social Insurance comprises about 80% of all social security expenditure. It is compulsory for all workers and consists of the employer's share (31% of salary) and a part paid by the employee himself (3% of salary). Pension Insurance comprises the greatest part of Social Insurance. There is also Sickness and Maternity Insurance as well as Unemployment Insurance. At the beginning of 2001 the official unemployment rate was close to 13%.

CONCLUSION: DOES MATHEMATICS INDICATE DEMOCRACY IN LITHUANIA?

In spite of a number of huge differences between India and Lithuania, nevertheless there are some similarities, particularly in political life, in mathematics of democracy[4]. Indeed, numerous contestants per seat in Parliament, small number of women elected, "first past the post" principle, most candidates win by minority not majority of votes, i.e., minority prevails over majority, coalition games inevitable - all these features are common for the both countries.

The main difference is that India had, during the several last decades, a stable economic and political system, whereas in Lithuania there was "revolution" in 1990 and during the last decade it had to qualitatively change its political and economical structures. Mainly due to this reason our paper differs from that of Mr. Sivaramakrishnan's[4], that is why we could not restrict ourselves only to the description of the mathematics of the elections of the Parliament.

Unlike in India, a country with stable economic system[4], there were drastic changes in all domains of the Lithuanian life after the restoration of independence (politics, economy, education, science, culture, health care and social security, etc.), occurrence of almost 40 political parties included. However, transition from state planned to free market economy, reorientation of economic relations from CIS to the West has led to an essential drop in production, jump in prices, which, in its turn, led to the decrease of the standards of living of the population, to differentiation of the society.

The middle class in Lithuania is rather weak. The gap between rich and poor people, as well as corruption, crime, alcoholism, prostitution, drug habit, emigration, "brain drain" increased dramatically. The hopes of a large part of the population of Lithuania that independence and democracy would bring about prosperity almost at once did not come true. Such a disappointment has reflected in the elections of the Seimas. Usually a majority of citizens voted not "in favour" but "against" parties in power (see Table 3). That is why after each elections parties in the opposition used to come to power (pendulum effect). During the last elections to the Seimas, consisting of 141 member, about 70% of members were elected for the first time. This also shows certain political instability.

Nevertheless, the signs of growing, improving economy, prospects of becoming a EU and NATO member indicate that there are good chances for Lithuania to become a full member of democratic Europe. There is no doubt that the future data on mathematics of democracy in Lithuania will confirm this conclusion.

REFERENCES

1. Lithuania. An Outline (Second Edition), "Akreta" Publishers, Vilnius, 2001.
2. Constitution of the Republic of Lithuania, Publishing House of the Seimas, Vilnius, 1993.
3. Lithuania After Entering the Third Millennium, "Algimantas" Publishers, Vilnius, 2001.
4. K.C. Sivaramakrishnan, Mathematics of Indian Democracy. In: International Seminar on Nuclear War and Planetary Emergencies (25th Session). Series editor and chairman A. Zichichi, edited by R. Ragaini, World Scientific, Singapore, 2001 (p. 373).

ANNEX

Mathematics of economic and social development
After the restoration of independence, Lithuania started massive transformation of its largely state run economy along free market principles. The initial period entailed many painful choices as price liberalisation triggered inflation exceeding 1,000%, and budget discipline resulted in the shedding of tens of thousands of redundant industrial jobs. In 1991-1992, the government launched a national wide privatization drive, which resulted in rapid growth of the small and medium-size enterprise (SME) sector and private agriculture.

Foreign trade began a painful reorientation from almost total dependence on the CIS market to a more balanced relationship between East and West. Today the EU is already Lithuania's largest trading partner. As Table 4 illustrates, the recovery gained momentum from 1996 to the summer of 1998, recording impressive annual growth rates of 7.3% in 1997 and 5.1% in 1998. The Russian financial crisis hurt firms which exported to the CIS, but the Litas remained stable and inflation continued declining. GDP growth, after a decrease by 4.1% in 1999, continued to grow reaching 3.3% in 2000.

Table 4. Yearly inflation rate and GDP growth in Lithuania.

Year	1993	1994	1995	1996	1997	1998	1999	2000
Inflation rate, %	189	45	36	13	8.4	2.4	0.3	1.4
GDP, %			3.3	4.7	7.3	5.1	-4.1	3.3

Estimated GDP growth will be al least 5% in 2001.

Lithuanian's state-owned banking sector is reformed, too. Private commercial banking boomed from 1992 to 1994, but in late 1995 a crisis was triggered. For more rigorous control international auditing standards became mandatory, and a system of deposit insurance was implemented. Trust in the banking system resulted in a big influx of foreign capital.

A National Stock Exchange was opened in 1993. This was the first stock exchange in the Baltic states open in the post-Soviet period.

Monetary stability has been achieved. The Litas-USD exchange rate has remained unchanged since its reintroduction in 1993, making it one of the most stable currencies in Central and East Europe. However, Lithuania is the last country in Europe to have its national currency pegged to the USD. Next year the Litas will be reoriented to Euro.

During Soviet times the Baltic states supplied an important share of the food products to Russia. Today many producers of food products have succeeded in upgrading production thanks to major foreign investment and may export the goods produced to the EU countries.

Another major industrial sector is Mazeikiai oil refinery, which accounts for 29% of the output of all Lithuanian industries. The refinery has been receiving all off its crude oil from Russia, but now it is able to import oil via an oil terminal on the Baltic coast.

Other significant sectors of the Lithuanian industry are textiles and clothing (11% market share), chemicals (8%), electronics (8.3%). About 52% of industrial production is exported.

51% of the land area is cultivated. 21.5% of the country's workers are employed in agriculture, hunting and forestry. Until 1990, the land mainly belonged to the state-run collective farms. Agricultural reform was aimed at the creation of a strong competitive agricultural sector, based on private farming. The land is going to be returned to the former owners or their successors. However now, the average size of a farm is only 12 hectares.

Forests covering 30.3% of the country are one of the largest natural resources. Approximately 60% of the forests are coniferous, mainly pine and fir.

Oil products constituted, in 2000, 34% of energy, natural gas 27%, nuclear power 29%, and other resources 10%. Lithuania's main source of electric power is the Ignalina Nuclear Power Plant, with its 3,000 MW capacity. In 2000, it produced 80% of the country's total electricity supply. With the help of foreign (mainly Sweden) assistance its safety standards were essentially upgraded. However, there is strong pressure from the EU to close the Plant.

Lithuania's privatization programme has been a crucial part of the transformation to the free market economy. The first stage of privatization for vouchers took place between 1991 and 1995. During that period 30% of the total state property, 94% of housing and most agricultural assets were privatized.

The second stage of privatization started in 1996 and involved sales of assets for cash to both local and foreign investors.

The economy is now based on private enterprise and ownership and a free-market system. Between 1991 and 1997 the contribution of the private sector to GDP reached 75%. It also employs about 70% of the workforce.

Lithuania's exports rose by 15% from 1996 to 2000 and import by 20%. From 1997 to 2000 exports to the EU rose by 46%. There are 6,583 registered foreign companies or joint ventures linking the Lithuanian economic entities with foreign capital.

Foreign investments have been growing permanently (Table 5).

Table 5. Foreign investment growth in Lithuania.

Year	1992	1993	1994	1995	1996	1997	1998	1999	2000
Mil.USD	19	149	310	352	700	1,041	1,975	2,413	2,650

Lithuania's geographical situation is highly favourable for the development of the transport sector. The Lithuanian sea transport system has been integrating itself into the international shipping and seaport services market. Railways are one of the most important means of transport in Lithuania for long-distance freight transport. They have good connections with the rail networks of the Baltic states and CIS countries. There is also a long tradition of civil aviation in Lithuania. Lithuania maintains a well-developed road network. The Via Baltica, linking Helsinki with Warsaw via the Baltic states is being constructed. The gas and oil pipeline network consists of a gas main with branches, and one oil pipeline. Gas and oil are imported mainly from Russia. Communications networks are being upgraded as well.

Table 6 summarizes the mathematics of economic and social development of Lithuania during 1994-2000. It demonstrates the growth of GDP, export and import as well as monthly salary. However, mainly due to economic crisis in Russia in 1998 and 1999, this growth was slowed down or even turned into decrease. The dynamics of minimal monthly salary, average monthly incomes and consumption per person illustrate clearly this effect.

Table 6.   Recent LITHUANIA   in Figures. Mathematics of economic and social development.

Year	1994	1995	1996	1997	1998	1999	2000
GRDp (10^9 USD)	4.2	6.0	7.9	9.6	10.7	10.7	11.2
Avg. monthly salary (USD)	81	120	155	195	233	247	252
Min. monthly salary (USD)	14	32	64	91	108	108	108
Avg. monthly incomes per person	-	-	82	92	106	107	104
Avg. consumpt. exp. per person	-	-	87	96	107	106	101
Export (10^9 USD)	2.0	2.7	3.4	3.9	3.7	3.0	3.8
Import (10^9 USD)	2.4	3.7	4.6	5.7	5.8	4.8	5.5
Nat. popul. growth (thousand)	-4.1	-4.1	-3.8	-3.3	-3.7	-3.6	-4.6
State jobs (thousand)	653 39%	608	547	544	528	524	495 31%
Private jobs (thousand)	1021 61%	1035	1074	1127	1129	1124	1090 69%

National population growth is negative, emigration is high. Preliminary evaluation shows that about 10% of the citizens have left Lithuania (let us hope that many of them - temporarily). This complicates essentially the demographic and social situation in Lithuania.

However, the further privatization of natural resources and state property, the increase of the efficiency of production, the growing role of private initiative, the permanent decrease of state jobs and increase of private jobs (see the last lines of Table 6) lead to the optimistic conclusion that a further strengthening of the role of democracy together with the consolidation of the free market economy will be the a solid foundation for the future prosperity of a democratic Lithuania.

5. TRANSMISSIBLE SPONGIFORM ENCEPHALOPATHY

BSE IN EUROPE

DAGMAR HEIM
Swiss Federal Veterinary Office, Schwarzenburgstrasse 161, 3097 Liebefeld, Switzerland, email: lukas.perler@bvet.admin.ch, Tél.: 031 322 01 56, Fax: 031 323 85 94

INTRODUCTION

The Transmissible Spongiform Encephalopathies (TSEs) are a group of diseases encompassing a wide variety of disorders in humans and animals. Of this group, the animal disease with the longest history is scrapie, which was first reported in the mid-18th century. For some decades, diseases in this group have also been reported in humans; the best known are classical Creutzfeld-Jakob Disease (CJD) and Kuru. Case reports of a variant of Creutzfeld-Jakob disease (vCJD) in humans, which were reported in UK in 1996, showed that a correlation between BSE and vCJD has to be assumed.

COURSE OF BOVINE SPONGIFORM ENCEPHALOPATHY

United Kingdom
The first cases of BSE were reported in the UK in 1986. Extensive epidemiological studies traced the cause of BSE back to animal feed containing inadequately treated ruminant meat and bone meal (MBM). The most important action to prevent new BSE infections in cattle was the ban on ruminant MBM for use in ruminants that came into force in the UK in July 1988. The use of mammalian MBM in ruminant feed was banned in November 1994. This eliminated the problem of distinguishing between feed of bovine origin and feed of mammalian origin, which in turn made it easier to monitor the feeding ban. Gradually, the measures were made increasingly stringent until in March 1996 the feeding of mammalian MBM to all farm animals was banned. The most important measure to protect consumers was the ban on the use of specified risk material (SRM) for human consumption (1989), namely brain, spinal cord, tonsils, thymus, spleen and intestines from cattle older than six months.

One of the problems of controlling BSE is that the effect of the measures taken cannot be evaluated until about five years have elapsed, which is the average incubation period of BSE. So the effect of the ban on feeding MBM to ruminants did not become clear until 1993. After a peak of 36,000 cases had been reached in 1992, the annual incidence fell. Following more than 180,000 cases in the UK until this year, it clearly can be stated that the measures taken, in particular the ban on feeding MBM to all farm animals, implemented in 1996, have been effective.

Bovine spongiform encephalopathy cases outside the United Kingdom
In 1989, the first cases outside the UK occurred in cattle imported from the UK. Not until the end of 1989 were the first indigenous cases reported in Ireland and the European Continent (France, Portugal and Switzerland). In the mid-90s, other countries reported cases of BSE (the Netherlands, Luxembourg, Belgium and Liechtenstein). In 2000 and 2001 cases of BSE were first diagnosed in six more countries: Denmark, Spain, Germany, Italy, the Czech Republic and Greece .

RISK ASSESSMENT FOR BSE

Risk of introduction of the BSE agent into a country
Two risk factors of introduction of the BSE agent in a country have to be considered: *The import of live cattle and the import of MBM.* As cattle born in the mid-1970s were affected in the UK, imports dating back to this period should ideally be included in the risk assessment investigation. Also, since 1990 other countries in addition to the UK have been affected by BSE, therefore imports from these countries also must be considered, unless adequate safeguards had been implemented. The risk of introduction of the BSE agent into a country should be evaluated with reference not only to a country's own import statistics but also to the export statistics of the UK and other at risk countries. This procedure allows an initial analysis to be performed to determine whether any potentially infectious material might have entered a country at any time.

In field cases of BSE in cattle tested, infectivity has not been recorded outside the CNS (brain, spinal cord and eye) in mouse tests. In experimental orally-induced BSE infectivity has been found in the distal ileum at intervals during the incubation period starting six months after exposure. Furthermore, central nervous tissues and dorsal root and trigeminal ganglia were found to be infective shortly before the onset of clinical signs.

Risk of propagating the BSE agent in a country
Any propagation of the BSE agent in a country must be prevented. The systems in place in the country must prohibit the introduction of the BSE agent into the feed chain and reduce the spread of the pathogen. A central issue is what happens with *specified risk material* (SRM) after slaughter. Some material, such as brain and spinal cord, may contain particularly high concentrations of the BSE agent. If these materials are removed at slaughter and then incinerated, the risk of recycling the pathogen is markedly reduced. If these materials are used for further processing to animal feed, there is a high risk of amplification of the BSE agent. If an SRM ban, including cadavers, is put in place at an early stage, this increases the stability of the system. The agent is extremely resistant to most physical and chemical inactivation methods. It is scientifically proved that even

treatment of infected material by 133°C at 3 bars for 20 minutes does not completely inactivate the agent if the initial infective load was high. Recent experiments have shown that residual infectivity can be present also when very high temperatures were used. Nevertheless, if the raw material is processed to MBM in a batch process at 133°C with 3 bars of pressure for 20 minutes, this decreases the risk.

The *feeding of MBM* to ruminants has to be investigated. In many countries, animals have traditionally never been fed MBM. But assumptions on this subject have to be looked at with circumspection. In the meantime, BSE cases have been diagnosed in many countries where it was not a customary practice to feed cattle with MBM. It has to be borne in mind that, even when no MBM has been fed to ruminants, a high risk still remains because of *cross-contamination and cross-feeding*. If MBM was allowed in feed rations for pigs and poultry, and these were manufactured in the same mills, and transported by the same vehicles, and if inappropriate feeding practices cannot be ruled out on farms, the risk remains high. It is lower than countries that have not prohibited feeding MBM to ruminants but it is still significant.

BSE-SURVEILLANCE

An important clue to the real BSE situation can be provided by the *surveillance system*. In most countries, BSE is listed as a notifiable disease, which is a basic requirement for a functioning surveillance system. Until a few years ago, BSE monitoring was confined to the notification of clinically suspected cases. It was assumed that this would allow early detecting of an outbreak and would be more effective than random sampling of all slaughter animals. However, a system of this kind, is dependent on many factors such as disease awareness, compensation practices and motivation to notify. In recent times, it has become increasingly obvious that a passive surveillance system alone, based on the notification of suspected cases, is not sufficient to provide evidence of freedom from BSE.

The availability of rapid BSE tests allowed the fast and uncomplicated testing of brain tissue on a large scale for BSE and to identify infected animals in the last stage of the incubation period. This made it possible to implement active surveillance programmes in populations at risk. Switzerland was the first country in the world that introduced active surveillance of BSE in 1999. Unfortunately these tests are not sensitive enough to identify animals that are infected but that do not yet have a high concentration of the BSE agent in the brain. Nevertheless, through the use of these tests, it is possible to get closer to the true incidence of BSE and they have proven to be a useful screening technique.

CURRENT SITUATION

In the past few years, there have been repeated allegations that some countries did not report BSE cases because the surveillance was not efficiently conducted. The absence of reported cases has to date always been automatically equated with freedom from BSE.

This can be a problem: a country has BSE, but it is not detected or reported, and export activities continue on the assumption that the country is free from BSE – then other countries become infected.

In many countries in Europe, an active surveillance system has now been implemented; tests for BSE have been carried out in populations at risk since 1999/2000 in some countries, and 2001 in others; the animals tested include cows that have died, or were emergency culled or emergency slaughtered. First results indicate clearly that active surveillance is a more objective approach and can help to assess the real BSE situation in a country. Consequently, some countries in Europe, that for years have been considered BSE free, are now shown to have BSE.

Although it is known that MBM and cattle have been exported all over the world, unfortunately, it is not yet possible to detect infection in these imported cattle or MBM. It will be some years before a country can determine whether or not the problem has been imported. Imported MBM is eaten by domestic animals and the SRM will be recycled and find its way into animal feed. This feed will in turn be eaten by domestic cattle – and the animals will not fall sick until an average of 5 years later, which could lead to a steady amplification of the BSE agent. If prophylactic measures are not taken in good time, and the surveillance is not efficient, it may take decades before an epidemic is seen.

REFERENCES

1. ANONYMOUS (2000). Final Opinion of the Scientific Steering Committee on the Geographical Risk of Bovine Spongiform Encephalopathy (GBR) (http://europa.eu.int/comm/food/fs/sc/ssc/out113_en.pdf).

2. ANONYMOUS (2000). Opinion of the Scientific Steering Committee on the Human Exposure Risk (HER) via Food with Respect to BSE (http://europa.eu.int/comm/food/fs/sc/ssc/out67_en.pdf).

3. Moynagh, J. and Schimmel, H. (1999). Tests for BSE evaluated. *Nature,* **400**, 105.

4. Prusiner, S.B. (1991). Molecular biology of prion diseases. *Science*, **252**, 1515-1522.

5. Taylor, D.M. (1993). Inactivation of SE agents. *B. Med. Bull.*, **49**, 810-821.

6. Wells, G.A.H., Hawkins, S.A.C., Green, R.B., Austin, A.R., Dexter, I., Spencer, Y.I., Chaplin, M.J., Stack, M.J., and Dawson, M. (1998). Preliminary observations on the pathogenesis of experimental bovine spongiform encaphalopathy (BSE): an update. *Vet. Rec.,* **142**, 103-106.

7. Wells, G.A.H., Scott, A.C., Johnson, C.T., Gunning, R.F., Hancock, R.D., Jeffrey, M., Dawson, M., and Bradley, R. (1987). A novel progressive spongiform encephalopathy in cattle. *Vet. Rec.*, **121**, 419-420.

8. Will, R.G., Ironside, J.W., Zeidler, M., Cousens, S.N., Estibeiro, K., Alperovitch, A., Poser, S., Pocchiari, M., Hofman, A., and Smith, P.G. (1996). A new variant of Creutzfeldt-Jakob disease in the UK. *Lancet*, **347**, 921-925.

TSE'S AND THE DENTAL TEAM

DR. ANDREW SMITH
Lecturer/Hon Sp Registrar in Microbiology, Infection Research Group, Glasgow
Dental Hospital & School, 378 Sauchiehall Street, Glasgow, Scotland, United
Kingdom.

INTRODUCTION

On the 20th March 1996, Stephen Dorrel (Secretary of State for Health, UK) announced to Parliament that the CJD surveillance unit had identified patients with a previously unrecognised and consistent disease pattern and that the most likely explanation was linked to exposure to BSE. These patients were younger than those with more classic forms of the disease, with prominent early psychiatric and behavioural manifestations and persistent paraesthesias and dysaesthesias. Pathological examination showed prominent and diffuse PrP^{sc} plaques similar to those found in other forms of Transmissible Spongiform Encephalopathies (TSE's). This presentation highlights some recent developments particularly in relation to infection control and the risks of iatrogenic transmission via dental treatment.

TRANSMISSIBLE SPONGIFORM ENCEPHALOPATHIES

Several TSE's have been described in animals and humans. All have incubation periods of months to years, and all gradually increase in severity and lead to death over a period of months or years. None evoke an immune response and all share a common noninflammatory pathologic process in the central nervous system. In all affected species, infectivity is greatest in brain tissue, is present in some peripheral tissues, but generally has been absent from all body fluids, such as saliva and urine[2]. The infectious agent associated with transmission events are variants of a host membrane sialoglycoprotein called prion protein (PrP). These transmissible agents appear to have a common mechanism of pathogenesis and possibly a common origin. Some have spread across species barriers (Transmissible mink encephalopathy and vCJD), some have reached epidemic proportions by entering the food chain (Transmissible mink encephalopathy, BSE and kuru), others have been transmitted by inheritance of mutations in the PrP gene (familial CJD, Gerstman-Straussler-Scheinker disease and Fatal Familial Insomnia) and others have been transmitted iatrogenically by implantation or injection of contaminated material.

NATURE OF THE INFECTIOUS AGENT

This will be dealt with more comprehensively by other speakers. However, some salient features of the infectious agent and perhaps the most controversial aspect of the agent that causes TSE's is the absence of conventional genetic material, such as DNA or RNA. An infectious protein hypothesis was first proposed by Griffith in 1967[15].

This was taken further by Stanley Prusiner who in 1982 coined the term "prion"[21]. Prions normally exist as protease-sensitive, cell surface proteins in many cells, such as B-lymphocytes and neurons, this form of the protein is designated prion protein cellular (PrP^c). PrP^c is continuously recycled through endocytosis, but its function is currently unknown. In humans a gene located on chromosome 20 called the prion protein gene codes for both the normal and abnormal form of the protein[12]. The influence of the host genetic background became apparent when scientists demonstrated that deletion of the prion protein gene protects mice from experimental scrapie on exposure to prions. Interestingly, deletion of the PrP^c gene did not affect development, viability or life expectancy in mice[9]. The normal prion protein is soluble and protease sensitive, the abnormal prion protein (PrP^{Sc}) is insoluble and resistant to proteinase. PrP^c is rich in a $\propto$-helical structures whereas PrP^{Sc} seems to be composed mainly of a (β-pleated sheet structure. PrP^{Sc} is postulated to act as a conformational template that promotes the conversion of PrP^c to further PrP^{Sc}. This triggers a chain reaction with the accumulation of further insoluble PrP^{Sc} in neural cells, disrupting function and leading to vacuolization and cell death.

Workers have identified at least 4 different types of prion protein with different clinical pictures and neuropathologies[11]. Prion replication with recruitment of PrP^c into the aggregated PrP^{Sc} isoform, may be initiated by a pathogenic mutation (resulting in a PrP^c predisposed to form PrP^{Sc}) in inherited prion diseases, by exposure to a seed of PrP^{Sc} in acquired disease, or as a result of the spontaneous conversion of PrP^c to PrP^{Sc} (and subsequent formation of aggregated material) as a rare event in sporadic prion disease. There is a body of opinion that the "protein-only" hypothesis does not adequately explain this variety of different strains and that another, as yet unknown factor, may be involved in the disease process.

KURU

Kuru was the first human TSE shown to be transmissible, this fatal disease of cerebella degeneration reached endemic proportions among the Fore ethnic group in a remote area of Papua New Guinea. The disease was spread by ritual cannibalism, with more cases in children and women who consumed more of the central nervous tissue, fewer cases were reported in males who consumed mostly muscle tissues. The disease has gradually disappeared over the last 40 years since the practice ceased.

CJD disease occurs naturally as both a sporadic and a familial disease. Its epidemiological and clinical patterns are different from kuru, but it produces similar spongiform changes in the nervous system.

SPORADIC CJD

This type of human TSE occurs world-wide with an incidence of approximately 1 case per million population per year. There is no seasonal variation, no evidence *of* changing incidence over the years, and no convincing evidence for geographic clustering, except for areas with large numbers of familial cases.

Many series and ease control studies have searched for risk factors, such as diet, exposure to animals, surgical treatment and occupational exposures. The incidence in Australia and New Zealand which are free of scrapie, is similar to that in the UK, where scrapie is endemic. Consumption of brains and offal nor lifetime vegetarianism alters the risk. Surgeons, pathologists, butchers, abattoir workers and cooks exposed to blood and uncooked animal products do not have increased risks. However, more recently, some workers[13] have reported that surgical procedures were significantly associated with the development of sCJD. This group also found a significant association between risk of sCJD and residence or employment on a farm or market garden for longer than 10 years. These workers found no significant risk associated with a history of major dental treatment. However another study[25] did not confirm the association of sCJD with a history of surgery or occupational exposure to animals or leather. A retrospective case controlled study[19] of 60 definite cases of sporadic CJD, which occurred in Japan between 1975 and 1977 found no association with extractions of maxillary or mandibular teeth. Epidemiological data to date then suggest that transmission of sCJD is a rare event. This documented evidence of lack of communicability via routine health care procedures should reassure the dental team who care for patients with these diseases.

IATROGENIC CJD

The human-human transmission of a TSE was demonstrated among the Fore people with kuru (incubation period 4-40 years). However, whether the original source was an index case of sCJD or some other event will probably never be known. Other cases of human-human transmission can be divided into 2 groups[7];

   a. Surgical transmission
       A number of procedures have inadvertently been associated with transmission of CJD.
          The first suspected human transmission was reported in 1974 when rapidly progressive neurological disease developed in a woman 18 months after a corneal transplant[4]. Contaminated neurosurgical instruments have also been suspected as modes of transmission in several cases, the most convincing reported after 2 patients underwent neurosurgery, using electrodes that had previously been implanted into a patient with CJD. The electrodes had been "decontaminated" with 70% alcohol and formaldehyde vapour, yet 2 years

later these electrodes were retrieved and implanted into a primate which subsequently developed CJD[28]. Over the past decade, more than 80 cases of CJD have been recognised 1.6-17 years after neurosurgical placement of grafts of human cadaver dura mater.

b. Transmission by pituitary hormones
In 1985, CJD developed in 4 patients who had received human growth hormone (all aged under 40 years). Injection of the hormone which was derived from pooled cadaver human pituitary glands, had been discontinued 4-15 years before the onset of disease. Recombinant growth hormone was licensed in 1985 and is now used. In the UK approximately 1% of recipients have been affected, with a mean incubation period of 12 years.

FAMILIAL CJD

Between 10-15% of individuals with CJD have a family history consistent with an autosomal dominant inheritance of the disease. More than 20 different mutations in the PrP gene have been described. On average, fCJD has an earlier age of onset and a more prolonged course than sCJD. The typical EEC changes are often missing, the prevalence of amyloid plaques varies but the essential changes in vacoulisation of neuronal cells and neuronal loss are generally present.

Several mutations lead to phenotypes that have been regarded as different diseases. Gerstman-Straussler-Seheinker disease (GSS) disease is an autosomal dominant illness that may have a prolonged course of 5-11 years, yet the mean age of death is only 48 years. The neuropathological findings are distinct, with many PrP^{sc} positive plaques throughout the brain. It must be emphasised that fCJD is extremely rare and only close blood line relatives are "at-risk" from the disease. sCJD is the most common type of CJD for which there is no risk of disease (either vertically or horizontally) to other family members.

VARIANT CJD

In July 2000, the Department of Health announced that the incidence of variant Creutzfeldt Jakob disease (vCJD) in the UK had reached statistical significance. Since the first case died in 1995, there have been 102 "definite" and "probable" vCJD cases (Figures correct to 29 June 2001) The incidence is rising by an estimated 20-30% per annum. However, the Spongiform Encephalopathy Advisory Committee (SEAC) has concluded that it is still too early to know whether this rising trend is likely to be sustained or to forecast the ultimate size of the epidemic.

Laboratory experiments provide compelling evidence that the causative agent of vCJD and BSE have a common origin[8,16,22]. Glycosylation patterns of PrP^{sc}, susceptibility studies in mice and patterns of disease in brain tissue from vCJD patients and BSE are similar but distinct from the patterns associated with scrapie, sCJD and fCJD. There is also a specific genetic predisposition to the disease (see Table 1). All patients with vCJD analysed to date have been homozygous for methionine at codon 129. Although this does not provide information on the route of exposure, it is widely believed vCJD entered the human population through consumption of bovine products from animals with BSE.

Table 1. Genetic Predispostion to human TSE's

Subject	Percentage of population with amino acid variation at codon 129		
	Met/Met	Met/Val	Val/Var
Healthy	37%	52%	11%
sporadic CJD (sCJD)	78	12	10
iatrogenic CJD (iCJD)	60	11	29
variant CJD (vCJD)	100	0	0

SCALE OF THE EPIDEMIC

Factors that are important in determining the probability of acquiring vCJD include incubation period, infective dose, route of exposure and genetic susceptibility. Data from the various types of human TSE's suggest that incubation periods of prion diseases in humans after peripheral or oral exposure, range from at least 4 years to 40 years, with a mean of about 10-15 years. Attempts by statistical modelling to predict the eventual scale of any vCJD epidemic only serve to emphasise the uncertainties, with estimates ranging from a few hundred to many hundreds of thousands.

INFECTIVITY AND INFECTIOUS DOSE

The pathogenesis of prion diseases varies with different host species and prion strain combinations. Evidence from BSE in cattle has demonstrated that infectivity is mainly concentrated in the central nervous system but is also present in lymphoreticular tissues including those in the gut. Experiments have shown that CNS tissues in infected cattle become infective in the few months before there is evidence of pathological changes in the CNS which themselves occur close to the onset of symptoms. Infectivity has also been detected in the distal ileum of orally challenged experimental cattle. However, it has never been detected in the lymphoreticular tissues of cattle that have acquired the disease naturally. In humans, however, post mortem studies have shown that abnormal prion is accumulated in CNS and lymporeticular tissues, including the tonsils and appendix in vCJD cases. It is likely then that infected human cases who are developing vCJD will have potentially infective tissues in the pre-clinical phase of their illness. This has highlighted the theoretical risk of iatrogenic transmission of the infective agent of vCJD. Currently it is assumed that CNS, followed by eye followed by lymphoreticular tissue are, in descending order, the tissues most likely to be infective[14].

In iatrogenic CJD cases where the infectious agent is introduced into or near the brain the incubation period is typically measured in months. Peripheral inoculation (IV,

IM or oral) produces incubation of years or decades[7,20]. In scrapie, it has been shown that inoculation through non-neural, peripheral routes involves an initial phase of replication in the lymphoreticular system LRS). This is followed by spread from the LRS to the thoracic spinal cord via visceral autonomic nerves and then to the brain[18]. The concern over iatrogenic transmission of vCJD arises from the fact that tissues from asymptomatic individuals incubating vCJD may be infective. The issue then is of the specific infectivity of various tissues at different stages of the incubation period and the transfer of infectivity by different routes. There is no current evidence to suggest that vCJD has been caused by iatrogenic transmission but this may be masked by the prolonged incubation period. The infective dose of prion protein required to cause vCJD is unknown. Much of the information is derived from other non-human TSE's. For example, an experiment has shown that 1 gram (this was the smallest amount given) of BSE infected bovine brain tissue can cause BSE when given to cattle by the oral route. A recent report has established some working levels of infectivity for a risk assessment of vCJD transmission by surgical treatment (see Table 2).

Table 2. Infectivity ranges for various tissues.

Transmission Route	Duration within incubation period	Range/logID$_{50}$/g
CNS to CNS (or posterior eye)	0-60%	0 (but sensitivity analysis around 40-60% of incubation period)
CNS to CNS (or posterior eye)	Remainder of incubation period	8 (but 9 as sensitivity analysis in final year)
Anterior eye to anterior eye	As above	0 then 5-6
LRS to LRS (or equivalent risk peripheral tissue)	All	5-6
Remaining tissues	All	0 (up to 4 in sensitivity analysis)

STERILISATION AND DISINFECTION

A consistent experimental finding is that PrP^{sc} is very resistant to the usual techniques for inactivating infectious agents. Ionizing, ultraviolet and microwave radiation have little effect on prion proteins. The current recommended process requires exposure to steam heat at 134°C for 18 minutes in a vacuum autoclave, although subsequent research has shown that this process does not guarantee complete inactivation[2]. In some strains of prion thermostability was increased as the temperature was increased from 134°C to 138°C[24]. Concentrated bleach appears to achieve inactivation of all strains, however, concentrated sodium hydroxide does not. A key feature of experimental work has been the recognition that the scrapie agent is more resistant to inactivation by autoclaving when infected tissue becomes dried onto glass or metal surfaces. Prior fixation of tissue in ethanol or formaldehyde has also been shown to considerably enhance the thermostability of the scrapie agent[24]. There is a further bleak hypothesis that theorises transmission could be caused by infective tissue coming into contact with that of another individual, rather than requiring any material to be transferred[14].

On a more practical level, a significant reduction in removing potentially infective material can be achieved by conventional washing and autoclaving processes. Workers[26] suggest that washing with detergents could reduce protein soiling plain flat surfaces by 5 logs, although this dropped to 3-4 logs with hydrophobic proteins such as fibrin. This has led to the suggestion that for most instruments, at least a 3 log reduction in mass adhering should be achievable by cleaning. Much may be dependent on the initial mass present and the design of the instrument to allow for easy cleaning. Although the effect of autoclaving may be variable[24] a reduction of 3-6 log has been used in risk assessment calculations. These findings emphasise the importance of cleaning equipment prior to sterilisation. Instruments that cannot be cleaned prior to sterilisation should either be dismantled prior to cleaning or disposed of. This has highlighted certain dental instruments such as endodontic files (see Fig 1) which by virtue of their small size and serrated surface are difficult to adequately clean.

INFECTION CONTROL ISSUES FOR PATIENTS CONFIRMED OR SUSPECTED CJD

Within the UK the Advisory Committee on Dangerous Pathogens (ACDP) and the Spongiform Encephalopathy Advisory Committee (SEAC) provide advice and guidelines for TSE's[2]. In general terms the advice for dental treatment on patients confirmed as suffering from any type of CJD is quite clear in its recommendations.

Instruments used for dental treatment should be disposed of after single use. Within dentistry, other than the use of disposable instruments no other special precautions are necessary over the routine universal precautions adopted for other patients undergoing routine dental treatment.

For patients with a suspected diagnosis of CSD which remains to be confirmed then instruments should be quarantined until a decision has been made about the diagnosis. If the diagnosis is confirmed the instruments should be destroyed.

INFECTION CONTROL ISSUES FOR PATIENTS "AT-RISK" FROM CJD

At present the situation for patients "at-risk" for CJD (see Table 3) is slightly more problematic. The first issue is identification of the ccat~risk~~ patient. Currently the British Dental Association (BDA) have revised their medical history forms to try and identify persons belonging to this group. If identified as "at-risk" and the procedure involves exposure to brain, spinal cord or eye then the ACDP/SEAC guidelines recommend use of disposable instruments. If however, the treatment (such as dentistry) is on tissues outside these categories then stringent decontamination processes are recommended (see Table 4). Since these are extremely difficult to comply with routinely this implies use of disposable dental instruments. Other than use of disposable instruments no other special precautions are necessary over the routine universal precautions adopted for other patients.

Table 3. Definition of patients "at-risk" to the development of human TSE's.

Definition of "at-risk" = Asymptomatic but potentially at-risk of developing the disease; • recipients of hormone derived from human pituitary glands, e.g., growth hormone. • recipients of human dura mater grafts. • people with a close (blood line) family history of CJD

ACDP/SEAC: Transmissible Spongiform Encephalopathy agents: Safe working and the prevention of infection. 1998.

The SEAC/ACDP guidelines for management of the "at-risk" group are not universally accepted. The WHO[27] recommendations consider that the risk of cross-infection from the CJD "at-risk" category during dentistry do not merit special precautions (except under conditions where there could be exposure to their high infectivity tissues). For close relatives of fCJD patients no consensus was achieved.

Table 4. Stringent disinfection and decontamination procedures for instruments used in clinical procedures on "at-risk" patients NOT involving brain, spinal cord or eye.

Process	Method
Cleaning	Items should be cleaned ASAP after use to minimise drying of blood and other body fluids. All instruments should be cleaned thoroughly at least twice prior to disinfection. Do not mix routine instruments with those used in TSE-related work. Following cleaning of instruments, the instruments washer should be run on an empty cycle. Any cleaning aids used, such as brushes, should be disposed of by incineration.
Sterilising	
Chemical agents	20,000 ppm available chlorine of sodium hypochlorite for 1 hour. 2 M sodium hydroxode for 1 hour.*
Physical processes	Porous load steam steriliser 134-137°C for a single cycle of 18 minutes or 6 successive cycles of 3 minutes each.*

(* but known not to be completely effective) ACDP/SEAC: Transmissible Spongiform Encephalopathy agents: Safe working and the prevention of infection. 1998.

PRION PROTEINS IN DENTAL TISSUES

Studies that have investigated the presence of prion proteins in oral tissues from human and animal models are summarised in Table 5. Of concern is the finding that prion protein can be detected in both gingival and pulpal tissue in the hamster scrapie model[17]. However, due to the differences in patterns of disease in animals models and strains of prion protein it is difficult to directly extrapolate these findings to humans but highlights the potential for transmission via the dental route. Data concerning the distribution of prion protein in the oral cavity is urgently required to make risk assessments of the risk of cross-infection from common dental procedures.

A preliminary analysis of previous dental treatment in vCJD cases (Ward, H. and Will, R.G. personal communication) has not revealed any consistent pattern suggesting past dental treatment as a risk factor for vCJD.

Table 5. Dental tissues, information on distribution of infectious agents.

Observation	Reference
Very low levels of scrapie infectivity from gingivae of infected mice. Failed to transmit scrapie by dental burs used on infected animals.	1
Infection of mice by scrapie agent via oral route is increased if gingival tissue is scarified.	10
Small clusters of sCJD cases possibly connected by dental procedures (3 cases UK and 3 cases Japan)	3, 28
Epidemiological evidence seems to exclude any correlation between tooth extraction or dental surgery and human TDE's.	13, 19
Gingival and pulpal tissues in hamster scrapie model have a high level of infectivity. Transmission of agent obtained by inoculation of tooth pulp.	17
No prion protein detected by western blotting in the dental pump of 8 sCJD patients.	6

OTHER DENTAL PROCEDURES

Concern has been expressed over the use of human dura mater grafts and bovine products for periodontal reconstruction[5,23]. Whilst the use of human dura graft material has now ceased and there are no documented reports of TSE transmission via this route it would seem prudent that appropriate records are kept of patients that have received these products.

CONCLUSION

The last five years has seen the emergence of a new disease termed variant CJD in humans. This has been accompanied by an explosion of scientific data on a novel infectious agent called prion protein. Many decisions concerning diagnosis, treatment and management of patients with CID are often based on incomplete knowledge on which to base important decisions. The main challenge to infection control of this agent is its extreme resistance to conventional methods of sterilisation and uncertainties on routes of infection. Surgical transmission of vCJD, including dental treatment cannot be ruled out as a risk to public health. Given current estimates of the infectivity of different tissues, operations on the central nervous system and posterior eye appear to carry the highest risk of transmitting vCJD. The effectiveness of instrument decontamination is the most important variable subject to influence. Recent work has demonstrated that current practice is extremely variable and this is being addressed as a matter of urgency. Since at this stage it is impossible to recognise patients in the pre-clinical phase of vCJD, this emphasises the importance of maintaining a high standard for routine cleaning and sterilising of reusable dental instruments.

REFERENCES

1.   Adams, D.H. and Edgar, W.M. (1978). Transmission of agent of Creutzfeldt-Jacob disease. *British Medical Journal* **1**, 987.

2.   Advisory Committee on Dangerous Pathogens and Spongiform Encephalopathy Advisory Committee (1998). Transmissible Spongiform Encephalopathy agents: Safe working and the prevention of infection. London, Stationery Office.

3.   Arakawa, K., Nagara, H., Itoyama, Y., *et al.* (1991). Clustering of three cases of Creutzfeldt-Jakob disease near Fukuoka City, Japan. *Ada Neurologica Scandinavica* **84**, 445-447.

4.   Duffy, P., Wolf, S., Collins, G., *et al.* (1974). Possible person-to-person transmission of Creutzfeldt-Jakob disease. *N Engl J Med* **290**: 692.

5.   Bartolucci, E.G. (1981). A clinical evaluation of freeze dried homologous dura mater as a periodontal free graft material. Study in humans. *J Periodontol* **52**, 354-361.

6.   Blanquet-Grossard, F., Sardovitch, V., Jean, A., *et al.* (2000). Prion protein is not detectable in dental pulp from patients with Creutzfeldt-Jakob disease. *Journal of Dental Research* **79**, 700.

7.   Brown, P., Preece, M., and Will, R. (1992). "Friendly fire" in medicine: hormones, homografts and Creutzfeldt-Jakob disease. *Lancet* **340**, 24-27.

8.   Bruce, M.E., Will, R.G., Ironside, J.W., et al., (1997). Transmissions to mice indicate that "new variant" CJD is caused by the BSE agent. *Nature* **389**,498-501.

9.   Büeler, H.R., Aguzzi, A., Sailer, A., *et al.* (1993). Mice devoid of PrP are resistant to scrapie. *Cell* **73**, 1339-1347.

10.   Carp, R.I. (1982). Transmission of scrapie agent by oral route: effect of gingival scarification. *Lancet* **16**, 170-171.

11.   Collinge, J., Sidle, K.C.L, Meads, J., Ironside, J., and Hill, A.F. (1996). Molecular analysis of prion strain variation and the aetiology of new variant CJD. *Nature* **383**, 685-90.

12. Collinge, J. (1999) variant Creutz:feldt-Jakob Disease. *Lancer* **354**, 317-323.

13. Collins, S., Law, M.O., Fletcher, A., Boyd, A., Kaldor, J. and Masters, C.L. (1999). Surgical treatment and risk of sporadic Creutzfeldt-Jacob disease: a case control study. *Lancet* **353**, 693-697.

14. Department of Health (2001). Risk assessment for transmission of vCJD via surgical instruments: a modelling approach and numerical scenarios.

15. Griffith, J.S. (1967). Self replication and scrapie. *Nature* **215**, 1043-44.

16. Hill, A.F., Desbruslais, M., Joiner, S., *et al.*, (1997). The same prion strain causes vCSD and BSE. *Nature* **389**; 448-450.

17. Ingrosso, L., Pisani, F. and Pocchiari, M. (1999). Transmission of the 263K scrapie strain by the dental route. *Journal of General Virology* **80**, 3043-3047.

18. Kimberlin, R.H. and Walker, C.A. (1988) Pathogenesis of experimental scrapie. In: Novel Infectious agents and the Central nervous system. Ciba Foundation Symposium No.135. edited by G. Bock and J. Marsh pp 37-62. Wiley, Chichester.

19. Kondo, K. and Kuroiwa, Y. (1982). A case control study of Creutzfeldt-Jakob disease: Association with Physical injuries. *AnnalNeurol* **11**,377-381.

20. Lueck, C.S., McIlwaine, G.G., Zeidler, M. (2000). CJD and the eye 1. Background and patient management. *Eye* **14**, 263-290.

21. Prusiner, S.B. (1982). Novel proteinaceous infectious particles cause scrapie. *Science* **216**, 136-144.

22. Scoff, M.R., Will, R., Ironside, J., *et al.* (1999). Compelling transgenic evidence for transmission of BSE prions to humans. *Proc Nat Acad Sci USA* **96**, 15137-15142.

23. Sogal, A. and Tofal, A..J (1999). Risk assessment of bovine spongiform encephalopathy transmission through bone graft material derived from bovine bone used for dental applications. *J Periodontol* **70**, 1053-1063.

24. Taylor, D.M. (1999). Inactivation of prions by physical and chemical means. J *Hosp Infict* **43**, (supplement) 569-576.

25. van Duijn, C.M., Delasnerie-Laupretre, N., Masullo, C., *et al.*, (1998). Case control study of risk factors of Creutzfeldt-Sakob disease in Europe during 1993-95. *Lancet* **351**, 1081-1085.

26. Verjat, D., Prognon, P., Darbord, J.C. (1999). Fluorescence-assay on traces of protein on re-usable medical devices: cleaning efficiency. *Int J Pharm* **179**, 267-271.

27. WHO/CDS/APH/2000.3 WHO infection control guidelines for transmissible spongiform encephalopathies.

28. Will, R.G. and Matthews, W.B. (1982). Evidence for case-to-case transmission of Creutzfeldt-Sakob disease. *Journal of Neurology, Neurosurgery and Psychiatry* **45**, 235-238.

6. FLOODS AND EXTREME WEATHER EVENTS — COASTAL ZONE PROBLEMS

FROM TOXIC ALGAE TO CLIMATE CHANGE: IMPACTS IN THE COASTAL ZONE

DONALD SCAVIA

Chief Scientist, National Ocean Service, National Oceanic and Atmospheric Administration, 1305 East West Highway, Silver Spring, Maryland, 20910, USA

INTRODUCTION

Human population has surpassed 6 billion, and almost 4.5 billion of those people live, work, and play within 100 km of the coast. While that simple statistic illustrates the economic, social, and aesthetic value people place on this vital "strip" of land and adjacent sea, it also underscores the growing suite of interactions among people and between people and the environment. Recent national and international assessments of the status of impacts in this coastal zone have documented a range of impacts, some clearly present, some emerging, and some on the horizon. This presentation highlights a number of those impacts.

The scope of this discussion runs from relatively well-documented trends (e.g., increases in harmful algal blooms and sea level), to future changes that are projected with relative certainty (e.g., increased oceanic CO_2 and temperature), to more speculative futures, such as changes in ocean circulation, patterns of precipitation, and subsequent water, chemical, and sediment runoff to the sea. The focus is on anticipated impacts and potential management, coping, and adaptation strategies for coastal development, shorelines, wetlands, coral reefs, marine fisheries, and estuaries. A significant portion of this summary is drawn from recent coastal assessments (e.g., Boesch et al.[1], Scavia et al.[2], CENR[3]).

HARMFUL ALGAL BLOOMS

Harmful algal blooms (HABs) are caused by microscopic algae that produce toxins that are subsequently transferred throughout the food web, released into the water, and in some cases volatilized and released into the air. These toxins can be harmful to both people and marine animals.

HAB species occur naturally. Their impacts are seen in coastal waters throughout the world, and evidence is growing that the blooms are becoming more common, more frequent, and more severe (Van Dolah[4]). While some of the increased prevalence can be explained by increased monitoring, detection technologies, and scientific attention, the current scientific consensus is that the increases are real, and potentially caused by increased exchange of ballast waters, increased nutrient loads and changing nutrient ratios, enhanced aquaculture and shellfish seeding, and changes in ocean temperature.

Impacts on People and Marine Animals

These biotoxins produce a series of poisoning syndromes in marine animals and humans. The human expression of these syndromes, induced through eating shellfish, include Paralytic Shellfish Poisoning (PSP), Amnesiac Shellfish Poisoning (ASP), Neurotoxic Shellfish Poisoning (NSP), Diuretic Shellfish Poisoning (DSP), and Azaspiracid Shellfish Poisoning (AZP). Human illnesses are also caused by eating fish containing Ciguatera toxins and by breathing air in areas where the biotoxins have volatilized. Impacts on humans from consuming shellfish (Shumway et al.[5]), as well as from breathing toxic aerosols (Baden[6]) and through physical contact (Carmichael[7]) have all been thoroughly documented.

In areas where shellfish consumption is not well managed, human illnesses can be common. It has been estimated that about 2000 cases of human poisoning have been reported, with fatalities numbering in the 100s. These are very conservative estimates because of the difficulty in distinguishing shellfish poisoning from other gastrointestinal illnesses, and the fact that mild cases are probably never seen (GEOHAB[8]). In most countries where strong monitoring programs are in place and the shellfish industries are managed conservatively, human death and illness is minimized, but at significant economic costs due to closures of shellfish beds when toxin thresholds are crossed or where monitoring programs are too expensive. A recent economic analysis of costs in the U.S. estimated annual losses averaging $40 M. Hard to quantify costs, such as lost opportunities from shellfish beds that are not able to open and losses in tourism and retail seafood industries would likely raise the loss estimates significantly.

These toxins have also been responsible for massive fish and shellfish kills and deaths of marine mammals and birds (GEOHAB, ECOHAB[9], CENR[10]). For example, harmful algal blooms killed 900 tons of fish in the Kattegat and Skagerrak areas of the North Sea, and caused massive fish kills in Japanese shellfish cultures, large-scale mortalities of manatees and sea lions in the US, and mass mortalities of sea birds.

Management, Adaptation. and Coping

Until the causes of the expansion and enhancement of HABs are discovered, diligent detection and monitoring programs provide the only practical way to mitigate impacts. Increasing our ability to detect, monitor, and forecast the blooms should improve protection. Emerging national and international research programs (ECOHAB, GEOHAB) should improve the understanding of the causes and consequences of HABs, and thus improve our ability to prevent, control, and mitigate blooms and their impacts. Also, increased awareness, in both the public and medical communities, of the symptoms and effects of HABs will help improve detecting and treating poisonings when they occur.

SEA-LEVEL CHANGE AND COASTAL STORMS

During the last 100 years, globally average sea level has risen approximately 10-20 cm, or about 1 to 2 mm per year (IPCC[11]), whereas local rates of relative sea level rise vary considerably due to regional differences in groundwater and oil withdrawal, compaction of muddy soils, subsidence, isostatic rebound, and tectonic uplift. Over the next 100 years, global warming is expected to accelerate the rate of sea level rise by expanding ocean water and melting alpine glaciers, resulting in increased sea level from 9 to 88 cm by 2100 (IPCC[12]); model averages from those analyses range more narrowly from 31 to 49 cm.

Trends in hurricane and tropical cyclone frequency cannot be attributed to current or projected climate change; however, there is a strong inter-decadal mode in North Atlantic hurricane variability, showing greater activity along the U.S. East Coast between 1941 and 1965, and in the 1990s (Landsea et al.[13]). Independent of their frequency, cyclone and hurricane wind strength could increase as a result of elevated sea surface temperatures (Knutson et al.[14]; Ken[15]). While there is no clear evidence for climate-induced changes in the frequency or intensity of hurricanes and coastal storms, there is little doubt that future waves and storms will be superimposed on a rising sea.

Impacts on Shorelines and Wetlands
Rising sea level can inundate low lands, erode beaches, cause barrier islands to migrate inland, increase coastal flooding, and increase the salinity of rivers, bays, and aquifers. Barrier islands illustrate the interactions between natural shoreline processes and human activities. These islands bear the brunt of hurricanes and winter storms, and protect the mainland from resulting damages. With a rising sea level, undeveloped islands slowly move toward the mainland through erosion on their seaward flank and over-washing of sediment to the waters of the bay. In contrast, as sea level rises on developed islands, the ocean sides erode, but buildings and other structures prevent washing sand toward the lagoons, leaving the islands and its development increasingly vulnerable to subsequent storms.

Storm-surge floods, waves, and coastal erosion cause some of the most visible and costly impacts to shorelines, particularly in developed areas. Hurricane Hugo caused an estimated $9 billion in insured losses in 1989; Hurricane Andrew damaged $27 billion in 1992; and Hurricane Georges damaged $5.9 billion in 1998). Winter storms also have significant impacts: The Halloween "nor'easter" of 1991 caused damages of over $1.5 billion along the Atlantic Coast, and a series of storms that battered the Pacific Coast during the 1997-98 El Niflo caused an estimated $500 million in damage in California alone (Griggs and Brown[16]).

In addition to increased erosion and flooding, a rising sea may also increase salinity of freshwater aquifers, alter tidal ranges in rivers and bays, change sediment and nutrient transport, and alter patterns of coastal chemical and microbiological contamination. Secondary impacts, including inundation of waste disposal sites and landfills, may introduce toxic materials into the environment, posing new threats to the health of coastal populations and ecosystems. Levees, seawalls, and other coastal structures are typically designed with the "100-year flood" as a basis. However increased sea level could mean that a future 50-year event may be more severe than today's 100-year flood (Pugh and Maul[17]).

Coastal wetlands are essential for sustaining the healthy ecosystems (wildlife, fishes, invertebrates, etc) that support commercial and recreational livelihoods of many coastal communities. The ability of these ecosystems to survive under a changing climate depends primarily on natural biological and geological processes that should allow responses (inland migration or vertical growth) to gradual changes. However, accelerated sea level rise and human development will likely confine wetlands to a decreasing margin between the rising sea and permanent structures, leading ultimately to their loss unless they are able to grow vertically. Changes in delivery of fresh water (driven both by human and climate alterations in hydrology) will also affect the sediment supply, which is essential for the needed vertical wetland growth.

Even if wetlands survive sea-level rise, they may become increasingly vulnerable to other storm surge impacts. During coastal storms, large quantities of sediment can move into marshes and smother vegetation, salt water can flow into fresh marsh systems and cause salt burn, organic marsh substrates can erode, and large quantities of underlying vegetation can be killed. Although plants can quickly reestablish themselves, changes in the frequency or magnitude of storm impacts, including the fact that storms will come in on a higher sea level, could additionally threaten the long-term sustainability of wetland ecosystems.

Management, Adaptation, Coping

Two general approaches are available to coastal communities for dealing with sea-level rise: holding back the sea, or allowing the shoreline to move inland. Holding back the sea with dikes, seawalls, bulkheads, and revetments generally sacrifices beach, wetland, and other intertidal zones but leaves dry land relatively unaffected. Two approaches to reduce impediments to the natural inland migration of shorelines as sea level rises are to prevent development in vulnerable areas, or use rolling easements, which allow coastal development, but prevent structures that lead to loss of wetlands and beaches. In both cases, setback lines would have to be based on both sea-level rise and subsidence projections.

Wetland survival depends on its ability to accrete soil or migrate inland to keep pace with the rising sea. In areas where upland migration is not feasible and wetlands have to rely on vertical adjustment, adjustments in controls of river discharge and sediment supply could be made to facilitate accumulation of substrate; however, that may be in conflict with social desires for flood control and water supply.

OCEAN TEMPERATURE, CO_2, CIRCULATION, AND ICE

There is strong evidence that ocean warming is due primarily to anthropogenic climate change (Levitus et al.[18,19]). In addition, arctic ice has declined in aerial extent by as much as 7% per decade over the last 20 years (Johannessen et al.[20]) and has thinned by as much as 15% per decade (Rothrock er al.[21]). While some of this change in ice cover may be related to decadal-scale atmospheric variability, comparisons with OCM outputs suggest the declines are related to anthropogenically induced global change (Vinnikov et al.[22]).

Ocean currents, fronts, and upwelling and downwelling zones that play significant roles in the distribution and production of marine ecosystems are likely to change in response to alterations in temperature, precipitation, runoff; salinity, and wind. While these changes are likely to influence biological productivity₅ region-specific changes are not yet predictable. However, one of the more dramatic projections is cessation of the North Atlantic portion of the deep-water "conveyer belt" circulation, with potentially dramatic feedback on large-scale climate patterns (Driscoll and Hau[23]; Broecker et al.[24]).

Air-sea exchange processes bring atmospheric and oceanic CO_2 concentrations into equilibrium. Therefore, as atmospheric CO_2 concentrations increase, so does the concentration of dissolved CO_2 in the ocean. Increased CO_2 dissolved in seawater decreases its alkalinity, which will impact coral reef development and repair as discussed below.

Impacts on Corals and Marine Fisheries

Marine, terrestrial, and atmospheric processes influence corals ecosystems over wide spatial and temporal scales. The physical boundaries for these reef communities can be defined by temporal and spatial distributions of temperature, calcium carbonate saturation state, salinity, light, sediment, nutrients, and physical energy regime (waves, currents, storms)—all potentially influenced by both natural and human processes. Human activities can also lead to physical destruction, over fishing, and toxic chemical contamination.

Sea surface temperatures (SST) might rise by 1 to 3°C over this century, although not uniformly (Pittock[25]). Reef ecosystems are susceptible to changes in the frequency or magnitude of temperature extremes because many coral species live near their upper limits of thermal tolerance. Coral bleaching can occur when these thermal tolerances are exceeded and the corals' symbiotic algae (zooxanthellae) are expelled, slowing or halting growth, skeletal accretion, and sexual reproduction, and increasing the susceptibility to pathogens (Glynn[26]). Warming events over the last several decades have led to extensive bleaching in the Florida Keys, the Caribbean, the Eastern Pacific, and elsewhere. More recently, unprecedented high sea surface temperatures and perhaps the most widespread coral bleaching ever observed have been associated with the 1998 El Niño (Wilkinson et. al.[27]; Hoegh-Guldberg[28]). If the high-frequency SST variation of the past 20 years continues, and is superimposed upon a warming surface waters, accelerated bleaching and mass mortality events could increase dramatically in many tropical reefs.

Corals build their reefs through calcification, a process that depends upon the ability to build their calcium carbonate structures. Calcification is sensitive to the carbonate saturation state of ambient surface water, which is depressed by rising concentrations of atmospheric CO_2. Carbon dioxide, dissolved in seawater, reduces alkalinity and calcium carbonate concentrations. This, in turn, decreases calcification rates of reef-building corals and coralline algae. Recent studies suggest that calcification rates are likely to decline between 17 and 35% by the. year 2100 (Gattuso et al.[29]; Kleypas et al.[30]) and as much as one-third of that decrease may already have occurred. The resulting reduced skeletal density and/or growth rate will increase reef vulnerability to physical damage, bioerosion, some forms of predation, and the corals' ability to compete for space. The effect of increased atmospheric CO_2 will be greatest at the northern and southern margins of coral distributions because CO_2 is more soluble in cooler waters. Therefore, these effects will be most severe at higher latitudes, reducing the ability of reefs to expand their ranges poleward as might otherwise be expected in response to ocean warming.

Altered temperature, salinity, precipitation, wind fields, and sea level will also affect the distribution, abundance, and production of marine organisms. Changes in ocean temperature are likely to drive poleward migrations of tropical and lower latitude organisms. For example, along the Northeast U.S. coast, cod, American plaice, haddock, Atlantic halibut redfish, and yellowtail flounder may all migrate northward, with mid-Atlantic species, such as butterfish, herring, mackerel, and menhaden extending their ranges into the Gulf of Maine (Murawski[31]). Many of these species provide an important forage base for other fishes, marine mammals, and sea birds; therefore, these migrations may have significant secondary effects on trophic interactions and relative distribution of predators and prey. Temperature shifts may also drive modifications in the distribution and abundance of intertidal and temperate reef species (Sagarin et al.[32], Hobrook et al.[33]) along the California Coast.

However, other habitat requirements may prevent or limit migration for some species, requiring them to accommodate higher temperatures (Colton[34]), and the energetic costs associated with living in sub-optimal temperatures may result in loss of regional populations. For example, based on fisheries oceanographic data, thermal limits of Pacific salmon, and temperature projections from the Canadian climate model, Welch et al.[35] suggest that by 2090 virtually none of the Pacific Ocean may lie within the thermal limits of sockeye salmon, thus restricting their distribution, and potentially other salmonids, to the Bering Sea and Sea of Okhotsk.

In addition to these direct temperature effects, changes in atmospheric dynamics, air temperature, wind stress, and freshwater runoff will likely alter water column stratification and regional circulation patterns. The effects of such climate variability is illustrated in the relationship among marine physical and biological changes and associated reoccurring patterns of interdecadal climate variability, referred to as the Pacific Decadal Oscillation, or PDO (Mantua et al.[36]). In the North Pacific, atmospherically driven warming of surface waters, decreased mixed-layer depths, and increased vertical stratification appear to have led to large-scale changes in productivity (Freeland, et al.[37]), including an apparent doubling of zooplankton biomass in the subarctic gyre between the 1950s and the 1980s (Brodeur and Ware[38]) and a contrasting 70% decline in zooplankton abundance in the California Current during roughly the same period (Roemmich and McGowan[39]). McGowan et al.[40] attributed these long-term declines in zooplankton populations in the California Current to the increased water temperatures and intensified stratification, and an overall lowering of mixing and nutrient regeneration in the upper water column.

Coastal freshwater discharge and the PDO are well correlated with salmon production. Within the Gulf of Alaska, decreases in the upper layer salinity from increased precipitation and glacial melt, are likely to increase the stratification and poleward baroclinic flow on the eastern boundary (Royer et al.[41]). A considerable volume of research indicates that this decadal scale climate variability (the PDO) has had substantial impacts throughout the North Pacific on the productivity and species composition of both lower and higher trophic levels, including many forage species, most populations of Pacific salmon, and many sea birds and marine mammal populations (McGowan et al.[42]; Anderson and Piatt[43]; Hare et al.[44]).

Increased temperature is also likely to continue to thin polar and subpolar ice and to change their spatial distributions. These changes will impact the marine mammals and seabirds that use ice shelves and flows as platforms for reproduction, pupping, resting, molting, and migration. Significant reduction in ice edge extent could have deleterious effects on marine mammals, such as walrus, ringed seals, polar bears, that depend upon these systems (Tynan and Demaster[45]). Ice edges also support highly productive ocean regions where physical and biological processes encourage substantial phytoplankton blooms and high levels of zooplankton and arctic cod production (Wheeler[46]; Niebaue[47]). Migration of belugas, narwhals, and harp seals to these ice edge regions have all been linked to surges in abundance of Arctic cod in these areas during summer blooms. Alterations in the timing, extent,, and location of these ice-edge dynamics may have significant impacts on these animals and the communities that depend on them.

Management, Adaptation, Coping

Almost all coral reefs that are in good condition are isolated from human populations (Miller and Crosby[48]). While degraded reefs typically suffer from combinations of natural and anthropogenic stresses, the latter are clearly the most significant shortterm drivers because coral ecosystems have been able to cope with natural change and variability in the past. Additional anthropogenic stresses are pushing individual reefs across critical thresholds; while at the same time, increasing CO_2 and ocean warming are providing a less hospitable environment at global scales. With increases in these longer-terms and less controllable climate pressures, the key strategy for corals is to reduce stresses from other, direct anthropogenic pressures such as nutrient over-enrichment, over-fishing, and sedimentation.

To cope with climate-induced change and variability, fisheries managers must take into account longer time-scale effects in their management strategies. Sustainable yields are tied directly to the state of the environment, and exploitation and environmental variability can interact to destabilize harvested populations. Environmental change that results in reduced productivity can lead to a decline or collapse of populations under levels of exploitation that are sustainable under more favorable conditions (Rice[49]). Rather than consider exploitation and environmental change separately, managers must recognize their interaction and adjust their strategies accordingly. Under periods of projected low productivity, allowable exploitation rates must be reduced to account for reduced recruitment rates. Under periods of projected high productivity, harvest limits can be increased.

PRECIPITATION AND WATER, NUTRIENT, AND SEDIMENT RUN-OFF

The hydrologic cycle controls the strength, timing, and volume of delivery of freshwater and its chemical and sediment load to coastal ecosystems. That cycle is likely to change under a changing climate; however, regional projections of precipitation vary considerably (NAST[50]). For example, the Hadley climate model projections result in a 34% increase in total runoff along U.S. Atlantic and Gulf coasts by the end of the century, while the drier and hotter Canadian climate model projections result in a 32% decrease. While these differences illustrate significant uncertainty regarding fixture rainfall and runoff patterns, one should not average them and assume that changes will be benign. Rather, these analyses suggest that significant changes in the delivery of freshwater into coastal environments may be significant, if highly uncertain. In addition, both models predict increases in extreme rainfall events, which can significantly increase the chemical and sediment load delivered to the coast. This increased "flashiness", which began during the 20th century, is likely to become more common (Karl et al.[51]).

Impacts on Estuaries

Estuarine impacts will be felt through intensification of current stresses; particularly those imposed by excess nutrients from significantly altered nitrogen cycles (Howarth et al.[52]; Vitousek Ct al.[53]). These impacts will be significant because nitrate-driven eutrophication is one of the greatest threats to many estuaries (NRC[54]; Cloern[55]). Eutrophication, defined as an increase in supply of organic matter due to flux from external sources or from nutrient-stimulated production within the system (Nixon[56]), can lead to greater phytoplankton biomass, decreased water clarity, hypoxic (low oxygen) bottom waters, more serious harmful algal blooms, degraded sea grasses and corals, altered planktonic and the benthic community structures, and loss of biotic diversity and fisheries (Jørgensen and Richardson[57]).

Changing precipitation patterns can significantly influence nutrient delivery and subsequent eutrophication. The load of nitrogen from the Mississippi River system to the Gulf of Mexico is a prime example. That load has increased approximately three-fold over the past four decades and is now the dominant factor in the massive hypoxic area off the coast of Louisiana (CENR[58]); however, while that loading rate has leveled off, there remains considerable inter-annual variability in driven by variation in precipitation and river flow. For example, the massive flood of 1993 produced the highest-recorded nitrogen delivery and an areal extent of hypoxia (17,600 km^2) twice the average of the preceding eight years. By contrast, the hypoxic area was only 40 km^2 during 1988, the 52-year low flow condition for the Mississippi River (Rabalais et al.[59]).

The effects of precipitation variability on nitrogen loss from the land may also be magnified by land use practices. Nitrate tends to build up in soils during dry years, largely as a result of reduced uptake of soil nutrients by crops' and is flushed into streams at much larger rates during subsequent wet years. Thus, wet years that follow dry years tend to produce the largest fluxes of nitrate. If future precipitation regimes are more variable and packaged into larger events, a trend suggested by many climate models, then larger nutrient loads and subsequent coastal over-enrichment and eutrophication are likely.

Estuaries also vary in their vulnerability to eutrophication, and climate change may influence that vulnerability in several ways, including changes in freshwater inflow, air temperatures, and precipitation patterns, and their influences on water residence time, dilution, vertical stratification, and control of phytoplankton growth rates (Malone[60]; Cloern[61]; Howarth et al[62]). For example, decreased freshwater discharge will increase estuarine water residence time, whereas increased runoff will decrease residence time (Moore et al[63]). Altered residence times can be significant because, even at their fastest growth rates, phytoplankton populations are only able to double once or twice per day. Consequently, in estuaries with water residence times of less than one day, phytoplankton are generally flushed from the system as fast as they can grow, reducing the estuary's susceptibility to eutrophication. However, if residence times increase as a result of altered freshwater delivery, susceptibility will also increase.

Increased air temperature may also lead to earlier snowmelt and peak in freshwater inflow. In those cases, summer flows may also be reduced as a result of higher temperatures and greater evapotransporation, resulting in increased salinity and modified stratification and mixing, thus influencing biotic distributions, life histories, and biogeochemistry.

Management, Adaptation, Coping

Estuaries are likely to respond to changes in sea level, temperature, and freshwater and nutrient delivery in different ways, based on their physical properties (e.g., flushing times). While many degraded estuaries already have commitments for restoration through pollution reduction strategies, for example, few of those plans take into account longer-term, climate-influenced changes in precipitation, runoff regimes, nutrient loads, and salinity. Water management and land-use policies should anticipate changes in the amount and seasonal distribution of water, human demand, and the needs of the estuarine ecosystems. Improved land use practices, such as more efficient nutrient management, and more extensive restoration and protection of riparian zones and wetlands may help meet longer term goals in a wetter future.

CONCLUSION

There are significant uncertainties in the forecasts of some important climate forces that will affect coastal and marine systems, particularly at regional scales. While forecasts of changes in atmospheric CO_2, air temperature, and sea level are becoming more reliable as model and data resolution increase, potential changes in the timing and strength of regional river runoff, coastal ocean and estuarine temperatures, and coastal circulation remain uncertain. However, enough is known to begin to prepare for such changes.

Climate changes that are likely to occur over the next century will have many consequences for coastal and marine ecosystems, and some of these may substantially alter human dependencies and interactions within these complex and linked systems. The climatic effects will be superimposed upon, and interact with, a wide array of current stresses, including excess nutrient loads, fishing, invasive species, habitat destruction, and toxic chemical contamination. While these ecosystem's ability to cope with or adapt to climate change is compromised by extant stresses, the inverse is also likely to be true; these ecosystems will be better able to adapt to climate change if other stresses are significantly reduced.

ACKNOWLEDGMENTS

I would like to thank the team of investigators who developed the original climate impacts assessment report that is summarized here, as well as Don Anderson, Fran Van Dolah, and Danielle Luttenberg for much of the information on harmful algal blooms.

REFERENCES

1. Boesch, D.F., J.C. Field and D. Scavia (eds.) 2000. The Potential Consequences of Climate Variability and Change on Coastal Areas and Marine Resources: Repon of the Coastal Areas and Marine Resources Sector Team, U.S. National Assessment of the Potential Consequences of Climate Variability and Change. U.S. Global Change Research Program. NOAA coastal Ocean Program Decision Analysis Series No. # 21. NOAA Coastal Ocean Program, Silver Spring MD. 163 pp.
2. Scavia, D., J.C. Field, D.F. Boesch, R.W. Buddemeier, V. Burkett, D.R. Cayan, M. Fogarty, M.A. Harwell, R.W. Howatth, C. Mason, D.J. Reed, T.C. Royer, A.H. Sallenger, J.G. Titus. (submitted) Climate Change Impacts on U.S. Coastal and Marine Ecosystems.
3. CENR. 2000. National Assessment of Harmful Algal Blooms in U.S. Waters. National Science and Technology Council Committee on Environment and Natural Resources, Washington DC. 38 pp.
4. Van Doiah, F. 2000. Marine Algal Toxins:. Origins, health effects, and their increased occurrence. Environmental Health Perspectives. 108(1): 133-141.
5. Shumway, S.E., H.P.V. Edmond, J.W. Hurst, and L.L Bean. Management of shellfish resources, in: G.M. Hallegraeff, D.M. Anderson, and A.D. Cembella (eds.) Manual on Harmful Marine Microalgae. UNESCO Paris. pp. 433-474.
6. Baden, D. 1983. Marine food-borne dinoflaggelate toxins. Int. Rev. Cytol. 82: 99-150.
7. Carmichael, W.W., 1997. The cyanotoxins. Adv. Bot. Res. 27:211-256.
8. GEHAB 2001. Global Ecology and Oceanography of Harmful Algal Blooms Science Plan. P. Gilbert and G. Pitcher (eds.) SCOR and IOC, Baltimore and

Paris. 86 pp.

9.   Anderson D.M. (ed) 1995. ECOHAB, Ecology and Oceanography of Harmful Algal Blooms. Woods Hole Oceanographic Institution, Woods Hole, MA. 66 pp.

10. CENR. 2000. National Assessment of Harmful Algal Blooms in U.S. Waters. National Science and Technology Council Committee on Environment and Natural Resources, Washington DC. 38 pp.

11. Intergovernmental Panel on Climate Change (IPCC). 1996. Climate Change 1995: Impacts, Adaptations and Mitigation of Climate Change: Scientific-Technical Analysis. New York: Cambridge University Press.

12. Intergovernmental Panel on Climate Change (IPCC). 2001. Climate Change 2001: The Scientific Basis Climate Change 1995: New York: Cambridge University Press.

13. Landsea, C.W., N. Nicholls, W.M. Gray and L.A. Avila. 1996. Downward trends in the frequency of intense Atlantic hurricanes during the past five decades. Geophysical Research Letters 23:1697-1700.

14. Knutson, T.R., and R.E. Tuleya. 1999. Increased hurricane intensities with CO_2 induced warming as simulated using the GFDL hurricane prediction system. Climate Dynamics 15: 503-519

15. Ken, E.A. 1999. Thermodynamic control of hurricane intensity. Nature 401:665-669.

16. Griggs, G.B., and K.M. Brown. 1999. Erosion and shoreline damage along the Central California coast: A comparison between the 1997-98 and 1982-83 ENSO winters. Shote and Beach 66:18-23.

17. Pugh, D.T. and G.A. Maul. 1999. Coastal sea-level prediction for climate change. C.N.K. Mooers, (ed.) Coastal and Estuarine Studies 56: Coastal Ocean Prediction. American Geophysical Union. pp. 377-404.

18. Levitus, S., J.I. Antonov, T.P. Boyer, C. Stephans. 2000. Warming of the world ocean. Science 287: 2225-2229.

19. Levitus, S., J.I. Antonov, J. Wang, T.L. Delworth, K.W. Dixon and A.J. Broccoli. 2001. Anthropogenic warming of earth's climate system. Science 292:267-270.

20. Johannessen, O.M., E.V. Shalina, M.W. Miles. 1999. Satellite evidence for an Arctic sea ice cover in transformation. Science 286:1937-1939.

21. Rothrock, D.A., Y. Yu and G.A. Maykut. 1999. Thinning of the Arctic Sea-Ice cover. Geophysical Research Letters 26: 3469-3472.

22. Vinnikov, K.Y., A. Robock, R.J. Stouffer, J.E. Walsh, C.L. Parkinson, D.J. Cavalieri, J.F.B. Mitchell, D. Garrett, and V.F. Zakharov. 1999. Global warming and Northern Hemisphere sea ice extent. Science 286:1934-1937.

23. Driscoll, N.W. and G.H. Haug. 1998. A short circuit in thermohaline circulation: A cause for Northern Hemisphere glaciation? Science 282: 436438.

24. Broecker, W.S., S. Sutherland, and T.H. Peng. 1999. A possible 20th century slowdown of Southern Ocean deep-water formation. Science 286:1132-1135

25. Pittock, A.B. 1999. Coral reefs and environmental change: Adaptation to what? American Zoologist 39:10-29.

26. Glynn, P.W. 1996. Coral reef bleaching: Facts, hypotheses and implications. Global Change Biology 2: 495-509.

27. Wilkinson, C., O. Linden, H. Cesar, G. Hodgson, J. Rubens, and A.E. Strong. 1999. Ecological and socioeconomic impacts of 1998 coral mortality in the Indian Ocean: An ENSO impact and a warning of future change? Ambio 28:188-196.

28. Hoegh-Guldberg, G. 1999. Climate change, coral bleaching, and the nature of the world's coral reefs. Marine and Freshwater Research. 50: 839-866.

29. Gattuso, J.P., D. Allemand, and M. Frankignoulle. 1999. Photosynthesis and calcification at cellular, organismal and community levels in coral reefs: A review on interactions and control by the carbonate chemistry. American Zoologist 39:160-183.

30. Kleypas, J.A., LW. Buddemeier, D. Archer, J.P. Gattuso, C. Langdon, and B.N. Opdyke. 1999a. Geochemical consequences of increased atmospheric carbon dioxide on coral reefs. Science 284:118-120. Klcypas, J.A., J.W. McManus, and L.A.B. Menez. 1999b. Environmental limits to coral reef development: Where do we draw the line? American Zoologist 39:146-159.

31. Murowski, S.A. 1993. Climate change and marine fish distributions: forecasting from historical analogy. Transactions of the American Fisheries Society 122: 657-658.

32. Sagarin, R.D., J.P. Barry, S.E. Gilman, and C.H. Barter. 1999. Climate-related change in an intertidal community over short and long-time scales. Ecological Monographs 69: 465-490.

33. Holbrook, S.J., R.J. Schmitt, and J.S. Stephens, Jr, 1997. Changes in an assemblage of temperate reef fishes associated with a climate shift. Ecological Applications 7:1299-1310.

34. Colton, J.B. Jr. 1972. Temperature trends and the distribution of groundfish in continental shelf waters, Nova Scotia to Cape Hatteras as determined from research vessel survey data. Fisheries Bulletin. 75:1-21.

35. Welch, D.W., Y. Ishida, and K. Nagasawa. 1998. Thermal limits and ocean migrations of sockeye salmon (Oncorhynchus nerka): long-term consequences of global warming. Canadian Journal of Fisheries and Aquatic Sciences 55: 937-948.

36. Mantua, N.J., S.R. Hare, Y. Shang, J.M. Wallace, and R.C. Francis. 1997. A Pacific interdecadal climate oscillation with impacts on salmon production. Bulletin of the American Meteorological Society 78:1069-1079.

37. Freeland, H.J., K. Denman, C.S. Wong, F. Whitney, and R. Jacques. 1997. Evidence of change in the winter mixed layer depth in the Northeast Pacific Ocean. Deep-Sea Research, 44:2117-2129.

38. Brodeur, RD. and D.M. Ware. 1992. Interannual and interdecadal changes in zooplankton biomass in the subarctic Pacific Ocean. Fisheries Oceanography 1: 32-38.

39. Roemmich, D. and J. McGowan, 1995. Climatic warming and the decline of zooplankton in the California Current. Science 267:1324-1326.

40. McGowan, J.A., D.R. Cayan and L.M. Dorman. 1998. Climate, ocean variability

and ecosystem response in the Northeast Pacific. Science 281:210-217.

41. Royer, T.C., C.E. Grosch, and L.A. Mysak. 2001. Interdecadal variability of Northeast Pacific coastal freshwater and its implications on biological productivity. Progress in Oceanography, (in press).

42. McGowan, J.A., D.R. Cayan and L.M. Dorman. 1998. Climate, ocean variability and ecosystem response in the Northeast Pacific. Science 281: 210-217.

43. Anderson, P.J. and J.F. Piatt. 1999. Community reorganization in the Gulf of Alaska following ocean climate regime shift. Marine Ecology Progress Series 89: 117-123.

44. Hare, S.R., N.J. Mantua, and R.C. Francis. 1999. Inverse production regimes: Alaska and West Coast Pacific Salmon. Fisheries 24: 6-14.

45. Tynan, C.T. and D.P. DeMaster. 1997. Observations and predictions of Arctic climatic change: Potential effects on marine mammals. Arctic 50:308-322.

46. Wheeler, P.A., M. Gosselin, E. Sherr, D. Thibault, D.L. Kirchmans, R. Benner, T.E. Whitledge. 1996. Active cycling of organic carbon in the Central Arctic Ocean. Nature 380: 697-699

47. Niebauer, H.J. 1991. Physical oceanographic interactions at the edge of the Arctic ice pack. Journal of Marine Systems 2: 209-232.

48. Miller, S.L. and M.P. Crosby. 1998. The Extent and Condition of U.S. Coral Reefs. NOAA's State of the Coast Report. National Oceanic and Atmospheric Administration. [Online]. Available: http://state-of-coast.noaa.gov/bulletins/html/crf_08/crf.html.

49. Rice, J. 1995. Food web theory, marine food webs, and what climate change may do to northern marine fish populations. In R.J. Beamish (ed.) Climate change and northern fish populations. Canadian Special Publicaton of Fisheries and Aquatic Sciences. 121: 561-568.

50. National Assessment Synthesis Team 2001. Climate Change Impacts on the United States: The Potential Consequences of Climate Variability and Change. Report for the U.S. Global Change Research Program, Cambridge University Irress, Cambridge UK. 620 pp.

51. Karl, T.R., R.W. Knight, D.R. Easterling, and R.G. Quayle. 1995. Indices of climate change for the United States. Bulletin of the American Meteorological Society 77: 279-292.

52. Howarth, R.W., G. Billen, D. Swaney, A. Townsend, N. Jaworski, K. Lajtha, J.A. Downing, R. Elmgren, N. Caraco, T. Jordan, F. Berendse, J. Freney, V. Kudeyarov, P. Murdoch, and Zhu Zhao-liang. 1996. Regional N budgets and riverine N & P fluxes for the drainages to the North Atlantic Ocean: Natural and human influences. Biogeochemistry 35: 75-139.

53. Vitousek, P.M., J. Aber, S.E. Bayley, R.W. Howarth, G.E. Likens, P.A. Matson, D.W. Shindler, W.H. Schlesinger, and G.D. Tilman. 1997. Human alteration of the global nitrogen cycle: Sources and consequences. Ecological Applications 7: 737-750.

54. National Research Council (NRC). 2000. Clean Coastal Waters: Understanding

and Reducing the Effects of Nutrient Pollution. Washington, DC. National Academy Press.

55.   Cloern, J.E. 2001. Our evolving conceptual model of the coastal eutrophication problem. Marine Ecology Progress Series 210: 223-253.

56.   Nixon, S.W. 1995. Coastal marine eutrophication: A definition, social causes and fliture concerns. Ophelia 41:199-219.

57.   Jørgensen, B.B. and K. Richardson. 1996. Eutrophication in Coastal Marine Systems. American Geophysical Union Washington, DC.

58.   CENR. 2000. Integrated assessment of hypoxia in the northern Gulf of Mexico. National Science and Technology Council Committee on Environment and Natural Resources, Washington DC. 58 pp.

59.   Rabalais, N.N., R.E. Turner, D. Justic, Q. Dortch, and W.J. Wiseman, Jr. 1999. Characterization of Hypoxia. NOAA Coastal Ocean Program Decision Analysis Series No. 15. Silver Spring, Maryland: National Oceanic and Atmospheric Administration.

60.   Malone, T.C. 1977. Environmental regulation of phytoplankton productivity in the lower Hudson estuary. Estuarine and Coastal Marine Science 5: 57-171.

61.   Cloern, J.E. 1991. Tidal stirring and phytoplankton bloom dynamics in an estuary. Journal of Marine Research 49: 203-221. Cloern, J.E. 1996. Phytoplankton bloom dynamics in coastal ecosystems: a review with some general lessons from sustained investigation of San Francisco Bay, California. Reviews of Geophysics 34:127-168.

62.   Howarth, R.W., G. Billen, D. Swaney, A. Townsend, N. Jaworski, K. Lajtha, J.A. Downing, R. Elmgren, N. Caraco, T. Jordan, F. Berendse, J. Freney, V. Kudeyarov, P. Murdoch, and Zhu Zhao-liang. 1996. Regional N budgets and riverine N & P fluxes for the drainages to the North Atlantic Ocean: Natural and human influences. Biogeochemistry 35: 75-139.

63.   Moore, M.V., M.L. Pace, J.R. Mather, P.S. Murdoch, R.W. Howaith, C.L. Folt, C.Y. Chen, H.F. Hemond, P.A. Flebbe, and C.T. Driscoll. 1997. Potential effects of climate change on freshwater ecosystems of the New England/mid-Atlantic region. Hydrological Processes 11: 925-947.

THE BLACK SEA ROMANIAN COASTAL ZONE—A GENERAL SURVEY OF THE EROSION PROCESS

DUMITRU DOROGAN
Ministry of Waters and Environment Protection, Bucharest, 75 061, Bd. Libertãtii 12, Romania. Email: <dsrc@mappm.ro>

DIACONEASA DANUT
National Institute for Marine Research and Development "Grigore Antipa" Bd. Mamaia # 300, Constanta 8700, Romania.

ABSTRACT

The erosion process of the Black Sea Romanian coastal zone is only a part of an entire Black Sea geomorphologic complex process. Consequently, this paper presents initially the general aspects of the Black Sea coastal zone, the international ICZM concepts with key elements, then, focusing on the Romanian Black Sea coastal zone, general presentation of the erosion aspects, and some proposed actions to be taken, for rehabilitation.

In general, it would be concluded that hydro technical constructions on the Romanian littoral had a negative impact, except some defense works initiated for the protection of the Mangalia, Jupiter and Constanta shoreline. At present, there are many urgent problems to be solved, but in a coordinated manner.

The main idea is that getting back to the initial status of the littoral is actually impossible, so the future efforts have to be oriented to preserve what we have now and to stop the destructive actions. For such a strategy it is necessary to initiate a coordinated coastal zone management system, possibly to be developed in the ongoing process for Romania to join the European Community.

BLACK SEA

Characteristics
Through its geographic coordinates (40°55' and 46°32' north latitude, 27°42' and 41°42' east longitude) the Black Sea, as the world's largest land-locked continental sea, is situated in the northern hemisphere.

The total shore length, not including the Azov Sea, is 4358 km (Ukraine 1628 km, Russian Federation 475 km, Georgia 310 km, Turkey, 1400 km, Bulgaria 300 km,

Romania *245* km (BSFP[1]).

From the geomorphologic standpoint the entire Black Sea shore can be regarded as complex. Alternating accumulative areas and bars with abrasion sections are characteristic of this type of shoreline. Lagoon like and abrasion landslide types of shore are most widespread in the north-western part of the Black Sea. The abrasion phenomenon is quite characteristic of the whole shoreline and in places takes complex forms.

There are no large islands in the Black Sea. The biggest one is Serpent's Island (1.5 sq. km) located in front of the Danube Delta, 45 km offshore of Sulina port, created in 1897 as a result of catastrophic floods and huge deposits of sediments transported with that occasion.

The Black Sea basin has similar geological properties to oceans and can be divided into four physiographic provinces (Ross[2]): continental shelf, continental slope, basin apron and abyssal plain (Fig. 1).

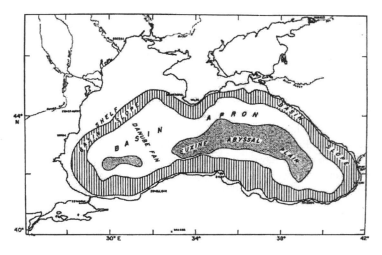

Fig.1. Main physiographic province of Black Sea.

- The shelf covers 29.9% of the Black Sea surface. In most areas the shelf edge is delineated by the 100 m isobath.
- The slope covers 27.3 % of the Black Sea territory and has a gradient typical of continental slope (1:40).
- The basin apron covers 30.6% of the body of the Black Sea with a gradient between 1:40 and 1:100, similar to that a continental rise, The Danube fan, a distinct depositional feature of the basin apron, extends across the Black Sea basin, dividing the abyssal plain into two unequal parts.
- The Euxine abyssal plain represents 12.2% of the Black Sea with a gradient of 1:1,000 and slopes to a maximum depth of 2,206m.

The origin of the Black Sea is still a matter of discussions. Many tectonic models have been proposed. One model presented it as a remnant ocean. Another one attributed its formation to the basification of the continental crust. A third model suggested that it resulted from a continuous uplift and erosion of continental area. A fourth model indicated that it resulted from strike slip fault movement. Later studies agreed on the back-arc rifting model. According to this model the West and the East Black Sea basins have separate origins. Speaking of timing, recent studies specified that the Black Sea formed during the Aptian to Cenomanian periods, as a result of a complex process between rifting, faulting and block rotation (Görür[3]).

A recent provocative thesis (Ryan[4]) argues that in about 7,150 B.C, with rising global sea levels, salt water from the Mediterranean and Aegean Seas apparently burst into the Black Sea, then a landlocked freshwater lake. The Black Sea rose (30-60 cm/day) inundating more than 100,000 sq. m of coastal plains (within a one to two year period) and giving the body of water its current size and configuration.

Coastline changes

Coastal erosion is a serious problem in the Black Sea. This is due to the destructive effects of the sea (wind, waves, currents, rising level, diminishing sedimentary stocks) as well as man-made activity. The building of port structures, concreting of coastal slopes, regulation of river run-off removal and dredging of beach and channel sediment amplifying thus the erosion forces, degradation by tourist activities and many other factors, disturb the equilibrium and change the living conditions of marine organisms in zone and landscape, facts which could affect the economical and ecological potential of a community.

As a biotope, the upper margin of the shore is determined by the limit waves reach during surges; the lower margin by the shoreline during offshore surges. In the Black Sea these borders are respectively located 2-4 m above average sea level on the shore and in the vicinity of the 1.5m isobaths in the sea.

The Black Sea coastal zone can be divided into three main morphodynamic categories (BSEP[2]):

- Low accumulative coasts (for Romania, the Danube Delta coastline is typical) usually associated with the mouths of big rivers and consisting of complex sandy barrier with strong long shore sediment drift systems. These coasts are most strongly influenced by global changes, specifically by sea level fluctuations and changes in the river sediments supply into the coastal zone. A rise in the sea level could under certain conditions result in an active and almost continuous retreat of the beach line (up to 20-30 m/yr in some sections of the Danube Delta).
- Erosive coasts within low-standing plateau and plains, with active cliffs (for Romania the coastline from Cape Midia to Vama Veche is typical); this category may also be affected by erosion processes, but the rates of coastline

retreat are smaller (only 1-2 m/yr.).
- Mountainous coasts, with cliffs, marine terraces, land slides, sometimes with sandy or gravely beaches (Romania has no such type of littoral, as Bulgaria and Georgia have); this category is the least affected by the erosion processes, being generally constituted of consolidated rocks, which are difficult to erode.

STATE OF THE ART

Erosion of the shoreline, caused by currents, tidal movements and wave, and wind action, is a common natural phenomenon. At present, approximately 70% of all sandy coasts in the world are subject to coastal erosion. An inventory of coastal evolution in the EU (CORINE program) showed 55% of the coastline (total length of 56000 km) to be stable, 19% to be suffering from erosion problems and 8% to be depositional. Coastal defense activity is following two main types of approaches[5]:

- hard engineering: structures which include breakwaters and seawalls designed to oppose wave energy inputs, groynes, to increase sediment storage on the shore, and flood barrages as water tight barriers;
- soft engineering: coastal processes regulation by manipulating natural systems, which envisage the reduction of anthropogenic loads on the coastal zone and stimulation of natural self-regulating processes.

A successful solution to the problems lies in a rational combining of both directions and types of engineering constructions.

The most used solutions against the coastal erosion and the longitudinal and transversal transport are to settle permanent solutions, like beach-groynes and reefs, which reduce and spread the wave intensity. In many situations this permanent solution was not established following sufficient studies of the coastal dynamics, and thus interfering with the natural erosion~deposition balance, which can create a knock off effects in neighboring localities, while not solving the problem in the site of intervention.

The Pan-European Biological and Landscape Diversity Strategy advocates a series of strategic principles. The Coastal Code of Conduct (XXX[5]) adopts these principles and defines some key elements, which relate to development and management in the coastal zone:

- principle of careful decision making: based on the best available information,
- principle of avoidance: appropriate procedures and projects to avoid adverse effects on biological and landscape diversity (establishing non-development zones);
- precautionary principle: for those activities whose impact has not been yet fully confirmed;
- principle of translocation: exceptional harmful activities which do not absolutely depend on the coastal environment, relocated to areas where they

will cause less impacts (outside of the coastal zone);

- principle of ecological balance and compensation: coastal protection measures should not deteriorate the ecological situation, (but promote increase in biological productivity and loss of coastal habitat) and must be balanced by compensatory conservation measures by the user;
- principle of ecological integrity: preservation of the natural character of dynamic systems (sand, dunes, beaches sea cliffs, deltas, can enhance the resilience of the coastline against coastal erosion and accelerated sea level rise);
- principle of restoration and re(creation): new habitats should be created prior to the destruction of existing ones. Restoration is the challenge for generations to come and it is recommendable to follow its directions:

    - natural development: rivers should be permitted to choose their own course, building banks freely, without being interfered with, restricted, or inhibited;
    - natural structures: during restoration, natural elements like trees, initial flora and fauna and sediments should be left where they are, whenever possible;
    - free-flowing: a river needs free flowing stretches (dams and barrage weirs block and prevent the natural development of rivers and their normal contribution of sediments);
    - more ecology in settlement areas: in densely populated areas ecological measures have to be taken with plants, and a diversified living environment is needed for the safety of the people who live there;
    - let nature grow: letting the dynamics of nature do their job is far better than any artificial scheme whenever possible.

- principle of best available technology and best environmental practice: materials used, which might enter marine or coastal ecosystems should not include contaminants; water conservation should be the primary concern in all developments;
- principle of complexity: technical solutions should combine coastal protection functions with a possibility to use them for recreational, transportation, biological, ichthyologic and other purposes;
- principle of correspondence to coastal landscapes: engineering constructions should be in harmony with coastal landscapes;
- principle of justified risk: decisions on coastal protection should be taken on the basis of scientifically justified measures to reduce anthropogenic transformation of a coastal regime;
- principle of functioning zoning of coasts: delimitation of a coastal zone territory according to priority of economic sectors;
- principle of universality: protection constructions should work effectively

under different conditions (sea level rise due to climate global changes);
- principle of international cooperation for monitoring, management and protection of the coastal zones;
- polluter pays principle: the cost of developments in coastal areas should be borne by development proponents.

ROMANIAN BLACK SEA COASTAL ZONE

Erosion general aspects.
The Romanian Black Sea coastal length is of 245 km meaning 6% from the entire Black Sea coastal length, with 80% beach and low altitude shores and 20% cliffs, or 84% natural shores and 16% constructed shores (ports, hydro technical works).

The Romanian Black Sea coastline (Fig. 2) consist of several geomorphologic areas grouped into two large zones (sectors):

- northern (delta), the creation of the Danube, with an approx. length of 160 km from the Ukraine border to Midia Cape, consisting of alluvial sediments with dunes, extensive lowlands marshes and lagoons, mostly at an altitude of 0.5-0.Sm;
- southern, from Midia Cape to Vama Veche- the Bulgarian border, with an approx. length of 85 km, a relatively higher zone consisting mainly of ground active cliffs (2-40 m high composed of anthropogene and neogene loess, clay and limestone), punctuated by short beaches from place to place and narrow strips (composed of sand) of old lagoons.

In the Danube Delta, under the influence of natural factors, the shoreline is mostly retreating. Here the beach lost in the last 35 years about 2400 hectares (about 80 ha/yr.). The shoreline has retreated in some points more than 400 m, most of it having retreated some 200m. At the same time, accumulation amounted to 169 hectares only, that is 6-7 ha/yr. During storm periods, in areas where the shore is lower, the coastal barrier is often damaged or submerged by the waves allowing seawater to flow into the freshwater lakes, breaking their ecosystem. In other areas, where the shore is higher, a layer of sand of among 80-100cm is removed[6].

The Danube Delta is the only delta in the world declared in 1990, with the support of J.Y. Cousteau, a biosphere reserve, joining another 352 reserves from 87 countries and is included, with the lake complex Razim-Sinoe, in UNESCO's Cultural Heritage List. According to the Ramsar Convention, the Danube delta includes 600 wet zones, and is in 8th position as the world's largest wet zone, and in 22nd position as the largest delta in the world and the third in Europe. The total area covered is 4178 sq. km of which: 82%-(3466 sq. km) is in Romania and 18%-(732 sq. km) is in Ukraine. As a real natural museum of biodiversity, it includes 30 types of ecosystems, with 5149 species of flora and fauna. The Danube Delta is the youngest landscape in Europe, growing annually due to the 40 million tons of alluvium dumped there by the Danube itself. With 15,000 people, the population of the delta is rather sparse on this 45th parallel where the main relief is the banks of alluvial deposited by the river's water, or by the Black Sea's water, and the area of plain and marsh which remain permanently above the floodwater level. The delta was formed around the three main channels of the Danube: Chilia (120km) the northern and the most active which borders the Ukrainian territory, Sulina (67.3 km), the shortest which represents a navigation channel, and St. George (69.7 km).

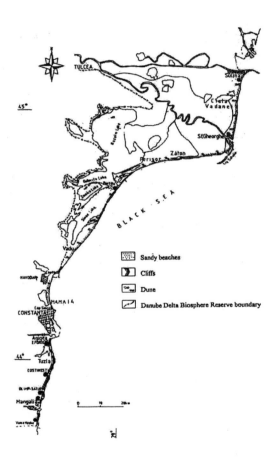

Fig. 2. The Romanian Black Sea coastline.

The coastal area, by its location, is characterized by the presence of unique resources (beaches, fishing resources, the great biodiversity, transport by water, therapeutically properties of water, recreation possibilities, as well as the Danube Delta, acknowledged internationally as a biosphere reserve with high ecological value), which differentiate it from the other parts of the country.

Beach segments can be found on 70.4% of the overall length of the Romanian shoreline, of which 62% is within the Danube Delta Reserve territory. The remaining 8% is distributed along the southern part of the Romanian shoreline, between Midia and Vama Veche in the form of beaches separated by cliffs. In the northern sector, the beach is of variable widths, from a couple of meters-where the erosion process is intensive, up to tens of meters- generally where cumulative processes exist or processes are have a relative equilibrium.

Littoral barriers represent mainly the Danube delta sector, with widths from less than 50-60 m up to medium values of 200-300 m.

In the southern sector Midia-Vama Veche, where the shore is relatively higher the shore-line has a different evolution. The destructive effects are reduced due to shore lithology (hard substrate).

Coastal protection has been built in the northern part to stabilize the coastal barrier. By closing Portita mouth, the lagoon type lake Razelm was separated by sea through a narrow strip of 6 km, and Sinoe lake by a 19,8 km protection barrier. But the majority were built in the southern part of the coastline, where most of the coastal activities involving industry and tourism occur. In spite of these protection works in some parts of this section (because of the degradation of 55% of the total littoral system of protection), the beach has retreated: 40m-Eforie (1981-1992); 24m-northern Neptune; 36m- Venus-Saturn (1983-1992).

The causes of these erosion aspects are:

- drastic reductions of the alluvial deposits outlet into the sea by the Danube and the rivers from the northwestern part of the Black Sea, because of the hydro technical dams built along these rivers;
- extension 8 km offshore of the main entrance into the sea of the Danube (the Sulina channel mouth), due to jetties built to protect Sulina's navigation channel (the majority of total sediments carried by the Danube are deposited north and at the end of these barriers);
- bringing the sediment transport further to the seashore as a result of Sulina's barriers, the development of Sakhalin Island which blocks the St. George-channel alluvial transport of sediments, and the extension of Constanta and Mangalia harbors, and Midia-Navodari port (a huge trap of sediments having as a result partially destroyed Mamaia beach),
- reduction of the bioorganic sand (diminishing of the Mya, Cardium and Mytilus shell stocks and illegal extracted sand from the shore for industrial purposes: construction, as raw material for heavy minerals extracted from Corbu beach, or as a chicken-feed ingredient)
- impact of the hydro technical structures for coast protection,
- the increased shipping traffic and pollution, which changed the ecosystem, thus reducing the organic source of sand formation.

In the last 55 years different protective hydro technical works have been built: drainage, consolidations, terraces and slopes to protect cliffs, and heavy-type solutions to protect beaches, such as dikes, barriers and break waves, groynes, jetties. After 1991 this activity decreased sharply.

During 1987-2000 in the southern littoral the coast-line varied according to Figure 3.

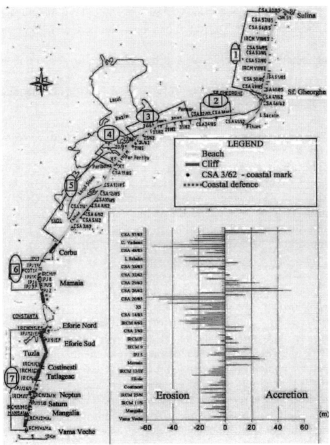

Fig. 3. Romanian shoreline evolution: national monitoring, 1987-2000.

National monitoring for the entire shore (Fig. 3) is based on 140 boundary marks mounted at 3-4 km distant from each other in the northern sector, where because of the erosion and destructive antropic actions, 75% of these marks has disappeared. In the southern sector, Midia-Vama Veche, coastal measurements are made on 98 profiles. On the tourist beaches, these marks are placed between 300-400 m.

In the last decades a variety of factors have had detrimental effects and degraded in many ways the Romanian Black Sea coastal environment. Studies revealed that the Romanian shore is in a critical stage, regarding erosion (about 60-70% from the entire shore length). This unstable equilibrium is the result of two actions with opposite effects. On one hand the drastic reduced Danube flow sand and consequently the transport of sediments on the beaches, and on the other hand the natural factors specific to the temperate area.

Natural factors: meteorological and hydrological, are essential in the erosion process.

Winds, northern predominance (40%-197]-94-XXX[8]), as the most important meteorological factor, have 2 actions:

- direct, blowing the sand from the beach to the sea, or in the opposite direction up on the beach;
- indirect, by waves, sea currents and temporary sea level rising (waves are erosion factors- the highest average waves recorded: l.3m height, 5.2 sec. period and 35 m length with a 28% max. frequency (XXX[8]); currents are transport agents for the eroded material -the existing surface current is north-south oriented with a width of approx. 6-10 miles and an opposite, compensatory one under this surface current, north oriented).

ACTIONS

The rapid coastal growth, the depletion and destruction of fragile resources, and the strategic importance of coastal zone environments, require the Black Sea's coastal nations to develop and implement coastal zone management programs that allow continued use of the resources for economic development. Integrated coastal zone management, in this context, is a dynamic process that develops a coordinated strategy and implements it for the allocation of environmental, socio-cultural, and institutional resources to achieve the conservation, rehabilitation and sustainable multiple use of the coastal zone. It shall be conducted to meet the needs of both present and future generations and to preserve essential ecological processes and biological, landscape and cultural diversity. ICZM according to the European Commission definition is "a continuous process of administration, the general aim of which is to put into practice sustainable development and conservation in coastal zones and to maintain their biodiversity."

Priority actions for the Black Sea coastal zone are: to create a system of protected areas within the coastal zone (biodiversity, fishing, tourism, health, archeology, geology areas), to form a legislative base for the Black Sea ICZM system (national regulations and policy and the alignment to European legislation), to reconstruct real coastal management system (involving industry, tourism, agriculture, urbanization) to set up an economically self-sustainable system of land use and nature use, to develop administrative economic methods as national priorities and to use public participation in developing policies and taking decisions.

An attempt to solve one aspect is a recent LIFE-Environment project (Coastal Erosion Recovery & Saving) proposed to the European Commission for funding (XXX[7]). Its result is to set up for the Eforie sector of the Romanian littoral an operation of coastal defense, taking care of all the factors concerned: hydrological, geo-morphological, natural, socio-economic and institutional, finally a sustainable development for the territory. The aim is to fight beach erosion, recovery of the beaches, slow down of human intervention in beach recovery, letting it evolve naturally, support the natural process, reconciling the protective demand with the conservative one. Considering the mentioned aspects, the proposal is to execute a plan for coastal protection, employing efficient and reliable techniques (reefs) combined with the innovative "beach-groynes" according to the mathematical model prepared by the Danish Hydraulic Institute. The innovative technique consists of combining the reef beach groynes (mainly submerged), with a diaphragm made of steel (or wood) boards (Larsen type), hinged between them and driven into the sand, or, as alternative with wood posts and joined one to the other, in order to make containers to be filled with sand. In the case of the wood posts, a geotextile coat will be placed to prevent the loss of sand. The shape of the diaphragm of the beach groyne will be similar to a Gauss curve with an extension on the seaside of about 50 meters (50% of the initial distance between the rocks and the sand shore) and width at the base of about 50 m. This natural solution has the advantage of restoring the initial condition without deteriorating the quality of the water, of the beach, and the coastal habitat in general. The beach will be naturally rebuilt, gaining, during the first year, 20/30 meters using "wind cuts" made of restored wood, stopping the damage suffered by the vegetation and dunes, with finally an important economic and environmental impact on social activities. A high quality sandy material will be restored, and the aquatic life will be promoted. In some parts of the coast, the reconstruction of dunes will encourage the development of vegetal species, generating a positive effect for the settlement and repopulation of autochthon animal species and the protection of the coast, improving also morphological stability conditions. The narrow strip beach reconstitution will provide protection from an attack by the sea on the naturalist oasis of Lake Techirghiol. The project aims to avoid the mistakes made in the past (very expensive, less effective, with a terrible impact on tourism) and adopts technical solutions of a low impact that can be moved further out to sea while recovering meters of sanded shores. The proposed project would be cost-saving compared to traditional works (rocks, barriers building, classical brushes and constant artificial refilling) estimated in a 40% reduction of costs. The final objective of the project is to rehabilitate part of the littoral, and also to recover its quality by trying to retain natural beaches and its autochthon botanical psammophille species, and protecting them in a restricted territory. In this particular area the results of the project might be used for future EU financed interventions aiming at solving the problem in the Black Sea region, therefore contributing to the employment growth in this specific sector.

156

CONCLUSIONS AND RECOMMENDATIONS

The need of the coastal zone law is evident in the context of the alignment of Romania to the European Community Aquis.

The main management problems and needs of the Romanian Black Sea coast (XXX[9]) that brought the necessity of the ICZM Law are the following:

- problems with the effective management, urbanization of the coast and the construction in the area (to agree on common aims);
- insufficient protection of historical and cultural sites;
- need for local self-financed administration and deconcentration of power to the regional and municipal levels of government; utilize financial resources efficiently;
- efficient instruments for public participation in the decision-making process for the coast development projects;
- need of new tools and procedures for the co-ordination of the conflicting sectoral interests and the conflicting interests of all the parties involved in the coastal development and preservation;
- conflicts in authority and functions of the government agencies responsible for the different sectors of the coast;
- create a legal framework for the implementation of the strategies and policies, establishment of a proper and effective institutional structure for the ICZM;
- insufficient and ineffective definition of the responsibilities of the state agencies and other authorities for different elements and activities at the coast–beaches and dunes, coastal lakes, fisheries, tourism;
- the implementation and enforcement of the legal context of the existing and well-defined environmental legislation for the area;
- improvement of the planning, development and management of an adequate environmental and technical infrastructure.

Short term solutions are soft engineering ones (mainly sand restoration with "wind cuts") and long term solutions are usually the hard engineering ones.

For the Romanian coastal zone, the national projected program of the hard engineering protective works was not entirely fulfilled, due to financial and complexity reasons.

Actually, there is no global program for maintaining and repairing the existing protective works that are in an advanced stage of degradation.

Protective works were only partially beneficial for the Romanian littoral.

Previous experience has revealed that to recover the original state of our littoral is impossible and all we can do from now on is to stop the destructive actions.

Consequently, it will be necessary to create a national coordination system for our littoral, a single concept which has to be on the Ministry's of Waters and Environment

Protection agenda for concerted cooperation on a national level. Particularly for the southern part Navodari-Vama Veche, the priorities are the Mamaia resort (photos 1, 2, 3, 4, 5, and 6), Eforie beach (photos 7, and 8), and the narrow barrier beach between Teehirghiol lake and the sea (photos 9, and 10).

REFERENCES

1. BSEP, (1997)-Biological Diversity in the Black Sea -A Study of Change and Decline (Ed. By Zaitsev, Y. and Mamaev, V.), United Nations Publications, New York. Black Sea Environmental Series. Vol.3 ISEN 92-1-126075-2.
2. Ross D.A., Uchupi, E., Prada, K.E., Macilvaine, J.C. -The Black Sea- Geology, Chemistry, and Biology; Memoir No. 20-The American Association Petroleum Geologists, 1974
3. Görür Naci -Origin of the Black Sea- in Environmental Degradation of the Black Sea: Challenges and Remedies, 1999 Kluwer Academic Publishers.
4. Ryan W.B.F., Pitmam, III, C.W., Major C.O., Shimkus, K., Maskalenko, V., Jones, G.A. Dimitrov, P., Görür, N., Sakinç, M., Seyir H.Y., 1997-An abrupt drowning of the Black Sea Shelf at 7~5 k. yr. BP. Geo-Eco-Marina, Bucharest.
5. XXX- Pan - European Code of Conduct for Coastal Zones, Council of Europe Document PE-S-CO (97)14 Add, Strasbourg, 1998.
6. XXX- The ICZM Report on the Romanian Coastal Zone- RMRI -focal point GEF BSEP. Constanta 1996
7. XXX- Life- Environment project CEReS- Coastal Erosion Recovery & Saving Constanta 2000.
8. XXX- National Institute for Marine Research and Development, Constanta, Romania- 1995 Study on the Erosion Aspects and Shore Protection for the Roman ian Littoral.
9. XXX- Coastal Zone Management Legislation for the Black Sea Region- proceedings-Bucharest, Romania, 20-21 September 2000.

Photo 1. December 1998. Mamaia beach, before a storm.

Photo 2. December 1998. Mamaia beach during a storm.

Photo 3. April 1998. Central Mamaia beach.

Photo 4. April 1998. Southern Mamaia beach.

Photo 5. Summer 1995. Mamaia Resort.

Photo 6. Summer 1995. South Mamaia. Hotel Park Protective Dyke.

Photo 7. April 1999. Eforie beach (northern part).

Photo 8. April 1999. Eforie beach (southern part).

Photo 9. 1998. Eforie beach hydrotechnical works. Technirghiol Lake.

164

Photo 10. 1998. Eforie beach/Techirghiol Lake.

CAUSES AND CONSEQUENCES OF NUTRIENT OVERENRICHMENT OF COASTAL WATERS

DONALD F. BOESCH
University of Maryland Center for Environmental Science, Cambridge, Maryland, 21613 USA

INTRODUCTION

While developed societies made significant progress in reducing the discharge of many wastes—including industrial wastes, organic matter in sewage, sludge, contaminated sediments and other material dumped at sea, toxic compounds, and oil-into coastal waters, nutrient over-enrichment intensified and spread widely among the planet's coastal environments during the last half of the 20th century. This has often had devastating effects on the fisheries, biodiversity and services provided by these ecosystems. Eutrophication, as used here, is the enrichment of aquatic environments by nutrients, especially compounds of nitrogen and/or phosphorus, causing an accelerated growth of algae and higher forms of plant life to produce an undesirable disturbance to the balance of organisms present in the water and to the quality of water concemed[1]. These nutrients come not only from treated waste streams, but also from land runoff, particularly from agriculture, and the atmospheric deposition of byproducts of fossil fuel combustion. These so-called nonpoint sources grew dramatically and arc particularly hard to control. Its pervasive extent, consequences, trends, and challenges qualify the eutrophication of coastal waters as a planetary emergency of the early 21st century.

This report provides an overview of the causes and consequences of coastal eutrophication, its global patterns and trends, sources of nutrients and efforts to control them, and efforts to restore degraded ecosystems.

CAUSES AND CONSEQUENCES

Effects of Overenrichment

Mineral nutrients are, of course, essential to life and to the productivity of coastal ecosystems. Because they receive the infusion of nitrogen, phosphorus, silicon, iron and other minerals both from land—via river discharges—and from the ocean-via upwelling of nutrient-rich, deep waters—coastal ecosystems are among the most productive on Earth and are valued for their rich fishery production.

The input of nutrients can, however, overwhelm the capacity and resilience of the ecosystem. Nutrients stimulate microbes, including phytoplankton, and macroscopic plants to grow. Biomass of planktonic organisms and macroalgae may accumulate and then decompose, consuming much of the oxygen available in the water column. This is especially the case where the water column is stratified, with saltier, cooler water near the bottom. Decaying organic matter accumulates in bottom waters below the density gradient, consuming oxygen, which is not replenished by mixing from the surface waters. The bottom waters become hypoxic (with dissolved oxygen levels below the requirements of metabolically active fish and invertebrates) or even anoxic (with essentially no dissolved oxygen). Hypoxia can result in mass mortality of benthic (bottom dwelling) species and tinder chronic stress can convert the seabed into a virtual biological desert devoid of animal life[2].

An increase in nutrient supply generally increases the production of phytoplankton, but has a selective effect, increasing certain rapidly growing species at the expense of others. Blooms of planktonic algae reduce water clarity, and thus light penetration, and, because of this selective effect, alter the quality as well as the quantity of food for secondary consumers. The nature of the phytoplankton response depends on the mix of nutrients available. Many coastal systems are receiving increasing supplies of nitrogen from human activities, while supply of silicon, which is required by diatoms, is declining because of trapping of this natural product of crustal erosion behind upstream dams[3]. Under eutrophication, the phytoplankton base of the food chain shifts, away from larger celled organisms toward flagellates and bacteria. Increased primary production coupled with qualitative changes in the producers have consequences throughout the food chain, leading to diminished success of fish larvae, for example, and increased production of jellyfish[4].

Sometimes, this also results in blooms of species of dinoflagellates and other organisms that produce toxins or have other harmful effects[4]. In recent decades there has been a worldwide explosion in the extent and frequency of these so-called harmful algal blooms, including those that produce paralytic, diarrhetic, neurotoxic, and amnesic shellfish poisoning in humans. Although eutrophication is thought to be a contributing factor in many of these cases, it is by no means responsible for all harmful algal blooms, some of which are occurring naturally and others may result from changes in climate and human introductions among other causes.

In shallow coastal waters with ample light penetration, eutrophication may cause proliferation of macroalgae growing on the seabed. These macroalgae smother benthic habitats, deplete dissolved oxygen when they decompose, and create nuisances when washed up on beaches. Nutrient enrichment can cause overgrowth or coral reefs by macroalgae, particularly where overfishing has reduced the populations of grazing animals[5,6]. Eutrophication has also resulted in extensive losses of seagrass meadows due to shading by increased phytoplankton biomass and overgrowth by epiphytic algae[7].

Consequences to Ecosystems and Society

The broader consequences of these effects of eutrophication-hypoxia, algal blooms, food chain shifts, and loss of seagrass and coral reef habitats - to the ecosystems and to human society are significant, but incompletely known. Eutrophic coastal ecosystems that are oxygen stressed and microbially dominated have reduced biodiversity and resilience in the face of other natural and anthropogenic perturbations[6,8]. The effects on living resources in coastal waters arc complex and not easy to quantify, even when the region affected is large and periodically affected by severe hypoxia[8]. To a certain degree eutrophication may increase fisheries production, particularly by pelagic species that feed on the increased supplies of plankton in the water column, but as hypoxia in bottom waters develops, supplies of benthic fish and shellfish are reduced[9]. However, many coastal ecosystems are also affected by overfishing, particularly of predaceous species that exercise top-down controls on the structure of ecosystems[6,10]. The interactive effects of eutrophication and harvesting can result in ecosystems prone to nonlinear response and collapse (Fig. 1). Dramatically altered coastal ecosystems, from decimated coral reefs of the Caribbean to the coastal regions of the Black Sea, may be the ultimate consequence.

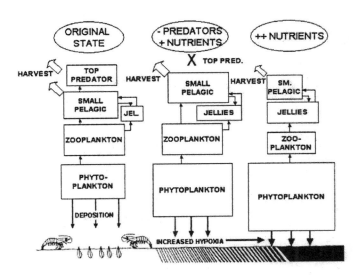

Figure 1. Simultaneous effects of eutrophication and fishery harvest on marine food chains (after Caddy[10]).

Species that are dependent on seagrass meadows or coral reefs are, of course, particularly susceptible to the loss or degradation of these habitats. In general, severe eutrophication is detrimental to the living resources of coastal ecosystems, however there remains considerable controversy about the degree to which eutrophication actually enhances production at higher trophic levels[11], whether the enrichment effects outweigh the detrimental effects[8], and about the consequences of reducing eutrophication on fisheries production[12]. In already productive coastal ecosystems, eutrophication tends to enhance the growth of forms of phytoplanklon that support microbial production rather than harvestable animal populations[13]. In addition to the effects on fisheries, eutrophication has other impacts on the economies of coastal communities, including diminished recreation and tourism[4]. For example, reduced water clarity, windrows of rotting algae on beaches, and health concerns about harmful algal blooms can deter beach goers and tourists, with devastating effects on local economies.

PATTERNS AND TRENDS

Worldwide Distribution
Many bays and estuaries and even some large regions of semi-enclosed seas exhibit the effects of severe eutrophication, including hypoxia, algal blooms, and loss of seagrasses. The problems are most severe in the Europe, North America and Japan, where there are large inputs of nutrients from land. Larger ecosystems affected include the Baltic Sea, the eastern North Sea, the northern Adriatic Sea, and the northwestern Black Sea in Europe; the Chesapeake Bay, northern Gulf of Mexico and Long Island Sound in the United States; the Seto Inland Sea in Japan; and the Gulf of Bohai-Yellow Sea in Asia (Fig. 2).

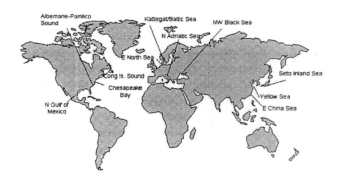

Figure 2. Semi-enclosed coastal seas experiencing severe eutrophication.

The deep basins of the Baltic Sea are continuously anoxia and many of the shallower embayments and the Kattegat at the entrance of the Baltic experience seasonal hypoxia[14,15]. Filamentous green algae have replaced brown seaweeds in shallow, rocky habitats and the depth at which attached algae can grow has diminished. Blooms of cyanobacteria and harmful algae occur with increased frequency and extent. The German Bight of the North Sea also experiences occasional hypoxia and extensive algal mats now cover the intertidal banks of the Wadden Sea as a result of the influx of North Sea water enriched by nutrients discharged from the 'mine and Elbe rivers'[6]. The northern Adriatic also undergoes seasonal hypoxia and produces noxious floating algal growth that fouls beaches[17]. Perhaps the most dramatic and extensive eutrophication was on the northwestern shelf of the Black Sea. Vast (10,000 km^2) meadows of attached red algae that once extended under clear water nearly disappeared as a result of decreased water clarity due to nutrient enrichment. Extensive hypoxia in bottom waters later developed, contributing, along with fishing pressure and introduced jellyfish, to the collapse of many fisheries

In North America, the most intensively studied coastal ecosystem infuenced by eutrophication is the Chesapeake Bay, a 300 km long estuary on the east coast. There seasonal hypoxia and anoxia in bottom waters have expanded and the majority of extensive seagrass beds that once covered many shallow water environments have been lost[19]. The deeper pans of Long Island Sound, east of New York City, experience seasonal hypoxia[20]. The continental shelf of the northern Gulf of Mexico off of the Mississippi River delta has the largest area of hypoxia, in some years covering 20,000 km2 of seabed[21]. Gulf of Mexico hypoxia has been the subject of a recent integrated assessment of its distribution, variability, causes, and history[22].

Several portions of the Seto Inland Sea in Japan witness seasonal hypoxia and harmful algal blooms as a result of eutrophication[23]. In addition, other large shallow embayments, including Tokyo Bay and Ise Bay experience severe hypoxia. Elsewhere in Asia, hypoxia has been observed off the mouth of the Chang Jiang (Yangtze) River and coastal waters in the Yellow Sea, particularly in the Gulf of Bohai, are experiencing increased frequency and severity of harmful algal blooms.

In addition to these larger coastal water bodies, ranging from large bays to the Baltic Sea, there are many smaller bays and estuaries in Europe, North America, Asia and Australia that are degraded by eutrophication. For example, 44 of 138 estuaries along the coast of the United States have high levels of eutrophication and an additional 40 have moderate symptoms[24]. These symptoms include increased phytoplankton growth, increased growth of macroalgae and epiphytes, low dissolved oxygen, harmful algal blooms, and loss of seagrass Similarly, eutrophication of coastal bays, fjords and lagoons is a widespread problem in Europe[25].

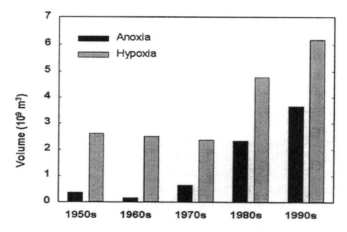

Figure 3. Changes in the volume of summertime
hypoxic and anoxic water in Chesapeake Bay[27].

Late 20th Century Phenomenon

Although some coastal ecosystems, particularly those off the mouths of major rivers, are naturally eutrophic, there is strong and parallel evidence that coastal eutrophication greatly intensified during the latter half of the 20th century. This evidence is based both on direct periodic observations of nutrient levels, water clarity, dissolved oxygen levels, seagrass distribution, and algal blooms and historic reconstructions using the chemical constituents and microfossils of buried sediments. For example, increasing trends in nitrogen or phosphorus have been demonstrated for the Black Sea, Baltic Sea and Irish Sea after the 1950s[1]. In one of the few areas where it has been consistently measured, primary production by phytoplankton doubled from the beginning of the 1960s to the 1990s in the southern Kattegat between Denmark and Sweden[26]. Using the same simple technology, the Secchi disk, first used there early in the century light penetration was shown to have declined dramatically in the northern Adriatic Sea after the middle of the century. The volume of summertime hypoxic and anoxic water in the Chesapeake Bay progressively increased during the decades of the 1970s, 1980s and 1990s[27] (Fig. 3).

Water transparency declined, extent of hypoxia increased, and harmful algal blooms became more frequent in Mikawa Bay, Japan after the late 1950s and into the 1980s[28] (Fig. 4).

The sediment record in the central basin of the Chesapeake Bay reveals a chronology of eutrophication that began with extensive land clearing by European colonists in the late 18th century[9]. More nutrients ran off the land, resulting in increased primary productivity and a shift in the type of dominant diatoms. Seasonal hypoxia was intermittent and modest until the 20th century and intensified during the during the 1950-1980 time period, as reflected in the degree of pyritization of iron, levels of reducible sulfur, and assemblages of microfossils.

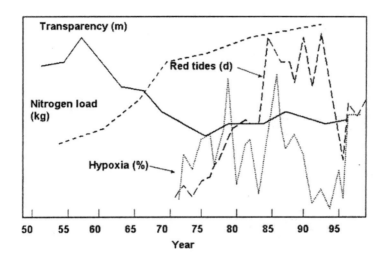

Figure 4. Relative changes in nitrogen loading, water transparency, extent of hypoxia, and frequency of red tides in Mikawa Bay, Japan (after Suzuki[28]). Scales of the variables are relative to their ranges: nitrogen to 14×10^6 kg, transparency to 6.3 m, frequency of red tides to 365 days, and hypoxia to 10 percent of area of bay with less than 30% hypoxia.

The dramatic eutrophication seen in many coastal waters around the world after the 1950s reflects substantial increases in the loading of nutrients, particularly nitrogen. Increased nutrient loading resulted from population growth, the collection and discharge of human wastes into surface waters, intensification of animal production, the combustion of fossil fuels (with the concomitant release of nitrogen oxides into the atmosphere), and, particularly, the widespread use of manufactured fertilizers after World War II. In addition, the capacity of river catchments to retain nutrients was reduced as a result of artificial agricultural drainage, the elimination of wetlands and riparian forests, and constriction of flood plains for flood protection and river training. As a result of the increased fixation of nitrogen and mobilization of phosphorus, the flux of these nutrients into coastal regions increased virtually exponentially. For example, it is estimated that the present flux of nitrogen into the North Sea is 13 times that when there was no human influence, flux along the northeastern United Sates nine times, and flux from the U.S. into the Gulf of Mexico five times[29]. The flux of nitrate, the principal form of inorganic nitrogen, from the Mississippi River basin into the Gulf of Mexico increased by a factor of three since the 1960s[30].

Sources of Nutrients
Fundamentally, the sources of human-mobilized phosphorus entering coastal ecosystems are soil erosion (as phosphorus tends to be bound to particles), agricultural fertilizers (largely mined from geological deposits), and industrial or domestic chemicals (e.g. detergents). These enter surface waters via diffuse sources of runoff from agriculture (from soil, fertilizers, and animal wastes) or urban and suburban areas and via point-source discharges of sewage or industrial wastes. Sewage and agricultural runoff typically dominate.

The fundamental anthropogenic sources of fixed nitrogen entering coastal waters are the combustion of fossil fuels (which release nitrogen oxides or NO_2 into the atmosphere), manufactured nitrogen fertilizers, biological fixation by crops, and mineralization of organic matter long sequestered in soils. Nitrogen has been called the most promiscuous element, because it can easily move from gaseous to dissolved to solid phases. Its many sources and routes make it more challenging to control than phosphorus. Atmospheric NO_2 can be deposited on land or surface waters dissolved in rain or in particulate form and is a significant cause of acid deposition as well as coastal eutrophication. Nitrogen may easily enter ground water (because of the high solubility of nitrate). Fertilizers are the original source of most of nitrogen in sewage discharges (from food), but because 70 percent of crops is fed to animals, much of the nitrogen in crops ends up in animal wastes, where some of it volatilizes as ammonia, to be deposited somewhere downwind. Because nitrogen typically limits primary productivity in marine as opposed to freshwater environments, most of the focus of abatement programs and the remaining discussion is on nitrogen.

The relative importance of anthropogenic nutrient sources varies greatly among affected coastal ecosystems. Using the U.S. Atlantic and Gulf coasts as an example (Fig. 4), the loadings of nitrogen into coastal waters with little agriculture in their catchments and receiving waste waters from urban centers, such as Massachusetts Bay (Boston) and Long Island Sound (New York City), are dominated by point source discharges[31]. Nitrogen loadings into coastal waters with heavily agricultural catchments, such as the Louisiana continental shelf off the Mississippi River or Pamilco-Abemarle Sound in North Carolina are dominated by agricultural sources (runoff; groundwater nitrogen losses, and deposited ammonia). Some coastal ecosystems, such as the Chesapeake Bay, receive a mix of inputs from agriculture, atmospheric deposition of NO_2, and point source discharges of sewage from its population centers. This means that pollution abatement strategies often have to be multi-pronged. Agricultural sources contribute half or more of the nitrogen loading in the larger eutrophic coastal systems that receive the discharges from large catchments, including the Baltic, Adriatic and Black Seas, the northern Gulf of

Mexico and Chesapeake Bay.

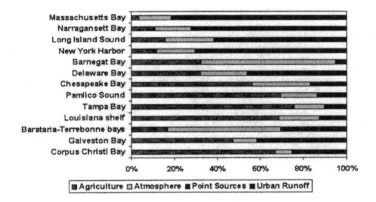

Figure 5. Proportion of sources of nitrogen loadings to selected bays and estuaries along the U.S. Atlantic and Gulf coasts[31].

Susceptible Coastal Ecosystems

Not all coastal waters are equally susceptible to nutrient over-enrichment. The most susceptible systems are those with limited tidal exchange and mixing, long retention time, vertically stratified water masses, and relatively low background concentrations of suspended sediments[4]. Highly turbid and tidally energetic systems like San Francisco Bay are not as susceptible to eutrophication as partially stratified systems with relatively little tidal resuspension of sediments such as the Chesapeake flay, even though their nutrient loading rates may be greater. Similarly, deep waters that lie close to the coast, such as off Southern California, provide rapid dilution of even large nutrient discharges without displaying eutrophication. This means that controlling nutrient inputs is of relatively little importance for achieving a healthy coastal environment in some areas, while it is of paramount importance in others.

THE FUTURE

Expanding Food Production and Energy Consumption

The widespread coastal eutrophication that developed during the later half of the 20th century is largely a manifestation of dramatic change in the nitrogen cycle of the biosphere, in which the world's supply of fixed nitrogen more than doubled in just a few decades[32]. This increase was much faster than the increase of CO_2 in the atmosphere and any other so-called global change. It has resulted in air quality problems (ozone formation), lake and soil acidification, reduced biodiversity in terrestrial ecosystems, and releases of N_2O (a potent greenhouse gas), as well as coastal eutrophication. The increase in nitrogen loading has been driven by increased consumption of fossil fuels (and associated release of NO_2) and the expanded use of chemical fertilizers.

Tilman et al.[33] demonstrated that the global application of nitrogen in fertilizers increased by eight-fold from 1960 to 1980 and the application of phosphorus in fertilizers grew by three-fold (figures exclude the former USSR). They further projected that, as the world population grows from 6 billion to 9 billion by 2050, the application of fertilizers will increase 2.7 times present values for nitrogen and 2.4 times present values for phosphorus. These increases will be driven not only by population growth but also by diets richer in meat. Fertilizer use in western Europe and North America has stabilized, so that most of this increased application of fertilizers, and thus additional releases of nutrients to coastal waters, will be in developing nations, where populations are growing and standards of living are increasing. As a result, one would expect the coastal eutrophication well in evidence in the developed world will expand to other coastal regions[34].

Controlling Sources of Nutrients

Significant reductions in nutrient inputs to coastal waters may be achieved by approaches that: (1) reduce the use of the nutrients in the first place; (2) control losses to the environment at the point of release (e.g. farm field, animal feeding operation, residential area, vehicle, power plant, or waste treatment plant); and (3) sequester or remove pollutants as they are transported to the sea[31,35].

A grand demonstration of the effectiveness of nutrient source reduction on eutrophication of coastal waters was presented in the northwestern Black Sea during the 1990s[36]. With the collapse of the centrally planned economics of eastern Europe, the applications of fertilizers rapidly declined to less than half the levels of the 1980s (Figure 6). Discharges of phosphorus and subsequently nitrogen declined several years later. In 1996 there was no hypoxic zone on the northwestern shelf of the North Sea for the first time in 23 years. Of course, the challenge we face is to reduce nutrient loadings from agricultural and other sources without having to suffer the economic collapse experienced in eastern Europe.

Phosphorus can be almost completely removed from wastewaters by additional chemical and biological treatment. This is now a widespread practice in Europe and North America and, along with bans of phosphate-based detergents, has resulted in significant reductions of phosphorus loading[25]. Significant nitrogen removal from wastewaters has been achieved in Scandinavia and for some U.S. estuaries by biological nutrient removal, a process in which one group of microorganisms convert wastewater ammonia to nitrate and another converts nitrate to dinitrogen gas[4].

Reductions in nitrogen oxide (NO_2) emissions to the atmosphere have been driven by air quality considerations generally outside the influence of water quality or coastal ecosystem managers. Goals have been to reduce ground-level ozone that poses human health risks and stresses forests and crops. Some reductions in emissions in NO_2 emissions from power plants and vehicles have been achieved in Europe and the U.S. and significant reductions should be achieved by additional regulations designed to meet air quality requirements.

Abatement of agricultural sources of nutrient pollution has been a more difficult challenge. To be practical, abatement of agricultural sources of nutrients must focus not only on reducing fertilizer use but also on plugging the many leaks in agricultural nutrient cycles. Although efficiencies in fertilizer use have been slowly but steadily increasing since the mid-1970s in western nations, about one-third of the nitrogen applied in the U.S. is not recovered in harvested crops[4]. Not all of the missing nitrogen contributes to eutrophication of coastal waters. Much is denitrified in soils or aquatic systems en route to the sea or is stored in soils or groundwater. In addition to increasing the efficiency of nitrogen uptake by crops, the return of nitrogen gas to the atmosphere can be enhanced through management practices.

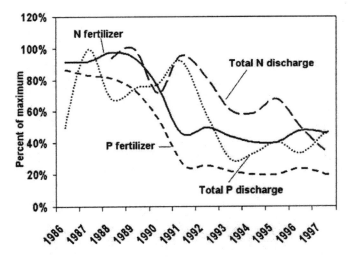

Figure. 6. Changes in the amount of nitrogen and phosphorus applied as fertilizers in the Danube River catchment and the loadings by the river to the Black Sea[36].

Various agricultural practices affect nitrogen and phosphorus runoff and losses to groundwater (which ultimately seeps into surface waters). Practices employed to reduce soil erosion also reduce nutrient pollution. Other practices are more specifically targeted to the efficient use and retention of nutrients: (1) soil testing to precisely match fertilizer applications to crop nutritional needs (many farmers still overapply to insure maximum crop yields); (2) applying fertilizer just at the time the crop needs it; (3) crop rotation; (4) planting cover crops in the fall; (5) using soil and manure amendments, and (6) specialized methods of application[4,31]. Landscape practices such as maintaining buffer strips between cultivated fields and nearby streams, moderating excessive drainage by ditches and tile lines, and maintaining wooded riparian areas can further reduce the leakage of agricultural nutrients to surface waters. By combining these approaches a significant portion of the edge-of-field nitrogen losses can be reduced.

Animal wastes are often the most significant source of nutrient pollution from agriculture[4,25]. Proper management of manures requires effective holding facilities and avoidance of overloading soils with manure applications. This is frequently difficult in regions of intensive animal production. Enclosures or trapping devices may eventually be required to stem ammonia emissions from animal wastes.

Urban runoff can also be an important diffuse source of nutrients. Reduction and control of urban and suburban diffuse sources can be achieved through: (1) reductions in horticultural fertilizer use; (2) effective and well-maintained stormwater collection systems (retention ponds can remove 30-40 percent of the total nitrogen and 50-60 percent of the total phosphorus); and (3) improved septic systems that promote denitrification[4]. Preservation and restoration of riparian zones and streams within urban and suburban areas is also an important aspect of effective nutrient control.

Removing or sequestering pollutants as they are transported downstream can also abate nutrient pollution. Often the majority of wetlands that once existed in a catchment have been drained and convened to other land uses. Also, floodplains have commonly been disconnected from their rivers by flood control projects or agricultural conversion and no longer serve as nutrient sinks. Because even the best land management practices will still release nutrients to surface or ground waters, reducing and controlling sources of land runoff must involve large-scale landscape management, including restoration of riparian zones and wetlands. For example, an integrated assessment of hypoxia in the Gulf of Mexico estimated that two million hectares of restored wetlands in the Mississippi River basin would reduce nitrogen loading to the Gulf of Mexico by 20 percent[22]. Coupled with feasible controls in agriculture, this would achieve a nearly 40 percent reduction in nitrogen delivered to the Gulf.

Coastal Seas Management

The various problems experienced by coastal ecosystems must be addressed by integrated approaches that address fishing activities and coastal zone development and related habitat modification, as well as pollution by nutrients and contaminants. But, to reduce the undesirable consequences of eutrophication such management must often reach beyond the coastal zone proper to extend to the entire catchment basin. Moreover, it may have to consider nitrogen originating outside of the catchment but transported into it through the atmosphere. These large and unconventional units for ocean and coastal resource management pose numerous challenges.

Concerted efforts to reverse nutrient over-enrichment are being undertaken for coastal ecosystems that range in size from small estuaries to the Baltic Sea. Once the seriousness and causes of coastal eutrophication began to be understood, multi-jurisdictional commitments were made in the late 1980s to reduce nutrient loadings of nitrogen and phosphorus to the Chesapeake Bay (by 40 percent of "controllable" loads), Baltic Sea (by 50 percent) and North Sea (by 50 percent). Multi-jurisdictional compacts-through the Chesapeake Bay Program, the Helsinki Commission, and the Paris Commission, respectively—established these commitments and guide and monitor their implementation[37]. Subsequent agreements were reached to stabilize the reduced loading levels to the Black Sea[36] and to reduce nitrogen loading from the Mississippi River basin by 30%. Other estuarine management plans[4], river basin action plans[25], or national laws (e.g. Denmark) have also established ambitious goals for reduction of nutrient loadings. These programs challenge intergovernmental commitments and actions. For example, for the Black Sea they must engage not only the six littoral nations, but also 11 other riparian nations in the catchment. Some 30 U.S. states fall within the Mississippi River basin.

Thus far, efforts to reduce coastal eutrophication have met with encouraging but limited success. Nutrient loadings from some European rivers have been reduced, particularly for phosphorus (as a result of discontinuing phosphate-based detergents and waste treatment), but reductions of nitrogen loadings from nonpoint sources have proven more difficult to achieve[25]. Improvements in coastal environmental quality have been achieved, for example in the Stockholm archipelago in the Baltic[15] and in Chesapeake Ray[19] and Tampa Bay[38] in the United States. Generally, these have come largely as a result of nitrogen and phosphorus removal from point sources. Greater success in reducing nonpoint sources, particularly from agriculture, will be required to achieve the restoration goals for most coastal ecosystems.

Strategies to reduce coastal eutrophication place a premium on environmental modeling, and monitoring[1,4] in an adaptive management framework[37]. Models are needed to track sources through the catchment, target abatement, and relate nutrient inputs to coastal ecosystem responses. Monitoring is critical in determining the effectiveness of abatement strategies, evaluating responses of the ecosystem, and placing these responses in the context of ecosystem variability.

REFERENCES

1.   J.E. Cloern. 2001. Our evolving conceptual model of the coastal eutrophication problem. Marine Ecology Progress Series 210:223-253.

2.   R.J. Diaz and R. Rosenberg. 1995. Marine benthic hypoxia: a review of its ecological effects on the hehavioural responses of benthic macrofauna. Oceanography and Marine Biology Annual Review 33:245-303.

3.   D. Justic, N.N. Rabalais, R.E. Turner, and Q Dortch. 1995. Changes in nutrient structure of riverdominated coastal waters: stoichiometric nutrient balance and its consequences. Estuarine, Coastal and Shelf Science 40:339-356.

4.   National Research Council. 2000. Clean Coastal Waters: Understanding and Reducing the Effects of Nutrient Pollution. National Academy Press,

Washington, D.C.

5. B.E. Lapointe. 1999. Simultaneous top-down and bottom-up forces control microalgal blooms on coral reefs. Limnology and Oceanography 44:1586-1592.

6. J.B.C. Jackson and 17 others. 2001. Historical overfishing and the recent collapse of coastal ecosystems. Science 293:629-638.

7. C.M. Duarte. 1995. Submerged aquatic vegetation in relation to different nutrient regimes. Ophelia 41:87-112.

8. N.N. Rabalais and R.E. Turner (eds). 2001 Coastal Hypoxia: Consequences for Living Resources and Ecosystems. American Geophysical Union, Washington, D.C.

9. J.F. Caddy. 1993. Towards a comparative evaluation of human impacts on fishery ecosystems of enclosed and semi-enclosed seas. Reviews in Fisheries Sciences 1:57-95.

10. J.F. Caddy. 2000. Marine catchment basin effects versus impacts of fisheries on semi-enclosed seas. ICES Journal of Marine Science 57:628-640.

11. F. Micheli. 1998. Eutrophication, fisheries, and consumer-resource dynamics in marine pelagic ecosystems. Science 285:1396-1398.

12. S.A. Ludsin, M.W. Kershner, K.A. Blocksom, R.I. Knight, and R.A. Stein. 2001. Life after death in Lake Erie: nutrient controls drive fish species richness, rehabilitation. Ecological Applications 11:731-746.

13. R.J. Livingston. 2001. Eutrophication Processes in Coastal Systems: Origin and Succession of Plankton Blooms and Effects on Secondary Production in Gulf Coast Estuaries. CRC Press, New York.

14. B.-O. Jansson and K. Dahlberg. 1999. The environmental status of the Baltic Sea in the 1940s, today and in the future. Ambio 28:312-319.

15. R. Elmgren and U. Larsson. 2001. Eutrophication in the Baltic Sea area: integrated coastal management issues, p. 15-35. In B. von Bodungen and R.K. Turner (eds). Science and Integrated Coastal Management Dahlem University Press, Berlin.

16. F. Colijn and K. Reise. 2001. Transboundary issues: consequences for the Wadden Sea, p. 51-70. In B. von Bodungen and R.K. Turner (eds). Science and Integrated Coastal Management. Dahlem University Press, Berlin.

17. D. Justit,. T. Legovic, L. Rottini-Sandrini. 1987. Trends in oxygen content 1911-1984 and occurrence of benthic mortality in the northern Adriatic Sea. Estuarine, Coastal, and Shelf Science 50:205-216.

18. Y.P. Zaitsev. 1999. Eutrophication of the Black Sea and its major consequences, p. 58-74. In L.D. Mee and G. Topping (eds). Black Sea Pollution Assessment. Black Sea Environmental Series Vol.10, UN Publications, New York, NY.

19. D.F. Boesch, R.B. Brinsfield, and R.E. Magnien. 2001. Chesapeake Bay eutrophication: scientific understanding, ecosystem restoration, and challenges for agriculture. Journal of Environmental Quality 30:303-320.

20. Long Island Sound Study. 1998. Phase III Actions for Hypoxia Management. EPA-902-R-98-002. U.S. Environmental Protection Agency, Stony Brook,

New York.

21.   N.N. Rabalais, R.E. Turner, D. Justic, Q. Dortch, W.J. Wiseman, Jr., and B.K. Sen Gupta. 1996. Nutrient changes in the Mississippi River and system responses on the adjacent continental shelf Estuaries 19:386-407

22.   CENR. 2000. Integrated Assessment of hypoxia in the Northern Gulf of Mexico. National Science and Technology Council, Committee on Environment and Natural Resources, Washington, D.C.

23.   T. Suzuki. 2001. Oxygen-deficient waters along the Japanese coast and their effects upon the estuarine ecosystem. Journal of Environmental Quality 30:291-302.

24.   S.B. Bricker, C.G. Clement, D.E. Pirhalla, S.P. Orlando, D.F.G. Farrow. 1999. National Estuarine Eutrophication Assessment: Effects of Nutrient Enrichment in the Nation's Istuaries. National Oceanic and Atmospheric Administration, Silver Spring, Maryland.

25.   P. Crouzet, and 11 authors. 1999. Nutrients in European Ecosystems. Environmental Assessment Report 4, European Environmental Agency, Copenhagen, Denmark,

26.   K. Richardson and J.P. Heilman. 1995. Primary production in the Kattegat: past and present. Ophelia 41:317-328.

27.   J. Hagy. unpublished.

28.   T. Suzuki. 2001. Oxygen-deficient waters along the Japanese coast and their effects upon the estuarine ecosystem. Journal of Environmental Quality 30:291-302.

29.   R.W. Howarth. and 14 others. 1996. Regional nitrogen budgets and riverine N & P fluxes for the drainages to the North Atlantic Ocean: natural and human influences. Biogeochemistry 35: 75-139

30.   D.A. Goolsby, W.A. Battaglin, B.T. Aulenbach, and R.P. Hooper. 2000. Nitrogen flux and sources in the Mississippi River Basin. The Science of the Total Environment 248:75-86.

31.   D.F. Boesch, R.H. Burroughs, J.E. Baker, R.P. Mason, C.L. Rowe, and R.L. Siefert. 2001. Marine Pollution in the United States: Significant Accomplishments, Future Challenges. Pew Oceans Commission, Arlington, Virginia.

32.   P.M. Vitousek, J.D. Aber, R.W. Howarth, G.E. Likens, P.A. Matson, D.W. Schindler, W.H. Schlesinger, and D.G. Tilman. 1997. Human alteration of the global nitrogen cycle: sources and consequences. Ecological Applications 7:737-750.

33.   D. Tilman, I. Fargione, B. Wolff, C. D'Antonio, A. Dobson, R. Howarth, D. Schindler, W.H. Schlesinger, D. Simberloff, and D. Swackhamer. 2001. Forecasting agriculturally driven global environmental change. Science 292:281-284.

34.   S.W. Nixon. 1995. Coastal marine eutrophication: a definition, social causes, and future concerns. Ophelia 41:199-219.

180

35. S.R. Carpenter, Caraco, N.F., Correll, D.L., Howaith, R.W., Sharpley, A.N. and Smith, V.H., 1998. Nonpoint pollution of surface waters with phosphorus and nitrogen. r.cological Applications 8:559-568.
36. L.D. Mee. 2001. Eutrophication in the Black Sea and a basin-wide approach to its control, p. 71-91. In B. von Bodungen and R.K. Turner (eds). Science and Integrated Coastal Management. Dahlem University Press, Berlin.
37. T.D. Jickells, D.F. Boesch, F. Coiijn, R. Elmgren, P. Frykblom, L.D. Mee, J.M. Pacyna, M. Voss, and F.Y. Wulff. 2001. Transboundary issues, p.93-112. In B. von Bodungen and R.K. Turner (eds.). Science and Integrated Coastal Management. Dahlem University Press, Berlin.
38. R.R. Lewis Ill, P.A. Clark, W.K. Fehring. H.S. Greening, R.O. Johansson, and R.T. Paul. 1998. The rehabilitation of the Tampa Bay Estuary, Florida, U.S.A., as an example of successful integrated coastal management. Marine Pollution Bulletin 37:468-473.

TROUBLE ON THE EDGE: THE IMPERATIVE OF INTEGRATED COASTAL ZONE MANAGEMENT

PETER M. DOUGLAS[*]
California Coastal Commission
45 Fremont Street, Suite 2000
San Francisco, CA 94105-2219
U.S.A.

INTRODUCTION

Coastal zones are among the most fragile and threatened ecosystems on the planet. Their economic and ecological importance is enormous. Nearly 60% of the world's population live within 150 kilometers of the coast. In island countries, a higher percentage lives only a few kilometers from the shore. In California, 80% of its more than 33 million residents are only an hour's drive away. Two-thirds of harvested fish depend on coastal habitat for survival and 95% of the globe's fish harvest comes from coastal waters. Development demand for transportation, commercial, industrial, tourist, residential, infrastructure, agriculture and other uses concentrated in coastal areas is growing exponentially. Global warming and sea level rise are impacting coastal areas by accelerating erosion, increasing storm frequency and severity, flooding, and saltwater intrusion of freshwater aquifers will render large reaches of coastal lowlands uninhabitable. Widespread marine pollution, over-exploitation and harmful extractive practices have driven many coastal eco-systems beyond thresholds of sustainability.

Unfortunately, to most people destruction and threats to coasts and oceans are phenomena out of sight and out of mind. Because the public is not clamoring for solutions to problems they do not see, many politicians lack motivation to spend more than lip service on the issue. Notwithstanding dire warnings from practitioners and scientists, policymakers find it easier to look the other way. Consequently, and similar to many complex environmental problems, responses from government will be reactive. It will take one or more crises, such as the collapse of entire natural systems (e.g., various fisheries around the world) or the major loss of life or economic health, to shock government into action, and then it may well be too little too late.

The worldwide destruction of marine eco-systems should be of grave concern to all nations. More must be done to effectively plan for environmentally sustainable use and preservation of surviving coastal habitats. In many places it is too late. In others, only restoration on a grand scale can ameliorate the losses. Government engagement, marked by leadership, vision, political will and strong, effective implementation initiatives is the only course of action that can inspire hope for a healthier environmental future that will slow and reverse the pace of destruction.

In many littoral nations, coastal protection efforts are either nonexistent or more symbolic than real (i.e., they do not make a difference on the ground). International efforts, while well intentioned, often fall short of the mark or go terribly wrong (e.g., promotion of the drift net fishery). Grand initiatives establishing integrated coastal zone management (ICZM) programs in a particular country are often announced with great fanfare only to wither away because the resources and tools to sustain local implementation capacity are rarely built into the equation. Today, it is generally understood that ICZM, like "environmentally sustainable development", is an accepted and expected part of a country's environmental agenda. But there is a big difference between ICZM as a process and ICZM as an adequately funded and effectively implemented program that regulates private and public uses of coastal and marine resources for the benefit of current and future generations.

This paper discusses why integrated coastal zone management is important, what it is, key terms and concepts,[1] how it can be implemented, and a description of and lessons learned from California's experience. Finally, this presentation is yet another urgent call for immediate action by coastal nations and the international community to strengthen timely, effective and sustainable protections for coastal resources.

WHY IS INTEGRATED COASTAL ZONE MANAGEMENT NEEDED?

Coastal zones of the world constitute geography of incalculable worth and value to current and future generations. Their relative importance to coastal nations is measured in terms of environmental, ecological, social, political, cultural, and economic parameters. Exploitative uses of coastal resources drive the engine of many local and regional economies and are the basis for livelihood and subsistence. Coasts support the most biologically productive and diverse eco-systems anywhere. They also have great historic, cultural, educational and social value as places where people live, recreate, conduct research and learn, and where many find solace, refuge, artistic expression and inspiration.

Coasts, like coveted landscape everywhere, are never finally saved. They are always being saved. The geography of dreams for explorers, adventurers, pioneers and entrepreneurs, coasts continue to suffer from massive, mindless human use and abuse. Vast reaches of coastal margins have been and are being destroyed at an alarming rate and at immeasurable cost to future generations of life. They are giant magnets continually attracting people in greater number who want to use the coast in ways that conflict with each other and, almost always, with nature. Commercial uses, such as fishing, agriculture, aquaculture, shipping, tourism, mineral extraction, energy production and manufacturing, compete for space. The competition for a piece of the action and a place in the sun is fierce. The stakes in the outcome, in terms of environmental despoliation, pecuniary profit and loss, and public health and welfare are enormous. Like waves beating on the shores of a rising sea, the surge of people to land's edge and their penchant to exploit its resources inexorably erode the capacity of coasts to survive as biologically productive and viable eco-systems. In the absence of deliberate and effective government intervention, the only winners in this struggle are the powerful exploiters who get in, pocket opportunistic profits and run, leaving to the disenfranchised, powerless, and unborn the task of living with the mess and arguing over diminishing returns.

A recent report by the World Resources Institute (*Pilot Analysis of Global Ecosystems (PAGE): Coastal Ecosystems [2001]*) clearly documents coastal destruction. 85% of mangrove forests have been lost in Thailand, the Philippines, Pakistan, Panama, and Mexico. Mangrove habitat has not fared better elsewhere. Much of the world lives on seafood derived from coral reefs, mangrove forests, estuaries, wetlands and seagrass habitats. Large-scale exploitation, destructive extractive practices and technological developments have decimated these natural systems threatening availability of seafood and exacerbating poverty in subsistence communities. The destruction of corral reefs by exploitation and by pollution and ocean warming is widespread. If not reversed, more than 70% of the world's 600,000 square kilometers of coral reefs will soon be lost. Pollution from urban, industrial and agricultural runoff has turned marine environments into dead or dying zones (e.g., Gulf of Mexico, Baltic Sea). Unregulated or ineffective management has led to the collapse of fisheries around the world and the extinction or near-extinction of once abundant species of marine life. The introduction of non-native species of marine organisms has drastically altered or destroyed native sea life and habitat.

More than 30% of uplands important to the health of coastal ecosystems have been destroyed or greatly altered by commercial, industrial, residential, infrastructure and tourist oriented uses. Shorelines around the world are eroding at an alarming rate as a result of unplanned and thoughtless land use practices and rising sea level. Global warming and climate changes threaten habitat and the lives and livelihood of countless millions of people living in low-lying coastal areas. Needless to say, eco-systems found at land's edge that are essential to life on the planet are in serious peril everywhere. What is needed to slow the pace of this devastating and senseless environmental destruction is strong leadership by politicians with the vision, understanding and courage to take effective, sustainable action at all levels of governance.

WHAT IS INTEGRATED COASTAL ZONE MANAGEMENT (ICZM)?

ICZM means many things to many people and it rarely means the same thing to any two of them. While practitioners agree on many features of the concept there are obvious differences of opinion about what specific elements should be in a good ICZM program. As a 30-year practitioner in this arena, I have reached several general conclusions.

General Principles

More than anything, ICZM is a way of *thinking* and a matter of *attitude*. It is an *approach* to the management of coastal resources and differs from the traditional way of *doing* the public's environmental stewardship business. ICZM is, or should be, about tangible results in the field, as well as process. It is both academic and practical, and, like saving a coast, it is never done only always being done.

ICZM is a continuing, dynamic, complex, multi-faceted process that incorporates many other concepts such as adaptive management, environmentally sustainable development, coastal management strategies, stakeholder involvement, precautionary principles, consensus decision-making, and multi-dimensional integration. There is no single model that fits all situations. Rather, ICZM must be adapted to the political, cultural, social, economic, environmental, institutional and legal characteristics and realities of the geo-political context in which it is being proposed. While undoubtedly an efficient and effective approach to sound stewardship of coastal resources, a meaningful program involves considerable costs. However, short-term costs are far outweighed by long-term benefits, and long-terms costs of not doing it. Indeed, a truism here is that it takes resources (e.g., money) to save resources.

Way of Thinking

ICZM is a way of thinking about *systems* and *interrelationships*. For example, how do land use and water use planning and management interact? What are the parameters, characteristics and interrelationships of particular eco-systems and how do they function? Is planning being conducted on a watershed or drainage basin level? Is policy decision-making linked to science and does the science include social as well as natural sciences? Is planning tied to regulation of coastal zone uses?

ICZM is not rocket science. It is the application of common sense. It requires a positive attitude, commitment to efficiencies in governance and to teamwork and partnerships among government agencies and public-private entities. It means putting aside narrow institutional jealousies and competition among agencies, and working toward a common purpose. In short, it means pulling on the oars together and in the same direction.

A good ICZM program is driven by a focused vision that informs program development and implementation. And vision is the horizon of hope for a better environmental future for coasts and seas toward which pragmatic efforts are directed, by which attitudinal and behavioral compasses are set, and against which achievement is measured. It is necessary to think vision in order to have vision, and to have vision is to see what needs to be done.

Way of doing

ICZM is, or should be, all about practical, positive results. To *do* it effectively requires understanding and buy in from the political leadership. Government leaders must acknowledge it as essential for the future environmental, economic and social well being of their country. Ideal is leadership buy-in at the national, regional and local levels. It is a top-down process only to the extent that the guiding coastal management policies and principles must be set at the national level. Successful implementation requires partnership with and buy-in at all levels of governance.

Doing ICZM requires sharing of fiscal, human and technical resources among agencies and meaningful coordination and communication among the various sectors having a stake in the program. It requires being flexible and adaptive in the management of the program as circumstances change. It is important to assess the capacity of existing institutions to carry out new responsibilities under the program. While creation of new government entities with new powers may be necessary to implement the program, existing divisions of institutional responsibilities must be taken into account to minimize counterproductive consequences of passive resistance from agencies whose cooperation is essential.

If a program is to succeed it must be given adequate support resources (i.e., fiscal, skilled staff, and technical) to do the job and it must be carefully and skillfully designed. Finally, it is important to structure and implement the program in manageable components. Doing everything at once is not possible. Depending on available resources and other circumstances, program components should be prioritized for phased implementation.

What is being integrated?

ICZM is a complex, multi-dimensional concept. One of its more challenging aspects is understanding what is actually being integrated in a particular coastal management program. Integration takes many forms and not all of them will be applicable in every situation.

1. *Inter-sectoral integration* refers to integration of planning and management among different types of non-governmental coastal resource users based on the resource used and where the resource is located. Examples include integration among coastal/marine sectors (e.g., recreational and commercial fisheries, recreational and commercial boating, marine transportation, offshore energy development, marine-oriented tourism) and integration between coastal/marine sectors and upland or land-based sectors such as development, mining, forestry and agriculture. There are also sectors of users that straddle upland and water areas such as aquaculture, marine research and educational institutions (i.e., aquaria). A "sector" refers to a type of use or the place where the use of a coastal resource occurs and refers to an activity or enterprise that tends to be focused on one or more related coastal/marine use or resource (i.e., offshore oil and gas).

2. *Areal integration* refers to the coordinated interaction of resource planning and/or regulation programs affecting the upland and seaward sides of the coastal zone (i.e., its dry and wet dimensions). It can also refer to the integration of resource management programs that are place-based and applicable to land or water areas with overlapping boundaries (e.g., special,

single purpose zones of management and broader coastal management zone). Another example is a coastal zone boundary that does not coincide with hydrological units (i.e., watersheds) for purposes of water quality control programs.

3. *Intergovernmental natural resources agency (vertical) integration* refers to the coordinated interaction among natural coastal zone resource management agencies at *different levels* of governance (i.e., local, state, regional, national and international). This integration can be vertical and horizontal (i.e., *among* agencies at the local, state, regional, national, international level and *between* agencies at the same level). Examples include local governments, wildlife agencies, park and recreation entities, air and water quality control agencies, and agencies that serve as landlord of other public lands. It refers to coordinated actions, effective communications and sharing of resources through such means as task forces, regular meetings, conferences and interagency staff assignments and staff sharing.

4. *Intersectoral governmental integration* refers to the coordination between governmental agencies at various levels with different programmatic responsibilities. Examples include coastal resources management and other environmental protection agencies and those responsible for economic development, housing, transportation, commerce and trade, energy facility siting and public works and infrastructure development.

5. *Intra-governmental integration* requires coordination among various bureaus and departments within a particular agency

6. *Science-policy integration* refers to the coordination and collaboration between and among disciplines (e.g., engineering, natural and social sciences) and policy decision-makers and policy-implementers. The integration of science and policy is one of the most important and challenging dimensions of ICZM.

7. *Planning and regulation* are essential elements of coastal management and must be effectively integrated in order for the overall program to succeed. This lesson has been learned time and again. Planning without regulation and regulation without planning simply does not work in land use management. These two aspects of CZM must be coordinated and incorporated in the same program.

ICZM is the contemporary concept of choice in the field of public sector coastal stewardship even though practitioners cannot agree on precisely what it means. Everyone can agree, however, on the notion that the essence of CZM is integration, *especially the integration of planning and regulation*. As a practical matter, integration is necessary to make best use of limited support resources, existing expertise, and to increase the likelihood of success in achieving program goals and objectives. How, when and to what extent integration occurs must be determined by customizing the approach to the circumstances and conditions of any given situation. The critical point is to recognize the wide range of benefits that can be derived from effective program integration and the need to design the particular coastal management program with this purpose in mind.

WHAT IS COASTAL ZONE MANAGEMENT (CZM)?

Like ICZM, coastal zone management means different things to many people and comes in many types, shapes and sizes. In a nutshell, CZM seeks to balance competing demands on finite coastal resources to meet social, economic and environmental needs in a manner that promotes environmentally sustainable economic development for the benefit of current and future generations. CZM attempts to achieve these goals by planning for and managing the use and conservation of coastal zone resources.

A coastal zone is generally defined to encompass coastal watersheds or drainages, other uplands that directly affect marine environments and a seaward extension to the outer limit of national jurisdiction. Precise coastal zone boundaries differ among states and are most often determined by local politics rather than geo-science. In any event, the territorial reach of coastal zones rarely coincides with political boundaries. It is this cross-jurisdictional dynamic that makes inter-governmental coordination and cooperation through integrated coastal zone management so important.

The perceived need for and feasibility of CZM in a coastal nation depends on many factors, including, status of economic development, population distribution and demographics, the level of public awareness and understanding, political leadership, governmental stability, degree of coastal environmental degradation, the threat of destruction related to coastal hazards, the intensity of conflict among coastal resource users, and the involvement of external organizations and forces. Much has been learned in the 35 years since CZM was first launched in California.

In the U.S., 34 of 35 coastal States, Territories and Commonwealths are participating in the national CZM program and 33 are more or less effectively implementing their own coastal programs. Estimates in 2000 of global participation in some form of ICZM shows a regional breakdown as follows:

Country	Number of Coastal Countries in Region	Number of Countries[2] doing ICZM in Region
Europe	33	30 (91%)
Near East	15	7 (47%)
Africa	37	13 (35%)
North America	3	3 (100%)
Central America	7	7 (100%)
South America	11	8 (73%)
Caribbean	13	8 (62%)
Asia	17	14 (82%)
Oceania	17	8 (47%)

These numbers have most likely changed as more countries initiate CZM programs.

Elements of a CZM program

Based on experience in the U.S., the essential elements of a sound CZM program are fairly well recognized. Generally, there must be an understanding of the problems and agreement on the need to address them. There must be strong leadership and general agreement on the goals and objectives of the program. A broad base of public and stakeholder support is important though not essential (in California many stakeholders vigorously opposed coastal management). The program must be adapted to meet the needs of and fit into the prevailing socio-political realities of the particular national setting (e.g., cultural, political, economic, legal, organizational). It must be sensibly structured and adequately funded and staffed with dedicated, knowledgeable personnel.

Enabling legislation at the national level is necessary and should contain a clear statement of vision, mission, goals, objectives, policies and procedures. The program should be comprehensive in terms of *who* it applies to, *what activities* are covered, and *where* it applies (i.e., geographic boundaries). It must be adaptive, and integrated, contain an effective governance structure, be clear about institutional divisions of duties and responsibilities, and it should encourage public participation. And it should incorporate the *precautionary principle*. More specifically, CZM enabling legislation should include, at a minimum:

1. a clear statement of purpose, goals and objectives;
2. definition of terms;
3. a concise statement of the governing policies that apply to planning and management decisions;
4. a feasible and effective governance structure;
5. a statement of the powers and duties delegated to the coastal management agency(ies);
6. delineation of planning responsibilities;
7. delineation of regulatory responsibilities;
8. description of the specific process and procedures to be followed for planning and regulation and the role of the public and other stakeholders;
9. setting out the role of regional and municipal governments;
10. meaningful enforcement provisions;
11. miscellaneous provisions (e.g., energy facilities siting, ports, education, public works, and other special situations); and
12. funding.

Building a CZM Program (CZMP)

Practical Considerations

1. <u>Recognizing realities:</u> It is important to acknowledge, at the beginning of the process, that it will take time to do what needs to be done and that not everything can be done at once. Another reality is that while environmentally sensitive coastal resources, like wetlands, *can* be permanently lost at any time, they are *never* permanently "saved", but only always in the process of being "saved" or protected.

2. <u>Pacing and expectations:</u> It is prudent not to do too much too fast and to avoid "extreme" strategies unless absolutely necessary (e.g., moratoria on new development). It is not feasible or realistic to expect that a new program will be able to address every existing and potential issue at the same time. A phased or incremental approach is usually appropriate and it is wise not to create expectations that cannot be met. This requires strategic thinking and will depend on variables uniquely applicable to the jurisdiction for which a CZMP is being developed (e.g., number and complexity of issues, the extent to which issues can be segmented and dealt with independently, the availability of fiscal and human resources for the job, the existence of needed legal tools).

3. <u>Stakeholders:</u> The "buy-in" (active support) of key stakeholders for the approach taken and the "solutions" identified is extremely valuable. This does not mean they must support every policy or solution deemed necessary or appropriate. Obviously, the likelihood of success is increased greatly if key stakeholders accept a particular approach to dealing with a problem and become active advocates for the identified solution.

4. <u>Costs and benefits:</u> Decision-makers, the public and stakeholders need to understand and appreciate that while there may be immediate "costs" associated with CZM, benefits will take time to become evident, many taking longer than others to materialize. In addition, many benefits, although real, are not quantifiable or visible. How does one measure the value of a spectacular landscape that has not been developed or of a wetland that has not been filled or public beach access that has not been lost? What is saved or not lost more often determines a program's success than what has been built.

   Most people do not accept the need for making sacrifices for benefits they cannot see or measure. It is important to identify benefits and <u>when</u> they are likely to be realized. When stakeholders (e.g., property owners, users of a resource) recognize that they will actually benefit in the long-term from a particular course of action, they will more likely be supportive. Experience shows that prevention of environmental degradation is less costly and, often, yields higher economic returns, than cleaning up the pollution or restoring degraded resources later.

5. <u>Alternatives:</u> The feasibility of different approaches to dealing with problems (e.g., an "incentive-based" versus "command and control" approach, voluntary compliance backed by enforceable policies, negotiated "rulemaking") should be evaluated and carefully considered. The approach selected will depend on factors such as costs (social as well as fiscal) and benefits (short-term and long-term), potential for effectiveness (relying solely on voluntary compliance may be unrealistic), and whether a specific approach is capable of being implemented within a reasonable period of time.

6. <u>One size does NOT fit all:</u> There is no single "best way" to do CZM. What "works" in one region may not work in another. For example, California relies primarily on a strong, independent commission while many states use a "networked" approach (i.e., coastal management responsibilities are divided among several agencies who are then expected to work together). The feasibility of a particular approach will depend on the political, economic, social, legal and institutional realities of each context. An effective CZMP must be "flexible" enough to be adaptable to local conditions while at the same time adhering to state-level goals and objectives (policies calling for "maximum" public access opportunities may be met in different ways in different localities - more accessways in urban areas than in rural places).

   CZM is a dynamic process and must be capable of adapting to changing conditions over time (social, economic and environmental needs, new information about the functioning and status of the natural environment, politics). The program must be open and clear about <u>what</u> is being done, <u>why</u> it is being done and <u>how</u> it is being done. Rules should be understandable and precise without immobilizing the system by details and inflexibility.

7. <u>People do the job:</u> It is vital the right people are selected to do the job. This means dedicated, skilled, knowledgeable people of integrity and good judgment must be found to make policy decisions. Highly competent and motivated staff are vital to the success of any program. Public service is often thankless work. Given that conflict comes with the territory, people who work in CZM programs must be committed to the job and able to embrace and work with conflict

8. <u>Politics:</u> Obviously, securing and maintaining strong political and public support for the CZMP is essential.

Scientific and Technical Considerations

1. <u>Linkages:</u> An important ecological planning principle is that no single part of an ecosystem operates independently of any other. Changes that are made in an ecosystem should be carefully evaluated for individual and cumulative impacts on other portions of that ecosystem. Linkages within systems and between human activities and adverse impacts on coastal resources (automobile oil dumped in the gutter ends up in the sea) need to be identified, explained and addressed.

2. Regulation is needed and requires good science: New development and uses of natural coastal resources, are most often not environmentally sustainable without some degree of planning and management of the use. Environmentally sustainable development recognizes that ecosystems have a limited capacity to survive human-caused stresses and that people must restrict their use of that system and its natural components in order to preserve options for future uses. But people do not voluntarily do the "right" thing even if they do understand the environmental impacts of their actions (which they most often do not or do not even think about), That is why some regulation is necessary. But to be effective and supportable, regulation must be based on sound science and technical analyses. In the U.S., regulatory actions can be reviewed by courts that are more and more looking for the technical and scientific basis for regulation.

3. Causes and effects: People are an important part of the environment and of many coastal ecosystems. While people can alter such systems in ways that improve human living conditions, more attention must be given to the avoidance of unintended negative secondary impacts.

    For example, opening a highly scenic coastal area to tourism may result in short-term economic benefits (jobs for construction). However, if tourist development is not carefully planned and managed, the very reasons tourists are attracted to the place (unspoiled scenery, seclusion, wildlife) may be destroyed and result in far greater economic and environmental losses than would have occurred had no development been permitted. Short-term gains often result in long-term "pains." Examples of the great social and environmental costs of such shortsightedness, most often driven by greed, are found everywhere. If carefully planned and managed, tourism can be sustained over time, resulting in much greater economic benefits to the region and certainly at less cost to the environment.

4. Mitigation and other concepts: In CZM, as in environmental protection programs elsewhere, there are important distinctions between harm prevention, mitigation, compensation, restoration and enhancement. Prevention seeks to avoid adverse environmental impacts in the first place. Mitigation is a way to avoid, lessen and minimize adverse environmental effects and often includes "in kind replacement", in equal or greater amounts (refers to physical area or habitat value) in the same general location where the impacted resource was. Compensation is used to "offset" or "pay" for the loss of a resource and takes various forms (money paid into an environmental protection fund, wildlife habitat or recreational improvements). Restoration refers to a process whereby a severely degraded natural resource area (wetland, woodland, grassland, riparian corridor, sanddune) is recovered or rehabilitated in a manner that restores, as nearly as possible, the biological functions of the area to its

condition before being degraded. Creation refers to those situations where the type of habitat lost is created in an area that did not previously support that kind of habitat. Environmental enhancement refers to the process whereby a resource that has been degraded but which retains some of its original environmental values is improved to increase biological productivity and diversity. Restoration, creation and enhancement measures are sometimes used as "mitigation" or "compensation" for adverse environmental impacts of new development.

It is important to keep in mind distinctions between the legal, scientific and technical aspects of these approaches to addressing adverse environmental impacts. For example, restoration projects and the creation of new habitat are usually more expensive and difficult to successfully carry out than are enhancement projects. However, if loss of a particular habitat (saltmarsh) is the primary concern, enhancement activities will not contribute to the overall recovery of that type of lost habitat. It only "improves" existing habitat and does not result in "new" habitat being created.

5. Information needs: Data and monitoring are important components of CZM. Informed decision-making requires good data (e.g., baseline information about the preexisting condition of an area or resource). Information relative to social and economic considerations is also important. To be useful, data must be reliable - it must be of good quality, gathered in a scientifically defensible manner, accurate, and it must be current. Data acquisition, storage and retrieval is expensive but a necessary cost of CZM.

However, there are limits to how much data can be collected or even how much information is needed to make decisions. The absence of "complete" data is often used as an excuse for inaction. A CZMP that incorporates a "precautionary" principle recognizes that action is often based on the best available information and need not wait until all possible information on a particular subject is in hand. Indeed, it can almost always be said that not enough scientific information on a particular issue is available to be absolutely certain about causal relationships and that what is proposed to be done is the least environmentally damaging alternative.

Monitoring is important to demonstrate the effectiveness of a particular approach and provides valuable information that can lead to necessary modifications. However, monitoring occurs on many levels and can take many forms and is expensive. Careful choices must be made regarding what will be monitored, how the monitoring will be conducted, who conducts the monitoring (should it be "independent" monitoring by someone other than the project/program proponent) and when and for how long. Monitoring can also be a key element in program evaluation and the application of adaptive management. For example, to determine if a habitat restoration or enhancement program has achieved its objectives, monitoring is essential. If monitoring identifies problems, appropriate remediations measure can then be required and implemented.

Meeting information needs does not require reinvention of the wheel. Knowing where to find existing information, building on the experience of others, and knowing where to find technical expertise and assistance is invaluable. This requires good information collection, storage, retrieval and sharing systems and networks.

6. Good science: Sound scientific (natural, physical and social) is essential. To be most productive however, the relative roles and limitations of science in CZM policy decision-making must be recognized and understood. Scientists come from and operate in a professional culture that differs significantly from that of policy makers and implementers. They speak a different language and have differing needs. For example, good science requires considerable lead-time and careful study before conclusions are reached. Policymakers usually need information in a short timeframe and often do not ask all the questions a scientist would in order to make a decision. Policy decision-making must be informed by science at nearly every stage of the CZM process.

7. Public involvement: Public education, understanding and participation in CZM is vital. Public support is usually critical to the ultimate success of a CZMP and public involvement should be actively encouraged. Effective CZM involves changing human behavior and requires that the public bear many implementation "costs". There are both practical and political reasons why public education and involvement should be key components of every CZMP. Public outreach and education efforts can take many forms. Public workshops and hearings, informational brochures, media outreach, special school programs, asking for public review and comment on proposed actions, and involving non-governmental organizations in the development of a CZMP are but some of the ways to achieve public support and understanding.

Steps to Establish and Implement a CZMP

Following is a listing of important steps and actions that should be taken to develop and carry out a CZMP.

1. Identify the full range of coastal problems and issues that should be addressed.
2. Identify the most significant of these problems and issues and prioritize them.
3. Identify key stakeholders and recognize the importance of involving them in certain phases of program development and implementation.

4. Identify, on a preliminary basis, the boundaries of a potential coastal zone, taking into account the value of a watershed and ecosystem-based approach to planning and management. In addition to the geographic reach of the program, factors relative to timing should be considered. (should all or some components of the program be temporary or permanent? What is to be accomplished within what timeframe?).

5. Identify and prioritize national and state goals and objectives for the coastal zone (description of the desirable future envisioned for the coast).

6. Examine the existing legal and institutional capacity of the State to address the issues that must be dealt with in order to achieve the identified goals and objectives. Identify any gaps in the existing governance system that need to be addressed. Evaluate potential changes in law should.

7. Identify real and potential impediments to the implementation of an effective CZMP and examine possible ways to overcome them without compromising key elements of the proposed program.

8. Using the broader CZMP goals and objectives for guidance, more specific policies should be developed and used to plan and regulate coastal land and water uses.

9. Identify the actions needed to implement the identified policies, including the establishment of new institutional arrangements, the implementation of planning and regulatory programs, initiation of public works projects such as the development of infrastructure (wastewater collection, treatment and disposal systems), and programs for the public acquisition of lands for public use and conservation. Innovative implementation approaches should be considered, including public-private partnerships, partnerships with NGOs (e.g., land trusts), and reliance on special public entities similar to governmental land and water Conservancies in the US.

10. Identify the data and information, technical or otherwise, that will be needed to influence and inform the decision-making process. This step should also include an examination of how additional information needs can be determined and possible ways to fund the acquisition of such information.

11. Identify the ways and means to finance the implementation of the CZM Program (special taxes and fees, bond measures, general fund revenues).

12. Evaluate, in a general way, the potential costs and benefits (economic, social and environmental) of implementing the CZMP. This step involves a public document that can be used to explain the program's purposes and what it may cost to carry out. The costs of not implementing a CZMP should also be taken into account.

13. Identify when and how the public and NGOs can be educated relative to the benefits of CZM and how they will be involved in the formulation and implementation of the program.

14. Develop a strategic action plan for moving forward with the development and implementation of a CZMP. Such a plan could include the empowerment of a governmental or quasi-governmental entity to prepare a proposed or draft CZMP for submittal to the appropriate Executive and/or Legislative decision-making authority for implementation.

CZM ISSUES

The issues that governments must address in their CZM program are numerous, complex and will differ from place to place. Examples include: Siting of land uses (residential, commercial, agricultural, industrial, visitor-serving, recreational, public services, coastal-dependent uses, multiple-use facilities, open space, wildlife refuges); protection of coastal water quality, coastal habitats, scenic resources, historic and cultural resources, and public access and recreational opportunities; coastal hazards; management of living resources (fisheries); managing renewable and non-renewable resources and public lands; ocean resources; public education; and restoration of habitats and urban waterfronts.

Not all these issues will be addressed in every CZMP. However, no matter how broad or limited the program is, there will always be significant challenges for decision-makers. The challenges described below exist in California and virtually every other coastal State in the US.

1. Lack of public understanding: The greatest threat to coasts and oceans is public ignorance and apathy. There exists an enormous lack of awareness and understanding by the public and decision-makers about the nature and extent of the forces that threaten the economic and environmental wellbeing of coasts and oceans. This challenge can be met, in part, through public education, special events, media attention, research, information-sharing, public meetings and increasing awareness among policy-makers in government.

    In California, a concerted public education campaign in 1972 resulted in the enactment of a citizens initiative that established its coastal protection program. Once the public and key legislators understand the nature of the problems facing their coast and ocean, an action plan can be promulgated and implemented. At the national level, the prestigious Stratton Commission Report, *Our Nation and the Sea* (1968), led to enactment of the federal Coastal Zone Management Act in 1972. Similar experiences are found in Great Britain (1970), Sweden (1971), Ireland (1972), France (1972), United Arab Emirates (1976), the Philippines (1978) and Ecuador (1983).

2. Too many and too few governmental agencies: In some areas, there are multiple and overlapping jurisdictions leading to fragmented decision-making. This can occur at all levels of government. For example, in California in 1972, over 150 local, regional, State, federal and special

district (school districts, recreation districts) governmental agencies had some degree of decision-making responsibility relating to coastal resources. Often, decisions made in one jurisdiction have substantial "spillover" effects in adjacent communities. In addition to the multiplicity of public agencies, there is, more often than not, no coordination or communication between them. This resulted in conflicts and poor decisions relative to the use of coastal resources. In California, the "solution" was not to reduce the size and number of agencies, but to create a new and more powerful entity with overarching authority that superceded that of others.

While the challenge in some regions is too many governmental entities, in others there are no agencies dealing with coastal environmental problems or those that do exist lack adequate legal authority to be effective. Then the challenge is creating the necessary legal and institutional structure to address the issues.

3. Lack of adequate funding and availability of technical expertise: Depending on the scope of a proposed CZMP, funding for research, support staff, information technology, and other operating costs is always a major challenge. In addition to the need for fiscal resources, finding the technical expertise (law, planning, biology, hydrology, water quality, geology, coastal engineering, economics, transportation) to devote to the effort is extremely difficult.

4. Cultural impediments: In California and other U.S. coastal States, the major "cultural" challenges to CZM implementation involved doctrines relating to "private property rights" and "home rule." An understandable tension arose from a clash between government imposed land use controls and a long tradition of private property owners believing they can use their land however they choose. In modern society with a burgeoning population and a declining supply of land area for new development, the broader common good, including sound environmental protection, requires reasonable controls over private use of land. While this concept has achieved some degree of acceptance relative to unique and special environments, such as coastal lands, it still generates controversy and hostility elsewhere in the country.

"Home rule" is the deeply ingrained concept that government closest to the "people" - local government - knows best. There exists in the U.S. great distrust of state and federal government involvement in local land use decision-making. State-level planning is often equated to "socialism" and resisted. However, in special areas of greater than local importance (coasts, lake basins, major bays such as San Francisco and Chesapeake bays, and other areas of unique biological value) the notion of State involvement in environmental protection through, among other means, some form of land use planning and regulation has been accepted.

Different regions will, of course, need to be sensitive and responsive to different cultural forces and values.

5. <u>Science-policy interaction:</u> Science-based decision-making is a fundamental element of CZM. The challenge is to bring these two cultures together in a timely and effective manner, to teach them to speak a language both understand, and to help them recognize each others unique needs and constraints.

6. <u>The politics of CZM:</u> The stakes involved in coastal land and water use decisions are enormous. Special interests have significant financial investments in coastal lands and freely use their political connections to influence development decisions. The politics of CZM are intense and relentless and present a unique challenge to coastal managers. Coastal politics will be a challenge everywhere.

7. <u>Special moneyed interests:</u> The coast is where big profits can be made and lost. Special moneyed interests have a huge stake in the outcome of CZM decisions. Many work hard to maintain the status quo because they benefit from it. Others seek to ensure any change that does occur is of special benefit to their particular interest, which often is not in the best public interest (e.g., resort development in remote, environmentally fragile and scenic places). Money speaks more loudly than those who speak for Nature all over the world. A sad, old story. Hopefully, this will change in more places of the world as public demand for effective coastal protection increases and becomes more politically sophisticated and effective.

CONCLUSION

Meaningful integrated coastal zone management is absolutely essential to the environmental wellbeing of the planet. Although most coastal nations now support some form of ICZM, many are weak, promote form over substance and lack the capacity to sustain themselves without external financial and technical support. There has been no systematic evaluation of the effectiveness of ICZM programs with only sporadic anecdotal evidence of success. The fundamental elements of sound ICZM programs are known even though there is no general agreement on precisely what it means. What is needed for the effective implementation of ICZM programs is bold political leadership at all levels of governance.

Because environmentally sensitive and coveted geography, like coasts, are never finally saved but always being saved, the work of coastal stewardship is never finished. This means that adequate support resources to sustain ICZMs in coastal nations must be dedicated in perpetuity as an ongoing cost of responsible world citizenship. In addition to political leadership and funding, public awareness and support for coastal and ocean protection programs is vital. There can be no doubt about the urgency and importance underlying the need to implement coastal conservation measures through ICZM around the globe. However, lip service is no substitute for effective programs that deliver good, on the ground, results.

It is the solemn responsibility of current generations of leaders in societies everywhere to advocate and work for meaningful environmental protections for coasts and seas. How well this challenge is met will be the measure of our legacy to future generations. We must worry less about the price of acting in defense of the Earth and more about the costs of inaction. Humans have the capacity as sentient, caring beings, to ensure a healthy environmental future for unborn life. By embracing commitment to making a difference and dedication to this resolve, we may yet step into the light shining just beyond the horizon of hope that inspires us to never give up.

REFERENCES

*   Mr. Douglas is the Executive Director of the California Coastal Commission (since 1985) and has been involved in coastal management issues at the state, national and international level since 1971. He was a member of the President's Panel on Ocean Exploration (2000) and serves on the U.S. National Oceanic and Atmospheric Administration Science Advisory Board and the U.S. – China Panel on Integrated Coastal Zone Management.
1.   See Glossary of terms.
2.   *Status and Prospects for Integrated Coastal Management: A Global Perspective,* Biliana Cicin-Sain, et al, Center for the Study of Marine Policy, University of Delaware (Summary of forthcoming article, May 2000)

GLOSSARY

UNDERSTANDING THE LANGUAGE AND CONCEPTS OF CZM

A. WHAT IS COASTAL ZONE MANAGEMENT?

"Coastal zone management" (CZM) is a government program carried out in a defined geographic area ("coastal zone") that is designed to plan for and to manage the use and conservation of coastal resources. Coastal zone management programs (CZMPs) may be broadly or narrowly defined and are structured to manage two or more coastal zone resources. CZM is contrasted to "sectoral" management which deals with a single area of use (e.g., agriculture, fisheries) or a highly focused area of economic development or importance (e.g., tourism, energy, transportation). A primary purpose of CZM is to deal with conflicting or competing sectoral uses. Another key feature of CZM is the integration of planning and management. The programmatic goal of CZM is to achieve the best possible balance between environmental protection, economic development and public and private land/water use. The society in which the program functions must define the meaning of "best possible".

There is no single institutional structure for the implementation of CZM. In some U.S. states, CZM is carried out by several agencies working more or less well together (so-called "networked" programs). In others (e.g., California), single agencies are designated as the "principal CZM" agency and given broad powers. (Note: Even in California, air and water quality standards are set by single-purpose agencies separate from the CZM agency.) In all States, municipalities have a key role (e.g., in California, cities and counties have "local coastal programs"). The federal government's CZM Program sets broad national goals. States voluntarily decide to participate (34 out of 35 are involved and have federally approved CZMPs) and receive federal money and the authority to regulate all federal agency activities that affect coastal zone resources.

B. WHAT IS ADAPTIVE COASTAL MANAGEMENT?

"Adaptive coastal management" is another contemporary concept derived from experience and necessity. The essence of this concept is that CZMPs should be improved, modified and adapted on a continuing basis as the result of experience, changed circumstances and lessons learned by "doing". Based on experience, program managers should modify program elements (e.g., goals, policies, procedures or practices) and adapt to new and changed conditions. Pursuant to this approach, management measures are always being evaluated and changes are made based on the information gathered. Adaptive coastal zone management should be viewed as both a strategy for the implementation of CZMPs and as a conceptual approach that is practical and necessary, especially in a subject area where so much uncertainty and complexity exists and where

200

the only constant is change. It allows innovation and experimentation and seeks to constantly improve the program and avoid stagnation.

C. WHAT ARE COASTAL MANAGEMENT STRATEGIES?

Countries and states use different strategies to manage the coastal resources. These include: (1) sectoral planning and management (e.g., fisheries, tourism); (2) statewide land use planning and regulation; (3) special area or regional planning and regulation programs; (4) shoreland exclusion zones; (5) critical area protections; (6) environmental impact assessments and review processes; (7) special coastal area guidelines for use and conservation; (8) coastal inventories or data banks; and (9) special coastal land acquisition programs. Often, a combination of strategies is used.

D. WHAT ARE COASTAL RESOURCES?

There is no fixed definition of coastal resources. CZMPs can be designed to cover a wide range of coastal resources. Examples of coastal resources are, renewable and non-renewable natural resources and commodities that only exist on the coast (e.g., fisheries, boat berthing spaces) or whose value is enhanced by its location on or near the coast (e.g., certain agricultural crops, visitor-serving uses, residential property), coastal environments, eco-systems and landscapes and the values associated with them (e.g., functioning wetlands, views and scenic values, beaches, dunes), human-based features and values (e.g., buildings, settlements, historical and archaeological resources), and certain uses or activities (e.g., recreation, education and research). The definition of "coastal resources" subject to a particular CZMP is determined by the responsible policy-making legislative organization in the state.

E. WHAT IS ENVIRONMENTALLY SUSTAINABLE ECONOMIC DEVELOPMENT?

"Environmentally sustainable economic development" describes a concept that is a principle goal of ICZM. CZM seeks to balance economic development with environmental protection in a careful, deliberative manner that allows people to use coastal resources today in a way that conserves them for use and enjoyment by future generations. The focus of primary concern is environmental protection taking into consideration economic and social factors. Some argue the concept must include all three factors by definition. If that were the case the concept, as a practical matter, looses meaning because it then means everything. The concept is often misapplied to government actions designed to protect, sustain and ensure profitability of a particular economic use. Another concept that is embodied in "environmentally sustainable development" is "sustained yield" which is the practice of harvesting a renewable natural resource (e.g., fisheries, trees) so that what is taken is replaced at the same rate so that overall supply is sustained over the long-term. It is possible however, to carry out a "sustained yield" program that is not "environmentally sustainable" because the focus of the former may be too narrow.

F. WHAT IS THE <u>PRECAUTIONARY PRINCIPLE?</u>

It is not always easy to determine, with any great degree of certainty, whether a particular use in a particular location is environmentally sustainable. The "precautionary principle" means that when there is substantial doubt about the environmental sustainability of new development or any other human activity regulated under the CZMP, the activity should be deferred or not permitted unless and until it can be shown that the activity will not have a significant adverse environmental effect. The principle applies to a wide range of human activities affecting the environment and the science used to assess or predict their consequences. Essentially, the idea is that when in doubt, decision-makers should follow a course of action most protective of important environmental qualities or resources. The rationale for this concept is that adverse environmental impacts are difficult if not impossible to reverse.

Other important elements of the "precautionary principle" are allocation of the "burden of proof" and fundamental assumptions built into the procedures used to implement the CZMP. For example, California's CZMP places the burden of proof on the proponent of any new coastal development who then must demonstrate to the satisfaction of the decision-making body (the Coastal Commission) that all applicable coastal resource protection policies have been met. In some programs, the decision-making body has the burden of proving why a project should <u>not</u> be approved. As a practical matter, the decision relative to who has the burden of proof is extremely important. When a public agency has the burden of proving that a proposed development <u>will</u> have an adverse environmental affect, the precautionary principle is not being applied. Similarly, California's CZMP contains explicit procedural requirements that any official decision by the Commission that would change the status quo relative to the use or protection of coastal resources requires a majority vote of the body (e.g., it takes a majority vote to approve a project unlike some programs that require a majority vote to deny a project).

G. WHAT IS <u>LAND USE AND WATER USE PLANNING?</u>

"Land use" planning is a purposeful governmental process through which information is gathered about natural resources, historic uses of land, social and economic needs, and natural and infrastructure systems. This information is then evaluated in the context of certain expressed public goals and policies and decisions are made about the future allocation of land for particular uses that are in conformity with the adopted goals and policies (e.g., that industrial uses will not be located in residential neighborhoods). The overarching goal of land use planning is to rationalize and harmonize human uses of land in order to further identified public purposes such as protecting environmental quality, ensuring compatibility of uses, and other purposes that promote "public health, safety and welfare."

Land use planning applies to urban and rural areas. Land use plans are not self-implementing and are separate and distinct from mechanisms to actually regulate and control land use. They merely show <u>how</u> land <u>should</u> be used and <u>where</u> certain uses should be sited. To make these plans enforceable, an implementation strategy must be devised and put in place. This usually involves the use of "zoning" ordinances and permitting systems. It is vital that land use planning be directly linked to regulatory controls of land use. Where the two are not coupled, land use policy plans quickly become obsolete as actual development that proceeds without regard to the plans overruns them and renders them moot. This is why under California's ICZMP land use planning and regulatory controls are essential and linked components of the program.

"Water use" planning, like land use planning, involves a rational process for making decisions about the siting of structures and the allocation of uses that occupy water space, change the intensity of water use and public access to water. This does not refer to the allocation of water rights among users for consumptive purposes uses (e.g., drinking, irrigation, cooling industrial plants). Of concern is how the surface area, submerged lands and the water column will be used, usually to the exclusion of other uses (e.g., ports, marinas, piers, aquaculture, recreational uses, landfills).

H. WHAT IS <u>MULTIPLE USE?</u>

"Multiple use" describes a condition of land use and is a planning tool to allow several different types of compatible uses to occur within the same structure or project area. For example, a destination resort usually includes a variety of uses that might not be allowed to co-locate in other settings (e.g., visitor-serving, residential units, retail, commercial office space). The co-location of permanent residential (apartments, condominiums, townhouses), visitor-serving commercial (hotels, recreational equipment sales and rentals, restaurants, art galleries), professional office space (medical, legal, consulting, architectural offices), and governmental uses (libraries, satellite offices for government agencies) are examples of what is being allowed in redeveloping urban waterfronts and congested coastal communities. Carefully planned multiple use plans are ways to revitalize urban areas and meet environmentally sustainable economic development goals.

I. WHAT ARE <u>CUMULATIVE IMPACTS</u>?

"Cumulative impact" refers to the incremental and additive effects of a development activity resulting from the interactions of many discrete activities, each of which independently may have an insignificant impact when considered individually, but which become significant when viewed in the aggregate. Cumulative effects may appear shortly after an activity has occurred or be delayed. They may have short-term or long-term consequences. They may occur in close proximity to the activity or in a more distant location, and they may interact in an additive or synergistic manner. Understanding what a "cumulative impact" is, evaluating them to help understand cause and effect linkages and their other features (e.g., spatial and temporal dimensions), and then managing them is a complex and challenging task at best.

J.   WHAT   IS   ECO-SYSTEM, WATERSHED, AND PLACE-BASED MANAGEMENT?

There is considerable overlap among these concepts and strategies for the management of coastal resources and environmental protection. A key feature of ICZM is to approach planning by looking at natural and human-based systems. Eco-systems are interrelated natural communities of living things and the physical and chemical characteristics on which the viability of the whole depends (e.g., wetlands). Principal features include the dynamic that when one part of the system is changed other parts will change as well and that such systems have a limited capacity to absorb external forces or impacts and if that capacity is exceeded, the system could collapse. Other natural systems include watersheds, hydrological units, littoral cells, estuarine water circulatory systems. Examples of human-based systems include infrastructure such as transportation, water supply, and sewage collection, treatment and disposal.

"Eco-systems" are determined by the interconnectedness of biological, chemical and physical features. "Watersheds" are hydrological units characterized by drainage into the same receiving waters. "Place-based" management refers to planning and management for a particular geographic area and could include watersheds, eco-systems, air basins, viewsheds, and areas that often overlap environmental systems such as parks, wildlife reserves, fishery zones, and specific units of forests. The defining feature is a specific physical place with clearly delineated borders.

A systems approach means planning for and managing the entire system or its key elements and evaluating human activities for their individual and cumulative impacts on that particular system.

K. WHAT IS NON-POINT SOURCE POLLUTION?

"Non-point source pollution" refers to pollution that does not come from a single point of discharge but rather from a diffuse source. Examples include polluted runoff from urban areas (e.g., stormwater), agricultural operations, forest practices, siltation from building sites, and mining. Because this source of pollution comes from multiple sources, it is difficult to prevent. Non-point source pollution is now the biggest cause of pollution of coastal waters around the US. These sources also pollute rivers, streams and lakes. Most coastal states in the US are now working to build programs to address non-point source pollution.

L. WHAT IS A STAKEHOLDER AND WHO ARE THEY?

A "stakeholder" is anyone who has a direct or indirect but expressed interest in a particular coastal resource use issue. The interest may be financial, material in another sense (e.g., health or quality of life are affected in a material way), academic, philosophical (concern about future generations) or by virtue of job responsibilities (e.g., a CZMP affects many governmental agencies such as the military). "Stakeholder" is a term used to describe those parties often asked to participate in the development and implementation of CZM programs. The term comes from the fact these people have a "stake" in the outcome. Their "buy-in" (acceptance of or support) to the approach and outcomes of the CZM process is vital to program success.

Stakeholders in CZM usually include coastal property owners, municipalities, environmental organizations, certain trade associations, certain industries, some labor organizations, certain ministries, marine transportation companies, energy companies, port authorities, fisheries (fisher groups, processing companies), tourist industry, marina operators, agricultural groups, certain groups of researchers and academics, certain non-governmental groups (NGOs), and various ad hoc public groups (e.g., civic organizations, neighborhood associations). Clearly, not all stakeholders have the same level of interest in CZM.

M. WHAT IS CONSENSUS DECISION-MAKING?

"Consensus decision-making" is a misunderstood and misused concept. Generally, it means working toward agreement among stakeholders that a particular problem exists and that an identified solution is preferred. While convening stakeholders and working toward consensus is laudable and desirable, requiring it as a precondition of decision-making is to sanction tyranny of the minority and simply perpetuates the status quo. By definition, stakeholders in an arena such as CZM have different interests that are often in irreconcilable conflict. While compromise is possible relative to many issues, when questions of livelihood or core interests are at stake decision by consensus is not possible and it is disingenuous to expect it. Effective coastal resource protection involves tough decisions with consequences that invariably mean hardship and loss to one interest group or another. A sensible and meaningful "consensus decision-making" process that does not sacrifice the natural resource on the alter of the politics of accommodation is one that strives for consensus but does not mandate it.

7. SCIENCE AND TECHNOLOGY
FOR DEVELOPING COUNTRIES

SCIENCE AND TECHNOLOGY: FRAMEWORK, GOVERNANCE, AND STRATEGIES TO BENEFIT SOCIETY IN DEVELOPING COUNTRIES

HOWARD ALPER

University of Ottawa, Ottawa, Ontario, Canada K1N 6N5

The economies of a number of industrialized nations are knowledge-based in nature, although the resources sector (mining, petrochemicals, and agriculture) is still of value in these countries. Generally speaking, developing countries have resource based economies, with manufacturing as an asset in certain cases. Given the different emphasis, globalization can have negative impacts in countries which are economically weak. However, developing countries can also seize opportunities provided by globalization to enhance the quality of life of its citizens. As C.N.R. Rao stated at the World Conference on Science in Budapest in June, 1999: "It is important that developing countries develop national development strategies where science and technology is properly integrated into the socio-economic plan of the nation". At the same conference, Ahmad Jalali stated that "what are lacking in most developing countries are the appropriate structures of research". To develop strategies and structures requires governments to have a viable policy framework for science and technology, and to work in partnership with their Ministries of Education, Environment, Health, and Justice to address societal needs.

Let us consider the basic elements of a science and technology policy. In some developing countries there is a Science and Technology Sector, usually in the Ministry of Science—if there is one—or the Industry Ministry. However, a number of developing countries have no section or division dedicated to science and technology.

Building expertise, with appropriate skill sets, is a basic element in the creation/operation of a solid science and technology policy. Thus, while the post of Minister is normally that of a politician, the person responsible for S&T policy must have the tools needed to lead this sector. He/she should be a person of considerable breadth in subject matter, and who is highly capable of informing and educating Ministers and other decision makers. Staff members should be well-versed in the protocols to establish, assess, and where required, modify or terminate programs. Selected staff should serve as interns, or be briefly (3 months) seconded to S&T policy organizations in countries with longstanding experience in this context.

Given the major challenges existent in everyday life in some of the developing countries, it is likely that basic research will be at maximum, a minor component of the S&T portfolio. Nevertheless, policies to build such research capabilities are to be encouraged. Environmental issues have to be a key component of a government S&T policy. Why? So many environmental matters are pivotal to our lives. Consider, for example, the quality of water, the use of pesticides, insecticides, etc. in food/agriculture, industrial pollution, and individual pollution (e.g. smoking) amongst others. The lifespan of citizens, let alone the quality of the life that they do lead, is subject to these and other factors. The S&T Policy sector thus must work with the Ministries of the Environment, Health, and Agriculture to develop policies on environmental issues, with the Ministry of Justice to draft and government to implement, legislation to improve the environment, and with the Ministry of Education, to inform the populace – especially those in schools, as to what actions can enhance the quality of their environment.

The importance of the development of talented people must be well-understood and championed by government. Education is the prerequisite to maximizing opportunities for young people. Access to elementary and high school education for both girls and boys leading, in its minimalist form, to high school graduation, is a fundamental goal in those countries, and there are too many, where such opportunities are for few rather than the norm. Education is not only a basic value of society, but interweaves with computer and internet technology, and thus with S&T policy. Wiring of the schools of a country should be a high priority, as democratization of access will markedly broaden the child's knowledge, and open up new vistas for the future. To do so means that the telecommunications system in a country must be at a level to assure success.

Conceptually, there are no borders to knowledge available on the internet. It is like dust blowing in the wind. In reality, governments can limit access to external sites (e.g. practiced by China). Nevertheless the benefit of the internet in terms of knowledge, literacy, etc. is clear. The internet has, in Canada, already had a phenomenal impact on society. Let me give you two examples. First, four years ago, I visited Baffin Island (above the Arctic Circle) which has spectacular scenery and, being there, adds a new dimension to the soul of a Canadian. Now I appreciate the beauty and diversity of Canada and its people. About 95% of the population of 25,000 on Baffin Island are Inuit who have lived principally by fishing and hunting. Their standard of living mirrors that of a more advanced developing country rather than a first world milieu. However, when I made an unannounced visit to an elementary and high schools in one of the small towns, and talked to some of the students, I was astonished. Here were children who had just excellent knowledge in using the computer and accessing the internet. They were proud about the Web Site for their community (e.g. Pond Inlet), how valuable it was. They talked about their dreams. One 13 year old told me how she must go to University "down south", complete a business degree, and start a quilt making firm which would market on the internet. She would also help develop tourism here via the Web and other means. What really was impressive was how, by the internet, she realized that girls/women can aspire to leadership as boys/men. I also was pleased that new bilingual textbooks explained, with sensitivity and understanding of the traditional cultural values of the Inuit, the changing role of women in Canadian Society.

While the internet is a key ingredient to economic development, it is important to be sensitive to traditional cultural values in advancing society. Larry Sampson of the Department of Sustainable Development, Government of Nunavut, Canada, said: "The wired world is here to stay, but hopefully, so too is the direct and sustainable relationship between Inuit and the natural world." He noted that "The subsistence economy has become a rare treasure. Hunting is about food on the table, but it is also about respect for the land, and building and maintaining ties with kin groups and with fellow residents. Nunavut will never see Ford plants and other big manufacturers. Yet it can be a place where people successfully straddle tradition and innovation – "the land" and the internet."

One other example of the impact of the internet on society is the greater access by the public in general, and patients in particular, to health information. This is so important in all societies, developing and developed. The internet can inform and educate in many areas. One example in Canada concerns cancer treatment. For instance, an oncologist told me that, as recently as 7-8 years ago, patients rarely asked about the chemotherapy or radiotherapy they were receiving to treat a particular form of cancer. Now they are so knowledgeable, as the internet has been valuable in providing and exchanging information on the different types of cancer and the options available to medically treat this disease. By being well-informed, not only do patients ask about the details of the proposed treatment, they also discuss with their physicians, in a meaningful way, the way forward thus leading to better decision making. Another example is the establishment of rural tech centers in India which provide current information on prenatal care and on other matters including farming techniques. As Mark Malloch Brown, administrator of the UN Development Program said: "low cost computers and low-literacy touch screens can further democratize education for the poorest of the poor". So the internet can transform society, its attitude to issues such as health, education, environment, etc. By having a S&T strategy which takes advantage of the internet, and by working with Ministries in different sectors, one can take steps to address environment, health, and other issues. One also expands the frontiers for the next generation in society. This is an example where, connecting globally, can make a favorable impact.

One must add a cautionary note: the internet is free-form and not peer reviewed. Consequently some of the information provided is inaccurate or wrong. So developing the ability to make wise judgments in this arena is important. To quote Mark Malloch Brown once again: "we should identify policies that can best accelerate the benefits of technological advances while carefully managing the risks that inevitably accompany change".

As we know, research is global. Participation in international research collaboration in science and technology can add value to a country. As Jalali noted: "even in the domain of subsistence research, the developing countries are obliged to enter into competition with more powerful rivals. All recent developments underline the necessity of defining and creating veritable research structures in development. The international community has a great responsibility for creating an international pattern of scientific cooperation to assist the developing countries to develop these structures."

In conclusion, the development of a framework for science and technology policy, and the tools necessary to assure responsible governance and networking across government sectors and the citizenry, can markedly benefit society in developing countries.

A PROJECT FOR SOLAR ENERGY AS A JOINT EFFORT OF WESTERN AND ISLAM COUNTRIES

GERHARD KNIES

Deutsches Elektronen-Synchrotron DESY, and Hamburger Klimaschutz-Fonds HKF, Hamburg, Germany, Stauffenbergstr. 15, D-22587 Hamburg

Email: gerhard.knies@desy.de

INTRODUCTION

This paper, unlike the others presented at this conference, is not reporting on the work of one of the Permanent Monitoring Panels (PMP) on global emergencies of World Federation of Scientists (WFS). It is reporting on two connected projects completely independent of WFS activities, that I am involved in, and which happen to deal with subjects related to the WFS activities. One is a political project, initiated by the President of Germany. In this project, 13 heads of state, from countries in the Western and in the Muslim world, have agreed that scientists from their countries conduct a dialog on common problems in the process of globalisation. The other project is the attempt to combine the potentials of European and Northern African/Arabian countries for a joint effort to use solar energy, in particular for generating of electricity for both regions. This latter project is emerging from the activities of the Hamburger Foundation for Climate Protection, and of the working group on global energy issues inside the German Physical Society. This South-North solar energy project has been adapted by the political dialogue project as a practical attempt to attack a problem common to the Muslim and the Western world.

GLOBALISATION

We are presently witnessing the process of globalisation–or better: many processes of globalisation. *Globalisation can be seen as the development of humans to humanity.* Humans so far have developed many forms of higher organization to social bodies, like families, neighborhoods, villages, peoples, nations, football clubs, jazz bands, industrial companies, Olympic Games, universities, world federation of scientists, religions, etc., with which individual humans identify. Humans now with increasing speed are becoming part of humanity—in supplement or in competition with peoples or with nation states. I believe that only when human beings identify themselves stronger with humanity than

212

with their nationalities or religions, they will become ready for solving global emergencies—across nations, military pacts and religions.

Globalisation as a move towards "mankind" has similarities with the phase transition of matter from gas to liquid or even to a solid state configuration. In gas, every unit is colliding against the others, in a confronting manner, independent and sovereign. A gas has no intrinsic ordering scheme—just collisions of its autonomous parts. In a crystal, however, all parts are coordinated. Shaping or structuring humanity on earth requires the setting up of organized and accepted interactions and communication between its constituents: is that possible?

After the collapse of the East-West conflict, which polarized or "ordered" the whole world, this process entered a new stage. Now the balance of growing human population on earth with the finiteness of earth's resources for human life is becoming a pressing issue. Sustainability will be among the global ordering principles. At the same time, scientific culture and its child, the technological civilization, have, in terms of satellites and electronics, established earth wide informational interconnections. The following map[1] shows those countries as gray from which institutions, located at the dots, are participating in the scientific data exchange network of the international research centres CERN[2] and DESY[3]. This map was shown at a congress on the subject: "One web, one world, one future". We observe two worlds. If one takes into consideration, that the countries Algeria, Iran, Pakistan and Uzbekistan are only minimally connected to this scientific network, this division in Eurasia is strikingly close to a separation into Muslim and non-Muslim countries.

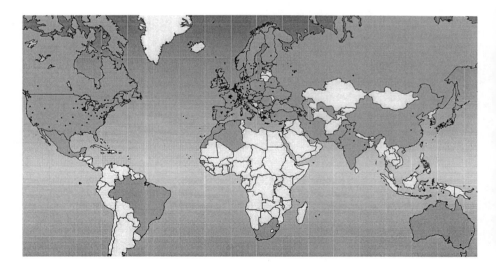

Fig. 1: One Earth – One World? Countries having scientific data exchange with CERN and DESY.

CLASH OF CIVILISATIONS

There are many conflicting interests and dividing lines between the peoples on this earth, like highly developed–little developed, rich–poor, democratic–non-democratic, oil exporting–oil importing etc. Mr. Samuel Huntington is expecting a clash of civilizations in the future[4]:

> *"It is my hypothesis that the fundamental source of conflict in this new world will not be primarily ideological or primarily economic. The great divisions among humankind and the dominating source of conflict will be cultural. Nation states will remain the most powerful actors in world affairs, but the principal conflicts of global politics will occur between nations and groups of different civilisations. The clash of civilisations will be the battle lines of the future."*

The place where we gather, Sicily, is located at the borderline between the Muslim and the Western world. The futures of these two worlds are going to be tightly connected, much tighter than in the past. The leading tie at present is oil and gas. Without supply of fossil fuels from Arab countries, European and other societies worldwide will collapse. Without the money from oil sales to Europe, some Arab countries will collapse. Here we have a symbiosis. The Western perception is: "They have our oil".

The oil resources are finite, with a reach of a few decades. What will happen then?

To me it seems meaningless to dispute whether cultural, civilizational, economical or any other source for clashes would be stronger. Most likely, they will amplify each other. In fact, this situation bears the potential for emergence of violent clashes—between the Western and the Arab World. Are they to come as an inevitable fate? Or are there more sustainable options than a fight for the largest share of natural resources?

PRESIDENTIAL PROJECT OF DIALOGUE BETWEEN MUSLIM AND WESTERN WORLD

In view of these conflictive perspectives, the former German President, Roman Herzog, in 1996 started out to initiate the enterprise: Dialogue between countries of the Western and Muslim world, for which he won the support by 12 heads of state. The 13 countries are:

- Egypt, Indonesia, Jordan, Malaysia, Morocco
- Austria, Chechnya, Finland, Germany, Italy, Norway, Spain, Switzerland

In 1999, his successor Johannes Rau has declared the *"Dialogue between Areas of Western and Islamic Culture"* to be a central theme of his foreign affairs activities.

214

Such an intercultural dialog cannot be carried by governments only. It requires the participation of broad societal circles. Neither is mutual assurance of good will without implications sufficient. To be taken serious it has to be driving at developing of co-operative solutions to common problems.

There are a lot of problems that could be subject to attempts for cooperative solutions, to "build cultural bridges" between Europe and the Islamic world. At the founding conference in Berlin in April 1999, four main areas have been identified for projects along these lines:

- Identity and discourse
- Politics and shaping the future
- Journalism and media
- Education and learning

The item "shaping the future" has a striking similarity to the goals of the World Federation of Scientists, and of the World Laboratory. I quote the World Laboratory presidents message: *"We believe that the new role of science is to promote a more effective collaboration between East and West, and to contribute in a direct way to fill the ever increasing gap between North and South."*

THE COMMON INTEREST IN CLIMATE STABILITY

One of the scientifically predicted consequences of the scientifically predicted climate change is that the dry zones in North Africa will expand to the north. In fact, the surface water, i.e. the water from rainfalls that does not evaporate before it flows down hill, has been monitored in Morocco annually since 1945. Until 1997, these data (Figure 2) show two distinct features:

- on average annual rainfall is going down
- the years with extreme low rainfall, dry years, are rapidly growing

Availability of fresh water is the limiting factor for expansion of food production and for the rise of living standard in all sunbelt countries. The measurements show a fatal trend. Limitation of climate change is of vital interest to the peoples living in North Africa.

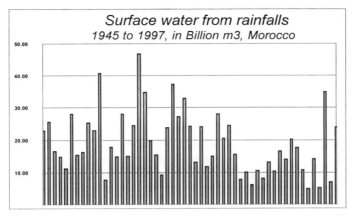

Fig. 2: Surface water from rainfalls, i.e. water left after evaporation[5], Morocco.

North African and Arabian (NA/A) countries have solar radiation in surplus to accommodate completely all their energy needs. Onto each square kilometer, the sun radiates the energy as of 1 Million barrels of oil, year by year. However, even a full replacement of fossil by solar energy in these countries would not help to curb climate change. The share of these countries in the global green house gas emissions is as low as 5%. The Islam countries can achieve their goal of climate stabilization only by a reduction of the "Western" emissions. In fact:

- the reserves of oil & gas are no longer too small, but too large.
- the production of any barrel of oil—wherever on earth—is an attack against the Arab world.
- NA/A countries cannot stop climate change by domestic measures.
- Climate change "turns off" national sovereignty and national security.

As seen from the West: "They have our oil, and we spoil their climate".

Also the European countries are facing severe problems from climate change. When heavy storms and intense rainfalls increase, when "century events" come at the pace of a few years, when they begin to repeat within the recovery time of damaged social or natural systems, their living conditions are at stake as well.

As seen from the West: "They have our oil, and we spoil their and our climate."

The NA/A and European countries are in the same "climate-boat." With the predicted and already measurable and visible phenomena of global warming as a result of anthropogenic production of green house gases, the assessment of conflicting interests between Western and Muslim world changes. The oil is a threat to the living conditions

216

of either side. Its unrestricted exploitation is in conflict with Western essentials like global and intergenerational justice and with universal human rights. With a co-operative development of solar energy potentials the Sunbelt countries and the "Western world" could serve their energy needs fully and economically, and reduce green-house gas emissions fast enough.

THE CLIMATE PROTECTION ALLIANCE

From the "demand side" this situation suggests a climate protection alliance. And in fact, by a solar energy co-operation—as if there were no boundaries—they can enhance their power for the necessary climate protection considerably, as shown in the following graph:

South – North Synergy Diagram on Climate Protection Power:

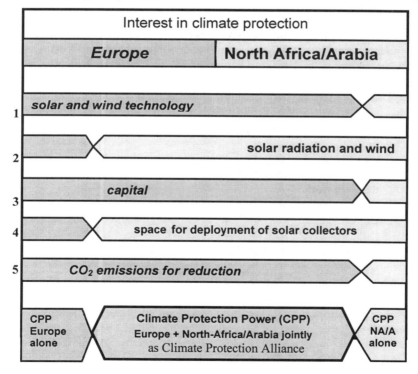

Fig. 3: The 5 factors required for climate protection power. Their relative strength in Europe and in North-Africa, qualitatively. The interest in climate protection is big in both regions (top bar). The separate regional climate protection powers are small (left, right bars in bottom row) as compared to the additional power from joining forces (bottom bar centre).

The European and the Northern African/Arabian potentials supplement each other to their common benefit. They have what we don't have—and vice versa. This situation calls for a climate protection alliance between Europe and Northern Africa/Arabia. Within the framework of the presidential enterprise "Dialogue between countries from the Western and Muslim world" a conference on these options is under preparation for the year 2002. A few questions arise:

- How can solar radiation be converted into usable energy?
- Is there enough space in North Africa to collect solar energy for the combined climate protection needs?
- Can solar energy be exported from North Africa to Europe?
- Cost of solar electricity?
- Benefits and problems from solar electricity co-operation—can the Muslim and Western worlds trust each other?

These questions will be addressed below.

ELECTRICITY AND DRINKING WATER FROM SUN AND SEA

In Sunbelt countries solar boilers can generate steam for conventional thermal power plants. If salty water is available, cooling of the power plant can be done by driving a desalination plant[6]. Typically, with 1 kWh of electricity 40 litres of drinking water can be co-generated.

A Solar-Thermal Power Plant can:

- generate electricity
- desalinate sea water as by-product (~40 litres / kilowatt hour)
- provide cooling as by-product
- grow vegetables in the shade under the collectors
- store heat for operation at night

Drinking water, salt, vegetables, cooling and electricity can be made from sand, sun & sea, with proven technology.

218

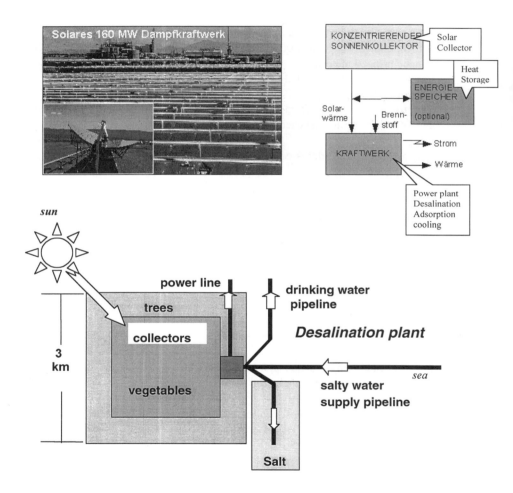

COLLECTOR SPACE FOR ELECTRICITY AT "CLIMATE SCALE"

Africa - Landcover

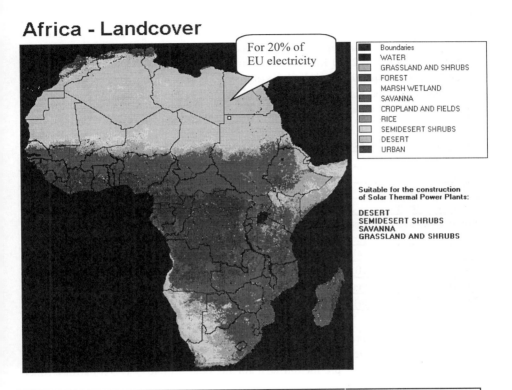

For 20% of
EU electricity

■	Boundaries
■	WATER
	GRASSLAND AND SHRUBS
■	FOREST
■	MARSH WETLAND
■	SAVANNA
■	CROPLAND AND FIELDS
	RICE
	SEMIDESERT SHRUBS
	DESERT
■	URBAN

Suitable for the construction
of Solar Thermal Power Plants:

DESERT
SEMIDESERT SHRUBS
SAVANNA
GRASSLAND AND SHRUBS

Typical strength of direct solar radiation of 2.5 MWh per m^2 and year,
  → thermal energy as in 1 barrel oil,   →electrical energy 0.25 MWh/m^{2}a.
For 500 TWh/a (= 20% of EU electricity)
  → collector area of ca. 50x50 km^2 (= 10% area of Sicily)

In addition to solar radiation, there are very good trade winds along the west coast of North Africa, available for electricity generation. The trade winds blow day and night, with remarkably constant speed. Lulls are extremely rare. They provide a good supplement to the solar radiation.

LONG DISTANCE TRANSMISSION TECHNOLOGY:

Power lines/sea cables for High Voltage Direct Current (HVDC, ~0.5 - 1 Million Volt).
Transmission losses: 2–3 % per 1000 km, 0.5% per converter station (AC → DC → AC)
Transmission losses from northern Africa to middle of Europe: ~ 10%.

THE COST OF CLEAN ELECTRICITY

The investments for solar power plants and for wind converters are at present quite similar, near 1000 \$/kW. These figures will continue to go down, as result of technological progress and of economies of scale. The annually available full load hours are between 2000 and 2800 for solar radiation, and between 3000 and 5000 for wind, respectively. Correspondingly, electricity from wind is cheaper, but the best wind regions are further south and less abound. The expected production and transmission costs are summarized in the following table:

Clean, Renewable Electricity Cost Estimates

	Wind[7] (Morocco)	Solarthermal[8] (Egypt)
Capacity	2 MW	400 MW
Full load hours	4000 h/a	2586 h/a
Economic lifetime	25 a	28 a
Maintenance, operation, insurance	3.5%	2.9%
Interest rate	5%	6.7 %
Investment cost (in 2010)	1000 $/kW$_{el. peak.}$	1000 $/kW$_{el. peak}$
Electricity production cost	0,025 $/kWh$_{el}$	0,044 $/kWh$_{el.}$
Transmission (4000 km) cost	0,012 $/kWh$_{el}$	0,012 $/kWh$_{el}$

Electricity generated from trade winds is already competitive with electricity generated from fossil fuel, at their present costs of 25 U.S.\$/barrel, and solar electricity is close. However, the bare production cost for oil is at around 5 U.S.\$/barrel. Since investment costs will go down and fossil fuel costs will go up with time, within the next 1-2 decades there could be a big market opening up for solar electricity export to Europe. It is to be noted, that the seasonal variations of sun and wind in North Africa are just opposite to the wind yield in northern Europe, providing an ideal supplement.

BENEFITS OF A SOLAR ELECTRICITY ALLIANCE: CO-OPERATION INSTEAD OF CONFRONTATION

There are not only costs connected with the electricity from a South-North partnership. With each kilowatt hour I see 4 important benefits that come across the Mediterranean for free, in the concept of a solar electricity symbiosis between Europe and North Africa/Arabia:

1. *Strengthening of the social situation in North African countries.*
   Oil makes a few countries or people very rich, solar energy will give jobs to many.
2. *Enforcing mutual peaceful behaviour.*

Violence between the solar electricity partners is not permitted–and not necessary.

3. *Dissolving the conflict potential of finite fossil energy resources.*
   An insufficient and exhaustible reserve is provoking violence.
4. *Solving the problem of global warming.*
   It may well turn out that the most efficient protection for the coast line of Bangladesh or worldwide would be power lines from North Africa to Europe.

COULD THESE BENEFITS COME?

Within the framework of *"Dialogue between Countries of Western and Islamic Culture"* a conference is being planned, at which the virtues, possibilities, the drawbacks and obstacles for a joint effort to exploit the synergetic potential of the respective regions for their common goals are to be examined. And to make these benefits real, the Western world could provide the transmission lines as "solar bridges" and the Muslim world desert space for collector fields as "solar acres".

REFERENCES

1. This map was shown by the member of DESY directorate, Dr. Hans von der Schmitt, at the Congress "One Web, One World, One Future", organized by the Social Democratic Party of Germany, 6th and 7th of July, 2001.
2. CERN, European Centre for Nuclear Research, Geneva, Switzerland.
3. DESY, Deutsches Elektronen-Synchrotron (German Electron Synchrotron), Hamburg, Germany.
4. Samuel P. Huntington, The Clash of Civilizations?, in Foreign Affairs, Vol. 72, No.3, (1993)22.
5. By courtesy of Abdelaziz Bennouna, Centre National de la Recherche, Rabat (Morocco).
6. Franz Trieb, Joachim Nitsch, Gerhard Knies, Energiewirtschaftliche Tagesfragen, 51. Jg. (2001)386.
7. Gregor Czisch, University of Kassel (2000), Study for the Hamburger Klimaschutz-Fonds HKF.
8. The Solarmundo Project, Exposee (June 2001), Solarmundo NV, Meir 44A, 2000 Antwerpen, Belgium.

AFTER GLOBALIZATION: FUTURE SECURITY IN A TECHNOLOGY RICH WORLD

THOMAS J. GILMARTIN
Center for Global Security Research, University of California
Lawrence Livermore National Laboratory, Livermore, California 94550

INTRODUCTION

Over the course of the year 2000, five workshops were conducted by the Center for Global Security Research at the Lawrence Livermore National Laboratory on threats to international security in the 2015 to 2020 timeframe due to the global availability of advanced technology. These workshops focused on threats that are enabled by nuclear, missile, and space technology; military technology; information technology; bio technology; and geo systems technology. The participants included U.S. national leaders and experts from the Department of Energy's National Laboratories; the Department of Defense: Army, Navy, Air Force, Office of the Secretary of Defense, Defense Threat Reduction Agency, and Defense Advanced Research Projects Agency; the Department of State, NASA, Congressional technical staff, the intelligence community, universities and university study centers, think tanks, consultants on security issues, and private industry.

For each workshop the process of analysis involved identification and prioritization of the participants' perceived most severe threat scenarios (worst nightmares), discussion of the technologies which enabled those threats, and ranking of the technologies' threat potentials. (Fig. 1) The threats ranged from local/regional to global, from intentional to unintended to natural, from merely economic to massively destructive, and from individual and group to state actions. We were not concerned in this exercise with defining responses to the threats, although our assessment of each threat's severity included consideration of the ease or difficulty with which it might be executed or countered.

At the concluding review, we brought the various workshops' participants together, added senior participant/reviewers with broad experience and national responsibility, and discussed the workshop findings to determine what is most certain or uncertain, and what might be needed to resolve our uncertainties. This paper summarizes the consensus and important variations of both the reviewers and the participants. The full report is available at: http://cgsr.llnl.gov/global/global.html

MAJOR CONCLUSIONS

In all, 45 threats over a wide range of impacts and probabilities of occurrence were identified, as were 60 enabling technology categories. Here we present the major conclusions, which each include consideration of several threats and their enabling technologies. The threats are listed in the order of highest concern.

Terrorist Nuclear Weapon

The danger that terrorists might use a crude or procured nuclear weapon to attack a city is non-negligible. The proliferation of nuclear weapons, the atrophying of huge cold war stockpiles, the global increase generally in nuclear technology, the rising tide of all non-nuclear, but related enabling technologies, from computing to robotics to remote control, and the ease of covert delivery, all increase the probability that a nuclear weapon will become available to and be used by a highly motivated agent. Attribution of such an attack may be difficult if the sponsoring group decides not to claim responsibility. Such extreme and potentially anonymous terrorism might be viewed as effective by hyper-zealous groups.

Natural And Engineered Disease

Unfortunately, diseases eliminated or controlled in public still exist in biological storage, are known to persist in relatively isolated populations, or are reemerging in drug resistant forms. Much of the once immunized population is again vulnerable, for example, to smallpox and to antibiotic resistant tuberculosis. In addition, new diseases are emerging, and biotechnology is now able to modify and combine disease elements to tailor their effects and potentially even to select their targets by genetic or habitual traits. Bio regulator technology, which can alter human function and performance, is increasingly sophisticated and available, for both beneficial and malicious uses. The technology, production means, and dispersal mechanisms needed to initiate a bio-attack are relatively simple and difficult to detect and attribute, and the knowledge of how to accomplish these ends is widespread. Even though the perpetrator might be the victim of his own attack, the potential for serious, widespread disease outbreak and global disruption is considerable.

Limited Nuclear War

Ironically, the technological obsolescence of legacy military technology and the revolution in military technology is favoring the nuclearization of emerging powers, which cannot afford and are unable to implement competitive sophisticated systems-of-systems forces. Nuclear weapons give immediate dominance over or nuclear peerage with local adversaries, deterrent capability in pre-conflict calculations and in conflict operations, and to-be-reckoned-with stature among world powers. Examples are Israel, India, and Pakistan, and potentially Iraq, Iran, North Korea, and others during the next two decades. Diverse asymmetries between these nations and their adversaries often make mutual understanding and stabilization difficult; the intensity of feelings impedes dialogue and restraint.

It is certainly possible, and may be even likely that some such situation will result in the use of nuclear weapons, out of desperation or vengeance, or in a low fatality (EMP or deep target) mode, this being less provocative, but militarily effective. Such localized use of nuclear weapons would reinforce the rationale for emerging nations to have such weapons and probably would increase proliferation and global risk of nuclear conflicts.

Major Nuclear War

While it is generally thought or hoped that the threat of global nuclear war has receded, massive arsenals and delivery capabilities still exist, are growing in some cases, and are now imbedded in a more complex geopolitical matrix. This situation might be more analogous to the multifaceted relations prior to World War I than to the bipolar Cold War stand-off, with now an array of powerfully armed nations and a second tier of emerging nuclear actors with intense animosities and a variety of alliances with each other and with the primary nuclear powers. This system is unpredictably unstable, very nonlinear, and far more difficult to crisis manage, if only because the scenarios are many, the interactors diverse, and the management protocols untried and undefined. This threat ranks high not based on any current tension, but because of the uncertainties and the potential for catastrophe.

Human Control Of Bio Forms

In addition to the malicious applications of biotechnology listed above, the fact that humanity is increasingly able to design and make new bio forms, from viruses and proteins (and prions) to bacteria and even to whole flora and fauna, is both wonderful and frightening. Evolution has constructed the microorganisms and biota of today over very long periods and has tested extremely complex interrelationships such that species are in dynamic equilibrium with their complex environments. In contrast, most of man's biological creations serve specific purposes, such as, the production of medicines and organs for human use, and the improvement of the productivity and performance of domesticated species, even of humans. These improvements will not be ecologically tested; such testing would be extremely complex, if possible at all. In fact, the increasing ease of biological creation will enable recreational genetics and bio hacking. The dangers of ecological and human disruptions will be great. In addition, nano technology and molecular scale information technology will blur the boundary between biology and heretofore inorganic technologies. To quote one project participant, "It will be a brave new world when man controls evolution and the worlds of carbon and silicon converge."

Spread of Advanced Military Technologies

Stealth, anti-aircraft IR (infrared) and radar counter measures, AWACS (airborne warning and control system), IR sensors and guidance, and precision ordnance have given the U.S. dominance and near impunity in projecting air power. An array of new air defense and air combat technologies threatens not only to compromise this dominance, but to push forward projection air and sea support systems back from target areas . These technologies include: IR focal plane array (IRFPA) sensors, which might defeat IR countermeasures; conformal IR missile dome optics, which give anti-aircraft missiles better sensors and greater speed and range; IR search and track (IRST) systems, and low frequency, multi-static, and expendable radar systems, which can lessen the effectiveness of stealth and anti-radar missiles; airborne or space based radar, IR, and visible sensors, specifically moving target indication (MTI) systems, which also lessen the effectiveness of stealth and of cruise missiles. Add to this that stealth technology is becoming widely available for aircraft, missiles, and ships, and that improved adversary IR systems will increase their night operational effectiveness. The result will be a diminishment of current modes of air dominance and forward strike capability, the need for greater standoff and other protections for aircraft carriers and AWACS, possibly the necessity of advanced forces to "share the night" with adversaries, and the need for a new generation of strike and countermeasure technologies. The global availability of advanced military technology is increasing and here-to-fore dominant technologies are no longer contained.

Control And Loss Of Control Of Nature
On the one hand, we are gaining greater control of natural phenomena. Models of global atmosphere/ocean/biosphere physics are being coupled to mesoscale and regional models, potentially enabling more accurate prediction and even, speculatively, some degree of control of weather, storms, and climate. This capability would be of great national advantage. Similarly, the understanding of earthquake initiation, of tzunami generation, and of methane release from metastable undersea clathrate formations are all potentially triggerable events.

On the other hand, human activities are changing atmospheric composition, adding green house gases and depleting ozone, which can change the global environment in ways that we are not as yet able to control. The effects of these changes will be predictably distributed, with much variation of benefit and harm among regions and nations. Our inability to control these effects is very troubling; and their actual distribution, when known, is certain to be a source of international antagonism. These effects on the U.S. homeland are, on balance, significantly negative.

Information Warfare
The number and variety of information operations that might be used against communication, communication-dependent infrastructure systems, and commerce occupied much of our discussion, from simple civil intrusion and denial of service to complex tapestries of financial, infrastructure, and military system attacks. Such attacks during the year 2000 disabled Internet services with undoubtedly large financial cost and inconvenience. However, although the frequency of lower level but costly mischief attacks will increase, and our information infrastructure will require constant defensive modifications to continue to function effectively, it was judged that defenses would evolve as needed and that such attacks would not ultimately threaten nations' sovereignty, economy, or military security. With adequate wariness and prudent

precautions, financial losses, disinformation, security breaches, system intrusions, and infrastructure attacks should be containable and preventable. In fact, the more complex the planned assault, the more probable its detection and avoidance.

Asymmetry

US military dominance is an example of asymmetry. It is highly unlikely during the next two decades that any adversary will defeat the U.S. in conventional conflict. In fact, the effectiveness of advanced military units might be increased by adopting techniques currently regarded as asymmetric (flexibility, adaptability, unpredicability).

But, concentrations of value (people, cities, infrastructure, industry, energy supplies, embassies, ships in port,...) are extremely vulnerable. It is repeated everyday in new reports that free and open societies are not properly organized, trained, equipped, or positioned to prevent devastating attacks on these sorts of targets.

While our discussions did repeatedly reveal new vulnerabilities to and new methods for such attacks, and did decry the commonly identified deficiencies of defense measures, it was also agreed that such attacks would not seriously threaten nations' military security or government, and that most perpetrators would eventually account for their actions. This is not to downplay the nature and difficulty of dealing with today's asymmetries, but to keep such potential actions in perspective.

Acute intelligence, mutual international commitments and collaborations, special forces, and clear responsibility for homeland defenses and emergency responses were all recognized as necessary to minimize the dangers from asymmetric threats.

GLOBALIZATION

Some of the characteristics of globalization relevant to security are:

Global Markets

Elements of large-scale economies including their needed resources, skills (education) and information, production, distribution, consumption, and finance will be increasingly transnationally organized and valued, and globally interdependent. Clearly this trend diminishes each nation's ability to control its assets, its enterprises, and its people. On the other hand, Francis Fukayama proposes (The End of History and the Last Man) that this will lead to greater similarity and cooperation among nations.

Economic Power

The ability to generate and control global markets has replaced military strength as a national strategy for and measure of dominance and security, for examples, by Germany and Japan, and as intended by China. We believe that the U.S. would much prefer a world in which it could compete economically without concern for military and physical security, but this does not seem to be an available option.

Private Capital

The scale of private money holdings is larger than the financial resources of many nations. More money flows across national borders every week than the value of the U.S. GDP. George Soros was personally able to affect a monetary crisis in south east Asia, which threatened even China; on the other hand, he is able to pour hundreds of millions of dollars into the rehabilitation of eastern European nations. Criminal cartels' money and thereby political control are threats to national survival in, for examples, Columbia and Russia.

Information Ubiquity

The Internet distributes information without regard to national borders, both intentionally for the purposes of collaboration and business, and unintentionally due to imperfect cyber security and malicious actors. With now wireless technology and wideband networks, almost no place on earth is not connected to the pool of human information and knowledge instantaneously. All the genies are slipping out of the bottles. In addition, the production and integration of information systems are global operations, which provide very sophisticated capabilities ubiquitously and prevent birth to implementation to death security assurance of even critical systems, especially with the increasing use of commercial products in military systems.

Disinfrastructurization

Global interconnectedness provides the ability to organize activities without localization and centralization constraints. Production resources are globally distributed, but coordinated as though they were in the same plant, using facilities wherever they are available, greatly increasing economic efficiency, but also enabling distributed production of weapon systems in a manner that eludes detection and control.

Ecology

The scale and reach of human activities exceed the boundaries and adaptability of most regional and even global ecological systems. The habitats of all species, except pathogens, are contracting. The future viability of most species is now requires a choice by man; only the selected and protected will remain. Such issues as atmospheric and ocean warming, pollution and eutrophication, and desertification are now transnational phenomena, with stakeholders motivated by widely varying needs, making cooperation difficult.

Migration

Populations are both mobile and motivated to move across borders and around the globe to improve their lives and to go to the centers of technology. They carry their culture, their genes, their diseases, and their politics, making the globe the new melting pot or boiling pot, as the case may be. In addition, ease of travel and increased wealth are making short term migration significant for recreation, work, and temporary residence.

Dynamic Socialization

Countering Fukayama's optimistic view of globalization, Samuel Huntington (*The Clash of Civilization's and the Remaking of the World Order*) writes that cultural differences will persist and cause conflict as globalization brings us all into closer contact and forces heretofore incompatible values into confrontation. In the same vane, the attraction of global action centers and the exposure of relative deprivations will drive migration and actions that relieve extreme social and economic gradients.

Conformed Governance; Standards, Law, Finance, Civil Rights, Environment

Fundamental to economic cooperation and the formation of global markets are developed and compatible systems of standardization, law, and finance, which enable business. Increasingly, civil rights and environmental standards are being tied into the terms for international transactions. The International Standards Organization, World Trade Organization, International Monetary Fund, and World Bank are examples meta-national organizations that require nations to bend to the global will and values.

Criminality

On the other hand, organized crime operates internationally on scales that threaten national sovereignty, for example, in drug trafficking in South America and South Asia, human smuggling and trade, and illegal currency trading and manipulation.

Haves and Have-Nots

Food and money once defined the principal distinctions between have and have-not populations. Today and increasingly in the future, the dimensions which divide haves and have-nots include additionally nuclear weapons, energy resources, water, information technology (the digital divide), and attitudes toward and the possession of genetic technology. Possession of nuclear weapons causes sharply different national stature and international risks. Genetically modified crops have benefited China in many ways, increasing crop yield and decreasing the use of pesticides; China is very supportive of many forms of genetic innovation. France is much more cautious. Both the advantages and problems of these extremes will accumulate and differentiate among nations, influencing nations' security and their ability to develop socially and economically

NUCLEAR, MISSILE, AND SPACE TECHNOLOGY

Ironically, advances in technology, which have given the U.S. global military dominance, have obsoleted much of the world's Cold War era systems and motivated emerging nations to acquire nuclear forces, which give them formidable weapons, the ability to deter superior conventional forces, and instant stature in international affairs. Even the U.S. can be deterred by nuclear arsenals much inferior to its own.

In addition, the general advance in global technology has decreased the cost to obtain nuclear weapons and their platforms. Computers, isotope enrichment, nuclear materials technology, robotics, precision machining, space launch and intermediate range missiles, cruise missiles, global positioning systems, and satellite imaging are all now commercially available.

The number of potential nuclear proliferators has increased, while international proliferation restraints and transparency have receded due to the waning of cooperative regimes, the emergence of barter economies, and the increasingly dual use of the relevant technologies. Manufacturing agility and virtual distributed industrial complexes (disinfrastructurization) discourage identification of proliferant activities.

Nuclear technology is one of the ships that is rising on the global technological tide. Global energy and environmental needs will support the continued spread of nuclear energy technology.

The web of nuclear threats is more complex now than it was during the Cold War (Fig. 2). The old actors remain; new actors have come on the scene or increased their capability; and the level of nuclear and platform technology has increased in regions of cultural and resource tensions, such as, South Asia, the Middle East, and the Caspian area. There are disturbingly many credible scenarios for the future use of nuclear weapons in regional conflicts.

If used for defense in a desperate attempt to prevent being overwhelmed in response to a massive attack, a nuclear response might be condoned. In fact, possession of "tactical', that is, short-range, limited-effect nuclear weapons for counter-force purposes, might be accepted internationally for national defense. This limited acceptance of tactical nuclear weapons might stimulate their proliferation, particularly in regions of tension, where they are most likely to be used.

This, in turn, increases the probability of further nuclear weapons development, and the potential for their preparation for other uses, such as in space or in other scenarios remote from the "defenders."

While it is reasonable to hope that the few massively armed nations will respect each others' capabilities and devise agreements and procedures that will continue to lessen, but probably never eliminate, the threat of massive nuclear war, it seems equally likely that nuclear weapons will gradually proliferate as the world becomes increasingly technological and nuclear energy is increasingly needed, and that a regional nuclear event or terrorist action will occur. Needless to say, these three possibilities should motivate extreme international cooperation in the effort to prevent them.

MILITARY TECHNOLOGY

The advanced industrial nations can dominate adversaries in symmetric engagements; they have the asymmetric advantage in a "fair fight." that is, in large scale force on force warfare. But, they are also asymmetrically vulnerable, having large concentrations of people and wealth, and globally distributed interests that are difficult to defend, particularly against a suicidal attacker. The global leaders generally respect each others laws, national sovereignty, people's natural and civil rights, property, and the environment, while asymmetric adversaries are not so constrained, and in fact use these foibles for advantage against nominally stronger foes.

The African embassy bombings killed a dozen U.S. citizens, while killing 200 Africans and wounding 4,000. The perpetrators were from six nations (apparently including the U.S.) which did not condone or know about their actions. It is likely that the all advanced nations, but the U.S. particularly, will have to deal with this kind of "asymmetric warfare" in the future.

As badly as advanced nations may be hurt by terrorism, they will probably not be destroyed by asymmetric adversaries. And as specialized counter forces evolve and adopt "virtuous" versions of asymmetric techniques, the adversaries' advantages and effectiveness might lessen.

Perhaps the greater fear is that current military advantages will diminish, as improved IR and radar systems diminish the effectiveness of stealth and countermeasures, as improved range, speed, and stealth of anti-aircraft and anti-ship missiles push support systems back from the theater, and as satellite systems improve adversaries' intelligence. In addition, adversaries' use of modern C4ISR (command, control, communication, computation, intelligence, surveillance, and reconnaissance) for their own purposes and to defeat advanced C4ISR systems might shift the information technology imbalance away from advanced forces. In general, we can expect the battlefield to become more transparent for adversaries and more opaque and dangerous for us.

The overall effects of adversaries' advanced technologies will be to increase the effectiveness and reach of their forces. It is probable that we will experience more casualties, even well behind the "front". The adversaries will be more difficult to suppress, with stealth, expendable and intermittent systems, and underground and disguised facilities. Both air and sea operations will be increasingly difficult due to the stealth and reach of advanced missiles.

Defense forces will be called upon to protect global interests, homeland cities, and civilian infrastructure, and to be prepared to deal with biological and chemical attacks. This will focus more defense resources on homeland defense.

The bottom line is that industrial nations might be less likely to intervene outside of their homelands, and will be more wary in protecting their homes.

INFORMATION TECHNOLOGY

Several general observations relevant to the prediction of the future the cyber revolution are in order. First, the rate of technological evolution and deployment in the cyber world is 2 to 4 times faster than the rates for traditional technologies; it is estimated that cyber innovation and technology improvement achieve in 3 months what more conventional technologies attain in a year. On this basis, the 15 to 20 year horizon for this study is the equivalent of more than 60 years in cyber time. The result, according to one knowledgeable participant, is that "whatever you can imagine for information technology will happen in this time frame."

Second, information and information technology are inherently global. Networks reach all parts of the globe instantaneously regardless of natural and man-defined boundaries, language, or social status.

Third, each information capability also defines a vulnerability. Information is also intelligence; interaction can easily be conflict; access, intrusion; exchange, theft; persuasion, propaganda; and so on.

Finally, cyber crime is still relatively unconstrained and retribution free. The perpetrators often act anonymously, from unknown sites, and, even when caught, are treated as white collar criminals, the loss of money or information being much less onerous than physical or personal harm.

Currently, the information defenders (individuals, businesses, institutions, and the national security entities) are not well coordinated for diagnosing attack, defending, or counter attacking. While the defense community is evolving toward better integration and defense, and the banking/financial community is reasonably secure, hackers, individuals, academics, and customer interaction businesses are by choice less constrained and less protected by security mechanisms and procedures. The consumer market places high value on freedom of choice and convenience, distributes the costs of attacker damage, and abhors the complexity of protection. As a result the Internet remains relatively unprotected, slow to diagnose malicious acts, and slower still to counter them or find and prosecute the malefactors, in spite of the year 2000's many and overall costly attacks.

Nations are deliberately developing the capabilities for more comprehensive and sophisticated information system attacks; undoubtedly, non-national groups are doing the same on smaller scales. The purposes range from surreptitious and intrusive intelligence gathering, to subtle bleeding of adversary's wealth and economic efficiency, to preemptive strikes in the opening phases of conflict and multi-facetted tapestry attacks during war, and to disruption of key tactical and strategic military operations. With national resources behind them, these methods could become more stealthy and effective, outpacing the development of defenses, which are being only weakly exercised in the absence of international information warfare. Gradually, these information weapons and criminal methods are becoming public and accessible to a wider range of disrupters.

Individuals, businesses, and national entities should link defense efforts, sharing data and analyses of information attacks. Commercial reticence creates asymmetry favoring the attacker, whose hacker technology is widely shared, while sharing among the defender communities of their attack experience, perceived vulnerabilities, and defensive measures is constrained by the desire for commercial access and by fear of the vulnerabilities created by revealing their weaknesses. Similarly, exchange between the open and national security communities is not yet productive. Clearly, the next decade will see major information operation events and the need for significantly greater cooperation and technology development in information system defense.

The consensus of our discussions, however, was that, while information technology and operations will be extremely important elements of both offense and defense, physical dominance is more likely to determine the outcome of conflicts during the next two decades.

BIOTECHNOLOGY

While the participants in the Biotechnology Workshop expressed considerable concern about the use of virulent pathogens by terrorists and warring nations, there was broader and deeper concern about the intentional and unintentional consequences of the genetic revolution that is just beginning.

Biotechnology has made tremendous advances in the last decades of the 20th Century. These advances have come about through the technologies of recombinant DNA research, genomics, and proteomics. The results have been rapid and have provided immediate and powerful tools to diagnose human disease and, in some cases, have provided clues to amelioration or cure. We fully anticipate this genetic and biotic revolution to continue in the 21st Century and further enhance the quality of human life.

However, these same technologies that provide enormous benefits to mankind can also be abused or misused. These future (and already emerging) capabilities will both increase the range and lethality of current bio threats, and will enable significant new types of accidental or intended bio threats.

Agricultural genetic modification has been underway for centuries (breeding) and for over a century on the basis of biological science (Mendel). Genetic manipulation has been publicly acceptable, and is accelerating, particularly in China. It is feared that surreptitious malicious modification might be used as a strategic weapon, or that well intended modification might have long-term and widespread undesirable side effects.

As we learn to genetically tailor antibiotics, anti-virals, anti-carcinogens, and anti-all-variety-of-undesirable-health-conditions, we strike the Faustian bargain that yields as well all of these same forms of malicious tools: super viruses and bacteria, subtle carcinogens, and attribute-selective (ethnic) and other forms of health and performance degrading pathogens. These pathogens might lie outside of the immuno/metabolic defenses of current life forms, and could cause wide destruction.

But the most profound threat is that we are approaching the era of human control of future bio forms. From genetic eugenics to animals with implanted human genes to the general ability to combine and modify species at the genetic level, we are learning that all life forms are assemblages of genetic parts that we might be able to mix and modify to suit our purposes.

This knowledge and the equipment needed to accomplish such changes will be ever more widely available and outside the control of authorities, analogous in some ways to the spread of information capabilities in the current information revolution. Indeed, biological commerce will globalize, enabling worldwide benefits, but also opportunities for bio hacking, bio crime, bio terrorism, and bio warfare.

These changes will come, are coming very quickly, within the two decades of this projection. Already many of the virulent bacteria and viruses have been genetically decoded and the first genetically designed antibiotic has been tested. These developments are driven by economics and by utility, not by any national plan or moral value. We are initiating a new biosphere in a very ad hoc fashion. Whereas nature has always tested its mutations against all other elements of the mutant's habitat, we have and will develop mutants that are optimized in one or a few dimensions, to produce more and better food or medicines or human compatible organs or function-specific organisms or even super humans, athletes and geniuses. These species will be out of equilibrium with their environments, thereby either dominant or fragile, but ecologically unstable.

The values, responsibilities, and mechanisms for guiding global bio activities are lagging far behind the scientific and technological advances. Even the immediacy of national defense has not motivated coherent action. The responsibility for bio agent detection and response within the U.S. lies with the Department of Health and Human Services, while the DOD, FBI, CIA, and FEMA all have overlapping responsibilities for anticipation and response to acts of bio crime, bio terrorism, and bio war. A more coordinated institutional arrangement is needed for prevention of and response to immediate, fast-acting threats. In the longer term, longer- and broader-view authorities might be commissioned for ongoing study and guidance of global biological R&D programs, policies, and regimes.

GEO SYSTEMS TECHNOLOGY

Geo systems, such as weather systems, ocean currents, crustal formations, and bodies of ice and water, are huge in scale and contain prodigious amounts of energy. For example, a large hurricane releases as much energy as a 1-megaton explosion roughly every 10 seconds (and the very largest, one megaton every second or so); a large earthquake releases the energy equivalent of 10 million megatons of explosive.

As models of and data on these systems improve, the ability to predict what will, or even might, happen will improve. Such knowledge could offer both a competitive and a self-defense advantage. The means may even emerge to modify, initiate, and redirect the energy contained in these systems by means of very high-gain trigger or boundary condition mechanisms. Myth has it that before these systems become mighty, the flutter of a butterfly's wing can set them in motion. Of course, it is also argued that many coherent mega-butterflies are needed, and that the chaotic nature of natural systems makes the consequences of a triggered natural event extremely unpredictable.

As society develops and becomes dependent on global intercouplings of products, infrastructure, information, and travel, natural events can cause significant disruptions of societies and economies (e.g., the drought, fires and economic collapse of economies in southeast Asia in the latter 1990s) that can have ripple or even tidal effects around the world. The potential to release huge natural energy and cause widespread disruption could be attractive to terrorists.

The atmosphere-ocean-land system is also the underpinning for the biosphere. Changes in the geophysical environment can determine the viability of living things and the local course of evolution. The ability to modify or corrupt these vast eco-systems or their local eddies could greatly impact our security.

The world is changing as a result of human actions: much of the world's land cover is changed, atmospheric composition is different and climatic change has begun, stratospheric ozone has been depleted, and more. We are not yet able to fully predict the consequences of these changes and are only starting to build the commitment to limit their influence. Over time, increasing information and insight will emerge. Having that information is likely to affect the balance of advantage among nations, and we need to be sure we are the well informed.

ACKNOWLEDGMENTS

This project was conceived and sponsored by Ronald Lehman, Director of the Center for Global Security Research, and led by Thomas Gilmartin, a senior fellow at the Center. The Workshop chairs were: William Schneider (Nuclear, Missiles, and Space Technology), Louis Marquet and Walter Sooy (Military Technology), Joseph Markowitz (Information Technology), Charles Cantor (Bio Technology), and Michael MacCracken (Geo Systems). For their service on the Senior Review Panel, we would like to thank George Shultz, Anthony Carrano, Arthur Cebrowski, Chris Chyba, Jeffrey Cooper, Jay Davis, James DeCorpo, Ted Gold, Donald Hicks, Robert Joseph, Steven Kornguth, James Lang, Michael May, John Nuckolls, James Richardson, Victor Reis, Paul Saffo, Wayne Shotts, Bruce Tarter, Richard Wagner, and Lowell Wood. We would like to express special thanks to Paul Saffo, Walt Sooy, Lowell Wood, Leonard Weiss, and Milt Finger for conceptual contributions. Thanks also for the participation of Eileen Vergino, Deputy Director of the Center, and for Karen Kimball's administration of and Tami Alberto's assistance with this project.

Fig. 1: Future Security in a Technology Rich World. The array of technologies that are increasingly available globally from a wide range of sources create wealth and give both offensive and defensive power to nations, sub-nationals, and even individuals.

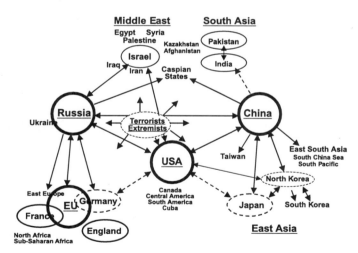

Fig. 2: Post-Cold-War Web of Global Nuclear Weapon Threat Relationships. The predominantly bi-polar nuclear weapon stand-off of the Cold War has given way to a set of multi-polar, interconnected relationships among nations and entities with dissimilar arsenals and technical capabilities, incompatible motivations, and political instabilities.

COMPACT NEUTRON SOURCES FOR PURE AND APPLIED SCIENCE

WILLIAM A. BARLETTA
E. O. Lawrence Berkeley National Laboratory, Berkeley CA 94720

The Accelerator and Fusion Research Division of the Lawrence Berkeley National Laboratory has been developing a new generation of neutron sources for university scale research laboratories, for clinical use and for portable field applications. While neutron tubes are not a substitute for fission reactors and accelerator driven spallation sources for the copious production of neutrons for the most demanding users, they can meet the operational needs (Table 1) of a wide range of researchers and clinicians especially in the University and small laboratory environment.

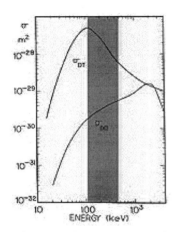

The new generation of compact neutron tubes being developed by the Berkeley Lab are based on D-D and D-T fusion reactions. The reaction cross sections shown in the figure suggests that the ideal neutron tube is a deuteron or triton accelerator operating at 100-200 keV.

The compact accelerator begins with a plasma source of deuterons (or mixed Deuterium-Tritium). The deuteron beam is extracted from the

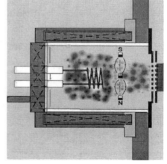

plasma volume through high-voltage electrodes to impact a deuterated or tritiated target plate. Neutrons are emitted from the target plate and pass through a moderator to produce a neutron beam of the desired energy. The heart of the tube is a sealed ion source. The source consists of a container lined with permanent magnets which contain a plasma that is generated by a radio frequency discharge from an rf-antenna. The figure shows the beam container, the antenna, the gas inlets and beam extraction electrodes.

An important part of the source is a magnetic beam filter which assures that only the desired species of ion (correct atomic or molecular state) is extracted. In our case we extract atomic hydrogen ions; the molecular ions which are extracted have only an energy per nucleon which results in a much lower fusion cross-section and therefore lower neutron yield per watt.

The measured yield for deutrons incident on a tritium- implanted, titaniam plated copper target is as shown. Ones sees that the optimum voltage is ~170 keV for an atomic tritium beam. (Sandia data). Further control of the beam species is also determined by the amount of rf-power fed into the discharge. Our measurements of this dependence suggest that a few kW of ~ 10 MHz rf-power yields nearly all monatomic hydrogen in the beam.

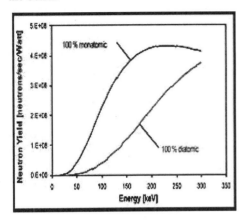 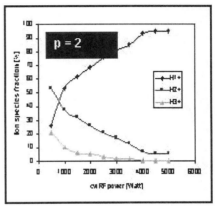

The first LBNL realization of the compact neutron generator is a sealed tube design that employs cross over beam optics to eliminate the erosion of the high voltage suppressor electrode. This design feature simultaneously improves the uniformity of the beam incident on the target.

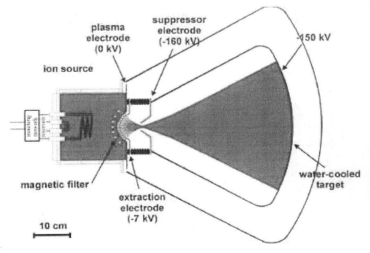

238

Unlike earlier versions of "neutron tubes" (such as the Livermore Rotating Target Neutron Source) which used a target made of titanium chemically bonded with tritium, the Berkeley tube uses a thin titanium layer bonded to a copper substrate with cooling channels. The tritium or deuterium is physically loaded onto the surface of the target by the beam. Thus the hydrogen cannot be depleted from the target as it is continually replenished by the beam. The target lifetime is limited by sputtering; the lifetime can be evaluated by using H ions so that no prompt radiation or activation is produced in the test components.

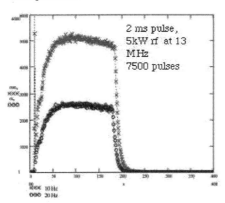

2 ms pulse,
5kW rf at 13 MHz
7500 pulses

Initial tests of neutron production have been conducted at low duty factor to minimize the shielding required at our experimental facility. The present shield is composed of 3 cm of borated polyethylene surrounded by 5 mm of lead sheet held in place between 1 cm plywood sheets. Tests of the tube operating at 150 keV demonstrate a z reproducibility especially in the sharp fall time of the 2 ms neutron pulse. The data taken by a Los Alamos group employ cadmium, shielded detectors to measure the non-thermal neutron flux.

The next series of neutron tubes employs a coaxial geometry to increase the neutron yield per unit volume by increasing the available target area. In a tube with dimensions - length = 26cm, diameter = 28cm and weight = 40lb - operating at 80 keV and 1 A peak current, we expect an output of ~ 1.2 x 10^{12} n/s for D-D neutrons and ~ 3.5 x 10^{14} n/s for D-T neutrons at a duty factor of 10%. The neutron yield per watt could be tripled by raising the tube

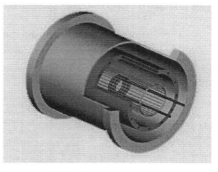

voltage to 170 keV at the cost of an increase in power supply cost per watt.

The limiting factor in tube performance is the power density on the target which we have set conservatively at ~650 W/cm^2. Tubes with as much as ten times higher average flux can be made by nesting the concentric targets and plasma regions as shown in the illustration. That

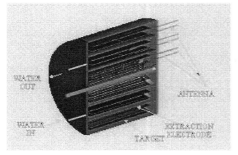

WATER OUT
ANTENNA
WATER IN
EXTRACTION ELECTRODE
TARGET

design would yield a neutron output of ~ 1.6 x 10^{13} n/s for a D-D tube and as much as ~ 4.5 x 10^{15} n/s for an 80 keV D-T neutron tube. Here again another factor of three can be obtained by raising the tube voltage to 170 keV.

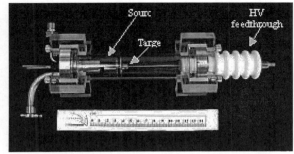

Tube geometry may be set by considerations other than maximum flux per unit volume. For example, for neutron radiography the geometry can be configured to yield a point source of neutrons. For applications such as down-hole well logging the tube must fit within a standard bore hole. For this purpose we have designed a 2" (o.d.) tube. This small tube is also ideal for moderator design studies and for pedagogic purposes in training nuclear engineering students about experimental neutronics on any of the many university campuses without a research reactor.

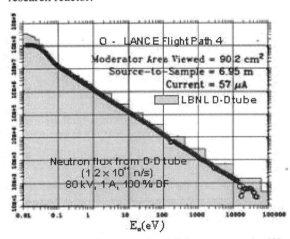

The utility of neutron tubes for research comes from the fact that the initial neutron energy is low unlike the case of high energy spallation sources using GeV protons on a high Z target. Importantly the neutron production is not accompanied by a powerful burst of gamma rays. This fact means that both the moderator and the shielding can be much thinner than those surrounding a spallation target which produces copious neutrons with energies well in excess of 100 MeV (actually up to the full beam energy). We can compare the output of a small (10 cm long, 5 cm diameter), 80 kW D-D tube surrounded by 4 cm of liquid methane moderator and 10 cm of concrete shielding with measurements. The spectrum of the neutron flux in neutron/(cm^2-s-eV) is shown as the shaded histogram in the figure. For comparison the circles show the measured neutron flux v. neutron energy (Ref: Paul E. Koehler, Nuclear Instruments & Methods A292, 541 (1990)) at LANSCE operating at a beam power of ~ 50 kW. One sees that due to the considerable distance (7 m) of the instrument from the neutron source, the flux of the high energy spallation source is nearly identical to the performance of the compact neutron tube. The overall electrical efficiency of a high power ~1 GeV proton accelerator is in range of 1 -10%, while the electrical efficiency of the neutron tube is ~90%. The consequence is that the fundamental energetic advantage of the spallation source (30 MeV per neutron produced) over the D-D or D-T generator (2000 MeV per neutron for D-T) is lost.

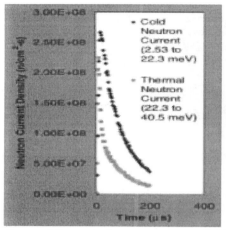

The calculations of the neutron spectrum also allow one to determine the time structure of the neutron pulse from a neutron tube with a thin moderator. In that case we have considered an initial pulse from the tube of duration 1 μs. As the figure indicates, the moderated pulse still has a very sharp time structure (17 μs FWHM) in the case of thermal neutrons.

Another comparison of the utility of a 1 A neutron tube can be seen in the application of providing source of neutrons suitable for the hospital environment for Boron Neutron Capture Therapy of glioblastoma–a cancer of the connective tissue of the central nervous system. In BNCT a tumor-seeking drug containing a large mass fraction of boron (a "neutron sponge") is administered to the patient. After a short waiting period the drug fixes in the tumor and clears from blood. At this point, the patient receives the radiation dose of roughly 30 Gray of epithermal neutrons. With the correct neutron spectrum the radiation effects will be concentrated in the cancer cells not in the

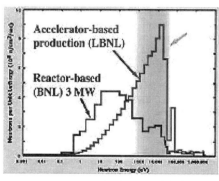

healthy tissue. For a 1 amp, 80 keV tube the spectrum can be moderated to be far more favorable to therapeutic requirements than a reactor with the same ~45 minute exposure time. The neutrons outside the shaded optimum region (i. e., less than 1 keV) will destroy healthy tissue rather than penetrate to the tumor and deliver a sufficient dose to kill the cancer cells. Since the actual dose that the patient receives must be set so as to spare the healthy tissue in front of the tumor, reactors have led to doses insufficient to control the tumor. In contrast a neutron tube can produce a nearly ideal spectrum leading to high tumor control probabilities even with available drugs such as borated phenoanylene (BPA).

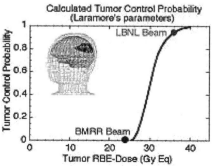

Calculated Tumor Control Probability
(Laramore's parameters)

In summary, the combination of efficient dc-power supplies and a new generation of plasma-based D-D and D-T neutron generators can lead to an expansion of university-based neutron science at a level which is comparable with all but the highest flux applications of reactors and spallation sources. These tubes would allow a renewal of nuclear engineering programs on university campuses and a spreading of neutron science and technology without the complications generated by nuclear proliferation concerns.

I would like to acknowledge the research efforts of Prof. Ka-ngo Leung and his group in my division at LBNL for their remarkable and rapid developments in plasma-based neutron generators. I am also grateful to Prof. Antonino Zichichi and the staff of the Ettore Majorana Center for Scientific Culture for their kind hospitality during this International Seminar. This work was supported by the Office of Science of the U.S. Department of Energy under Contract No. DEAC03-76SF00098.

Table 1. Applications of Compact Neutron Sources.

• Condensed Matter Physics (10^7 n/cm^2/s) Scattering of slow neutrons in condensed matter (solids or liquids) can determine structure on the atomic or molecular level. Neutrons penetrate deeply into matter enabling study of new materials in real T, P, and other ambient conditions
• Material Science (10^7 n/cm^2/s) Study point defects, dislocations, inter-phase boundaries, intrinsic junctions with micro-cracks, pores, etc.
• Molecular Compounds (10^7 n/cm^2/s) Small-angle neutron scattering (SANS) is a powerful method to investigate polymer systems and surface-active substances (SAS). Specular reflection provides information about the structure along the surface of the material.
• Biology (10^7 n/cm^2/s) Neutrons can "see" hydrogen better than photons, which enables determining details of the structure and function of biological systems
• Engineering Analysis (10^6 - 10^7 n/cm^2/s) Neutron diffraction probes internal stresses deep in multiphase materials
• Earth Sciences (10^6 - 10^7 n/cm^2/s) Neutrons probe study the texture of rock materials and minerals and effects of the external pressure on the structure of samples
• Medicine (10^9 n/cm^2/s) Neutron capture therapy
• "Forensics" (10^6 - 10^7 n/cm^2/s) Identify concealed materials

BUILDING INTEGRATED DECISION TOOLS FOR PLANETARY EMERGENCIES

DR. DAVID W. MULENEX

U.S. Embassy Counselor for Environment, Science and Technology, Rome, Italy

INTRODUCTION

Our goal today is to understand a little better the national and international interactions among the information systems which underpin decision making about planetary emergencies. For reasons that I hope will become clear, I include within planetary emergencies emergencies both immediate, which require a coordinated international response due to their severity, and potential emergencies which, because of the likelihood that they will affect large geographic areas and large numbers of people over long periods of time, could threaten significantly the global environment and humankind.

My first job will be to outline the salient characteristics of the three information systems in widest use by decision makers today: political, economic, and scientific information systems. I would not argue with anyone who wished to add other information systems to this list, for example religious or social/cultural systems, but our time is too limited for a comprehensive discussion. We will be especially interested in the spatial and temporal characteristics of information systems. Next, we will explore by examples the ways in which decisions are typically made about the information these systems produce, and the consequences of using information derived from one system in the decision making typical of another system.

My second excursion will be to present examples of hypothetical or ideal decision making structures for planetary emergencies: large scale natural and manmade disasters, focusing on the Global Disaster Information Network; climate change; and biosafety. Our job together is to deconstruct these ideal systems to better understand why it is so difficult to achieve the integrated planning and decision making necessary to effectively avoid and mitigate potential planetary harm.

Last, drawing on the previous two efforts, we will suggest some criteria for new tools we will need to improve the likelihood of managing planetary emergencies, and evaluate the likelihood of deploying these tools in our information and decision making systems. I'll suggest that our information and decision making processes are badly outdated, and need revamping if they are to cope with globalisation.

THREE INFORMATION AND DECISIONMAKING SYSTEMS

The August 4 issue of *The Economist* included an article by a Danish statistician, in which the author, Bjorn Lomborg, attacks what he claims as four pillars of environmentalism: (1) exhaustion of natural resources; (2) global population growth; (3) the threat of biodiversity loss; and (4) growing pollution of air, sea, soil, and water. Mr. Lomborg argues that environmentalists look on the dark side of things, and have not given due credit to human ingenuity in finding new resources and increasing productivity, or tallied up accurately the reduced pollution in highly developed countries, or looked at the apparently gentler rate of population growth. Some of you doubtless read Lutz, Sanderson, and Scherbov's letter in the August 2 issue of *Nature*, in which the authors argue that about mid-century the probability of a decline in global population is at least 50-50. Peak population, before fertility decreases and a rising percentage of over 60 individuals sets in, would reach about 9 billion around 2075. To their credit, Lutz *et.al.* describe their model in some detail, and discuss the problem of variance, and produce data for different subregions. It is beyond the scope of their paper to analyze the social and economic effects which will be causes by the dramatic shift of the world's population to include almost one in three over the age of sixty.

By contrast, it's hard to get a handle on Lomborg, because he is delving into various information systems without enough discrimination. He produces risk analysis statistics linking costs of environmental control measures to an additional year of human life, without sharing with us the basis of the data—he provides a citation, but doesn't tell us anything about the population included in the sample, the countries involved, or how the calculations were made. The effectiveness of his challenge to the greens is diminished by his lack of rigor. Most damaging of all is his tendency to generalize without an accurate recognition of the variance across countries and regions of the planet, a pretty clear orientation to the societies of wealthy, culturally coherent countries like Denmark, where hard choices between investments in economies and investments in pollution reduction are still a little easier than in most parts of the developing world. If all countries were Denmark, Mr. Lomborg might be more right than wrong.

Lomborg's virtue is his attempt to bring a lot of data from experts into the more accessible format of a broadly read magazine. Consider by contrast this statement from the report of the National Academy of Science to the White House, released at the time of President Bush's announcement that the U.S. would not pursue further the Kyoto Protocol:

Climate change simulations for the period of 1990 to 2100 based on the IPCC emissions scenarios yield a globally-averaged surface temperature increase by the end of the century of 1.4 to 5.8 degrees centigrade, relative to 1990. (p.4)

This is pretty dense stuff for a non-scientist: why 1990 as a starting point? If the global average predictions show such a wide range of results, what's the range of regional predictions? The choice of surface temperature for the data might not be enough; we might ask about ocean temperatures, or changes at different altitudes, say 2,000 to

5,000 meters, where most of our mountain ranges occur. Nonscientists will not possess the qualified perspective necessary to address these questions.

Let's try a different approach: which questions did the White House ask the Academy to address? Here's the letter (exhibit 1) sent to Bruce Alberts by Gary Edson, the President's advisor on climate change. The White House asked the academy to (1) identify areas "in the science of climate change where there are the greatest certainties and uncertainties", and (2) express "whether there are any substantive differences between the IPCC Reports and the IPCC summaries." The White House did not ask whether our current knowledge provided an adequate basis for national or international action to slow climate change or implement mitigation strategies, and the Academy studiously avoided entering into such discussions in its report.

Now let's look in detail at the Academy's reply to possible differences between the IPCC summary and the IPCC reports.

The committee finds that the full IPCC Working Group I report is an admirable summary of research activities in climate science, and the full report is adequately summarized in the Technical Summary. The full WG I report and its Technical Summary are not specifically directed at policy. The Summary for Policymakers reflects less emphasis on communicating the basis for uncertainty, and a stronger emphasis on areas of major concern associated with human-induced climate change. *This change in emphasis appears to be the result of a summary process in which scientists work with policymakers on the document.* (page 5, italics mine)

> After analysis, the committee finds that the conclusions presented in the Summary for Policy Makers (SPM) and the Technical Summary (TS) are consistent with the main body of the report. There are, however differences. The primary differences reflect the way in which uncertainties are communicated in the SPM. The SPM frequently uses terms (e.g., likely, very likely, unlikely) that convey levels of uncertainty; however, the text less frequently includes either their basis or caveats. (p. 22.)

The Academy Report suggests that in the process of making scientific information useful to policymakers, the effort at simplification—which takes the characteristic form of *translating quantitative data to qualitative statements* may have inadvertently made the scientific picture clearer than warranted by current knowledge. I'd suggest that this act of translating scientific information into politically and economically relevant information creates an additional problem, that the relevant decisionmaking frameworks cannot accommodate easily exogenous, in this case scientific, information.

Exhibit II represents an attempt to summarize some salient characteristics of economic, political, and scientific information systems and the kinds of interests and criteria included in decisions made on the basis of the information systems. If we ask about the source of basic data for the three information systems, we recognize that economics is based on ownership, and the ability to transfer ownership, whether what is owned is physical property like an automobile, or intangible property like knowledge and technology. Our political system, and I am consciously describing representative models to the exclusion of dictatorships, and religious autocracies, takes as its most salient data voters, constituents and special interests, parties and elites, and large intragovernmental interests such as defense and social services, whose size and impacts on large numbers of voters provides a kind of voter feedback on changes, and thereby an additional source of information. Our scientific information system comes from experimental observation and analysis of the physical and biological world and the repeated testing and verification of causal relationships using quantitative means.

A quick look at prevailing quantitative units of measurement is instructive. While economic information always comes down to money as the medium for conveying ownership, political systems look internally to voter support, to party loyalty, and to the quantitatively measured support for policy decisions in opinion polls, part congresses and key parliamentary votes. The scientific information uses quantitative measures derived from the scale of the phenomena under study. Because quantitative data is essential for verification of results, scientific information systems employ a large range of measuring tools capable of greater precision, and hence better able to characterize uncertainty and variance, than either economic and political information systems. Last, if we look at the orders of magnitude typically included within the three systems, the political and economic systems operate within 12-15 orders (up to 1,000 trillion, but usually much less, in the ranges of 9-10 orders of magnitude, say 10 billion, the projected population mid century), while scientific observations span at least thirty orders of magnitude.

What time scales are typically included in our three systems? The economic system, especially enabled by electronic global markets and exchanges, operates on a scale from minutes to a usual outward boundary of thirty years or so, the period of time for a large plant investment to be amortized by its productive capacity. As all of us who have tried our hands at environmental economics know, the intergenerational problem of allocating benefits to the future while paying costs in the present is a formidable obstacle. And, of course, our problem is compounded by the time scales included within typical representative governments, which may range on the short side to two years, the tenure of a first term U.S. Congressional representative to the tenure of a "founding father" politician whose influence might extend over two generations or more. More typically, the political time frame is bounded by term limits, whether parliamentary or presidential, and rarely extend beyond eight years or so. In the case of scientific information, time scales for critical findings can range from nanoseconds to eons, and in the case of those problems of greatest topical interest, like climate change, biodiversity, or severe natural catastrophes, include several decades at a minimum.

This is as good a place as any to introduce another key, if obvious idea: the phenomena under which all three information systems operate are dynamic, but this dynamism is usually underrated by an actor in one system who looks in non the others and regularly perceives his own system to be more dynamic, and more affected by exogenous factors. A political decision to expropriate foreign-owned assets disrupts the fundamental principle of secure ownership on which economically rational decisions are made. Less dramatically, new environmental laws and regulations based on scientific

246

findings may disrupt the valuation of economic assets. A political decision to relax targets on atmospheric emissions may change the baseline used by scientists to project longer term emissions, produce new results and raise questions about scientific reliability and uncertainty.

Another important difference among our systems is the yardstick used to measure success. The economic system is basically black/red: increased net wealth, for an individual, a company, or a nation, is success. Our representative political system measures success by staying in control of the government, and getting reelected or reappointed. The scientific system measures success in terms of leadership and contribution to specific fields of research. The operative principles of the three systems are competition for economics; influence for the political systems, and peer review and recognition for science. And no, I am not unaware that competition and influence affect the rise to leadership in every field.

Having outlined, if very broadly and incompletely, some of the salient characteristics of the three information systems, I want to look briefly at the formation of decisions using the three information systems, and then look at biosafety as an important example for understanding what happens when political, economic, and scientific information mix without clear rules and markers. Let's consider the internal processes which shape political decisions first. Since the measure for success is staying in government, actions taken which produce short-term benefits (which may be political or economic, but are rarely scientific) are preferable to short-term costs, and short-term benefits which incur costs only beyond a politically relevant time frame are generally acceptable. Second, decisions which generate a political benefit for the decision maker but may generate a cost for a geographically distant and politically distinct neighbor are acceptable if the costs are sufficiently diffused and spread over a long period of time, i.e. unlikely to produce retaliation, a particularly negative kind of feedback. Last, factual understanding may be an inadequate basis for action if the facts are not salient or accepted by political constituencies, either because of uncertainty or because costs will be paid only in the future. In short, when costs can be displaced in time or space, or made doubtful by uncertainty of facts and opinions, and political benefits accrued, this will usually happen.

Spatial and temporal displacements also affect economic decision making, and a global economy increasingly based on knowledge may exacerbate these displacements. For example, when a new technology or new product is developed in one country with potential application to many, there will be a natural tendency recognized in international practice to protect the invention through patents or IPR agreements in order to charge the highest rent for the longest period of time in order to maximize the return on investment. Some object, as did litigants in South Africa over HIV/AIDS medications, that access to life saving drugs should better calculate ability to pay. Others argue that IPR protections must be limited when they tend to produce monopolies which can determine supply and prices. Two more ordinary examples: first, the debate on trade and environment, and claims that the transfer of production from highly regulated countries to countries with lower environmental standards is unfair to competition in the more developed countries which must meet stiffer standards. The second, a subset of the first, involves the exploitation of natural resources from metals to timber without paying the full environmental costs or following sustainability guidelines—which are simply political rules to reshape economic behavior—n order maximum exports and their profitability.

Scientific decision making is sometimes almost schizophrenic in its simultaneous wish to be ideally independent and objective, and to be employed usefully and in the service of society. The responsiveness of the National Academy of Science report to the White House on climate change science is a good example of the difficulty in striking a balance. The Academy has a long and distinguished history of preparing expert reports at the request of the U.S. government, and doubtless wanted to start off its relations with a new administration in the most positive way. At the same time, President Bush's predilections about the Kyoto Protocol had been known since February 2001, when he withdrew carbon dioxide from the Multi-pollutant Strategy, to the embarrassment of his new EPA administrator, Governor Christine Whitman, who had a few days earlier at a meeting of the G-8 environment ministers used the inclusion of carbon dioxide in the strategy as evidence of the seriousness with which the President viewed climate change. The Academy could not have avoided seeing in the White House request a hope for scientific support for his rejection of the Kyoto Protocol as "fatally flawed." The Academy, many of whose members were represented in IPCC working groups, had to weigh its own credibility. By choosing politically to follow the narrowest of mandates, the Academy was able to produce a politic but credible report, and to avoid charges of politicism.

The Academy report, by pointing to IPCC problems with translating uncertainty from qualified quantitative ranges to strictly qualitative ranges, probably gave support to economic and political interests who argued that any uncertainty was too much where their interests were concerned, and where, in any case, the entire scientific community could not agree on the best course forward on the scale between mitigation and no action (or full adaptive measures, if you prefer.) The President's subsequent call for more research in climate change science and in new technologies to create more options to respond to climate change was a positive response to the Academy's report, but was as narrow as the mandate originally given to the Academy. Both the White House and the Academy effectively disengaged themselves from the most relevant question, i.e., whether we know enough to take precautionary measures to protect ourselves from the most potentially damaging effects of climate change at the top end of the projected temperature range.

DECONSTRUCTING EXEMPLARY SYSTEMS INTO REALITY

Climate Change has not given us much to work with in terms of tools for planetary emergencies. The United States has questioned whether an emergency is really on its way, and expressed a confidence that it can be avoided or managed, while the advocates of Kyoto as the only available tool, which must be adopted to avoid an emergency, have dropped the emissions targets to 1.8 percent below 1990 levels from a previously concluded 5.6 percent drop—some 75 percent—and have offered special provisions on sinks to Canada, Japan, and Russia which reduce their obligations further. While the European Commission is expected to announce a package at COP-7 at Marrakesh which will encourage member state ratification of the Kyoto Protocol by the time of the Johannesburg summit in 2002, this is a clearly political move, and some member states are likely to balk because no serious work has been down on coordinated policies and measures within the EU that would deal with very different costs of Kyoto compliance across member states. The commitments of rapidly growing economies—China, India, Brazil, Indonesia—have not been resolved. Nor have monitoring and compliance regimes.

Let's look now at a new subject, but which will soon yield to some of the perspectives we have been developing. The subject is Biosafety, or as it appears in some recent FAO documents, biosecurity. Consider these three statements:

An effective global biosecurity system must have consistent and measurable international standards across the entire food chain, including laboratory and field testing for new products before commercial use.

An effective global biosecurity system must include the impact of new technologies and organisms which may affect biodiversity, the environment, traditional agricultural practices, and international trade in agricultural products.

An effective global biosecurity system must not only provide perfectly safe and nutritional food all the time, but must also meet the demands of consumers for clear information about the origin of foods and their processing, and make the most advanced agricultural biotechnologies available affordable to the developing world.

These three statements represent a progressive order of difficulty, but also represent an exemplary system that would satisfy most international health authorities- —those responsible for phyto-sanitary agreements and the *Codex Alimentarius* for example—, most national food safety authorities, most environmental and consumer organizations, most farmers and traders, most scientists, and most companies which have developed biotechnology food products.

The reality is very different, as we know. International standards are in place for microbiological and other contaminants, detection of parasites and vermin, post-harvest treatments, storage and shipping, and for ongoing inspections and laboratory testing. There are in fact more foods in international trade than ever before, and that food is overwhelmingly safe. And despite genuinely desperate cases of famine in areas under extended drought, floods, or political unrest, the recent FMD epidemic in the U.K., and the emergence of BSE and a new variant of CJD, biosecurity does not seem to meet the criteria for a planetary emergency.

Nonetheless, food safety is one of the hottest issues under discussion around the planet. Cause number one is the dramatic increase in agricultural trade, and the increased demands placed on national food safety authorities. The food chain is longer and more diverse, and represents many more unknowns to national authorities regarding production sources and methods, and the mixing of commodities from different producers in bulk shipments. In the European Union, the lack of a EU wide food safety authority has eroded consumer confidence in food safety. Beyond the major crises created by BSE and FMD, there have been numerous smaller crises ranging from dioxin in feed to elevated dioxin levels in fish from terrestrial sources. These developments have been accompanied by an almost Garden of Eden approach by some environmental and green groups; in Germany, for example the Minister of Agriculture announced upon her selection that her first constituency would be consumers, not farmers. Despite ample evidence that organic farming in Europe is not as chemical free as many Europeans believe, it continues to expand and organic products, many now being produced under tighter rules, are gaining more markets because the products satisfy the cultural and ideological demands of consumers who are putting beliefs, not food safety facts, into practice.

In the United States, the food safety system regulated by USDA, U.S. EPA, FDA has largely retained the confidence of American consumers. Products controversial in Europe, like U.S. hormone treated beef, remain unlabelled in the United States. Genetically modified corn and soy varieties dominate U.S. farm production, and products made from GMO grains remain on U.S. grocery shelves undifferentiated from non GMO products. In the last year, as a reaction to serious allergic reactions attributed to the unauthorized and unintended introduction of Starlink corn into products intended for human consumption, U.S. agencies have begun to reexamine labeling for GMO products.

Insecurity about the safety of foods in the global marketplace is being addressed by a variety of measures. The creation of a European Food Safety Agency, which may employ initially about one thousand scientists and technicians is one response. From our perspective, however, the European Agency has a potential problem. The decision taken in the design of the Agency gave it no authority to approve new food products, or to withdraw others; such decisions are reserved to European political decision makers. While the EU based its approach on a laudable desire to protect scientific research, analysis, and recommendations from political influence, the net effect may be in economically or politically important cases to decouple scientific information from other forms of information in decision making.

Agricultural policy is one area in which science has often been overridden as the basis for decisions. For example, the United States has tried for years to bring the widest variety of agricultural exports to European markets, and has placed its research and product safety testing at the disposal of European authorities. When almost twenty years of testing data on hormones used in beef production were repeatedly rejected without clear grounds by the European Commission, the United States took the matter to the Dispute Resolution Body of the World Trade Organization. The United States won its case on appeal, but the European Union chose to pay compensation rather than open its markets to American beef. In its written arguments to the DSB, the European Union argued that although all available studies of the effects of hormones on beef had shown no measurable adverse effect on human health, further studies needed to be done, and that it was the right of the European Union to adopt standards higher than those prescribed in the *Codex Alimentarius* under its science-based precautionary principle until any remaining uncertainties were resolved. In the same argument, the EU introduced the first time its view, which has been refined and restated over the last four years, i.e. that the precautionary principle could be used to block imports based on social values and norms, including consumer demands, in short arguing *that science should not be the sole basis on which decisions about biosecurity should be made.*

It was unsurprising therefore, that when U.S. products containing genetically modified organisms encountered resistance from European consumers, farmers, and governments, that the precautionary principle would be invoked in several forms. First, there were questions about product identity and safety. Second, there were questions about the possible impacts on the environment and biodiversity of planting GMO varieties in Europe, in particular potential gene flows from GMO plants that could create unintended and potentially harmful organisms such as superweeds. Third, there were questions about an economic American invasion of European agriculture, since the Round-Up Ready varieties developed by Monsanto required the use of Monsanto's patented herbicide. Some objected to the "terminator gene" found in certain varieties as a market ploy. Others just objected in principle to varieties they concluded had been unnaturally produced and inadequately tested. Many more believed that GMOs products were just the first wave of applications along a scientific and technological continuum which would lead inevitably to manipulation of the human genome, to which they were ethically or religiously opposed.

We could treat this discussion as a chapter of U.S.-EU trade relations of little planetary interest but for three factors. First, a good deal of scientific work was directed toward providing answers to questions such as geneflows from transgenic plants, for which we have no field evidence and a laboratory calculated probability of less than 10 minus 12, with a very low probability of survival rate even in soil bacteria capable of rapid mutation. The studies have pointed out in detail the importance of homology between the transgenic DNA and the DNA of the potential incorporating organism. This work is relevant to every country considering the use of GMO varieties. Second, the differing American and European points of view were reflected in trade disputes which affected agricultural production in other parts of the planet. Last, the EU and he United States took their trade views and the science "supporting" their respective views into the broader international discussion of the Biosafety Protocol to the Convention on Biological Diversity (CBD). The choice of forum gave the EU a distinct political advantage, since the U.S. has not ratified the CBD and consequently could not participate directly in the Protocol negotiations.

By anyone's tally more than eighty percent of the European Union budget is spent on agricultural price supports. As a result, a high volume of exports to dispose of overproduction, often on subsidized terms, and higher food prices within the EU, have characterized European policy. At the same time, access to European Union markets has been an important trade policy tool of the EU in its dealings with non- EU producers, particularly in Africa. The EU moratorium on new approvals for GMO varieties under the old 90/220 now replaced, on environmental safety grounds and consumer demands for traceability and labeling, led the EU, for consistency's sake, to enforce those same rules not only for the American imports the rules were designed to control, but also for all agricultural products destined for the European Union. By this time, some successful trials with GMO varieties had taken place in Africa and other non-EU countries, and substantial quantities of GMO seeds were being used in Canada, in Argentina (for soybeans), for corn, and for cotton. Other developing nations, such as the Philippines, were engaged in laboratory work and confined field trials. In a particularly striking case of EU action, the EU told an African beef producer with access to the EU market, that the adoption of GMO cotton varieties, with the potential to use seedcake as feed, would result in loss of the EU as a beef export market. The country in question blocked the use of the GMO cotton, with dire consequences for parts of its domestic economy.

While trade actions were taking place, the negotiations on the Biosafety Protocol to the Convention on Biological Diversity were proceeding. Some of you will know that the United States decided not to join the CBD in part over unresolved discussions of so-called "breeder's rights" and the apportionment of rights of countries whose economically valuable plants had been used to create new products. The Biosafety Protocol had a different but related purpose; to help countries maintain and protect biodiversity by devising an advance consent system for all living modified organisms crossing national boundaries. This would include GMO seeds, and as originally written, could have included bulk unprocessed commodity shipments of soybeans and corns.

Under EU leadership, discussions came to center on that broad application of the precautionary principle detailed legally in the Commission's reply to the dispute over U.S. beef treated with hormones response Application of a "zero-uncertainty" precautionary principle would have applied to GMOs an unprecedented series of costly risk assessments and would have imposed an equally costly trace and label system on exporters and importers. Many developing countries, including some with advanced scientific capabilities like South Africa, admitted they lacked the technical and administrative capacity to evaluate all the varieties which would be included under the Protocol. One positive result is the Biosafety Clearing House now being created under UNEP, a data bank intended to help all countries assess information about varieties intended for import by providing available lab and field test results.

The debate over acceptance of GMOs and GMO products has shifted, in the absence of any solid research demonstrating gene flow from GMOs, or harmful effects on human health—even the celebrated Starlink corn case ultimately traced the allergic reactions of affected individuals to other characteristics of the corn other than the Cry 9A gene—to consumer wishes and cultural preservation in Europe, and in the developing world to the potential contributions of GMOs to boost nutritional values- the case of golden rice-, productivity, to reduce the need for herbicides and pesticides and human exposure to these often toxic chemicals, and the long term possibility of reducing that two-thirds of the world's fresh water budget now spent on crop irrigation. It appears that the failure of first generation GMOs, mostly targeted at farmers and offering little direct benefit for consumers, can be overcome by offering such benefits in the form of taste, cost, appearance, and shelf life, and by offering consumers, at least those in Europe, a clearly labeled choice. In the developing world, good science and sound administrative capacity will be needed to evaluate the usefulness of GMOs for particular countries, and the environmental impacts, positive and negative. For example, a decision to change rice cultivars from wet to dry varieties may save water, but may also affect the extent, biodiversity and productivity of former wetlands. The introduction of GMO varieties will require changes in traditional practice, and require farmer education, costs to be included in the overall benefits analysis.

The biosafety example suggests some ways in which political and economic systems response to rapid changes in science and to expectations about science. At the outset, existing EU economic arrangements and interests protected by a complex political system were exposed to new challenges by the expansion of the economic system under free trade rules of the World Trade Organization. New products from the U.S., mostly generated through industrial rather than public research, sought market entry under the new rules. Politicians used scientific evaluation by expert bodies as a legitimate tactic initially, but later as a stalling tactic to protect markets from erosion. They also used the uncertainties included in scientific risk assessments to build public support against new products, and in some cases, as in Italy, bought into a green agenda that linked opposition to new products on grounds of health and environmental uncertainties to a perceived positive agenda of preserving cultural identity through traditional food varieties, even if uneconomical to produce. The net affect was to use scientific information decoupled from scientific reasoning and decision making in political decision making alongside information "naturally occurring" in that system. This sort of "transgenic information flow" has built something quite new and unforeseen by the scientific studies. i.e., a value-dominated process governed by democratic equality (all opinions have equal weight at the ballot box) which has made the body of scientific evidence confirming the safety of GMO products largely irrelevant. Although the survey numbers have fallen since 1999, about a third of Europeans still reject GMO foods even if conclusively demonstrated by experts to be safe and approved by governments. The consequences for research turn out not to be negligible. In the case of Italy, the EU ranked Italy at the top of biotechnology research at the beginning of the 1990s, but it had fallen to near the bottom by the end of the decade. Only an outright ban on agricultural biotechnology research by the Agriculture Minister in early 2001 finally provoked the Italian research community, led by a number of Nobel laureates, into public declarations and meetings to assert the need for continued research.

Fortunately, the stubbornness of scientific evidence, its repeatability and its ability to progressively confirm earlier results in likely to enable more rational decisions to be made about GMOs in the coming years. The Biosafety Protocol is an important step forward, because it made more international, and fairly so, what had been previously a market based dispute between the United States and the European Union.

LINKING INTERNATIONAL INFRASTRUCTURES: THE GLOBAL DISASTER
INFORMATION NETWORK (GDIN)

"The right information, in the right format, to the right person, in time to make the right
decision." This is the goals statement of the Global Disaster Information Network as
announced at the GDIN annual conference held in Ankara in April of 2000.

GDIN originated from a meeting held in Washington in July of 1998 in which two
facts became apparent. The range, quality, timeliness, and number of sources of remote
sensing data relevant to disaster mitigation and response was growing rapidly, while the
ability of potential users in to access and pay for maps and data which could save lives
and money was not. GDIN was founded as an international organization of volunteers
committed to creating the links among existing infrastructures to reduce the cost and
improve access to key data. From the beginning, GDIN decided it would not produce its
own data or enter into direct disaster mitigation and relief activities.

GDIN is a small scale enterprise compared with climate change or biosecurity. Its
annual budget is in the single millions of dollars, mostly contributions in kind. By
comparison, damage estimates from natural disasters in Latin America in the last decade
alone reached twenty billion dollars. Argentina, Belize, Brazil, Colombia and the
Dominican Republic will spend over 600 million dollars in 2001 on mitigation.

To be successful, GDIN has to build its niche among existing organizations
concerned with disaster relief and mitigation. With a natural disaster of international
significance occurring about every three days, there appears to be plenty of opportunity.
GDIN decided on an open membership structure for its working committees, and offered
membership on its funding committee to any organization or group of organizations
pooling resources that contributed five percent or more of GDIN's core funding.

There are some GDIN principles which will be difficult to live up to. The Ankara
Declaration commits GDIN to the respect of copyright, intellectual property, proprietary
information, and national security concerns, and at the same time providing tailored and
timely information with "no or minimal costs and restrictions." The GDIN approach is
to negotiate a set of MOUs with the owners of relevant data, often the agencies of
national governments but in some cases private sources, to provide data to GDIN, rather
than end users. Some agreements are in fact already in place.

Operationally, GDIN relies on a network of more than 1,000 volunteers organized
in seventeen working groups held together by an executive director and three person
staff. Information collection and sharing and analyses to determine best practices are
accordingly first priorities.

GDIN's five year plan, starting from 2001, focuses on outreach to new partners
and donors, and on regional networks in the Pacific, the Americas, and the
Mediterranean. The last two years of the plan include the development of real disaster
products, i.e., maps and reports useful in terms of resolution and parameters, and
supported by real time communication with experts associated with the GDIN network.
However, the placement of product development after organizational infrastructure and
internal funding requirements probably will push product development further out.

Still, GDIN has some early successes. After the Turkish earthquakes in 1999, GDIN delivered daily maps showing street-by-street damage. In 2000, GDIN provided maps to assist with flood relief in Vietnam and Mozambique. This year, GDIN linked data from ESA, Space imagery, and SPOT to assist El Salvador with earthquake relief.

If you want to know more about GDIN, look at their website: it's www.gdin.org. You'll also see a draft agenda for the 2002 GDIN annual conference which will be held in Rome next year, and will feature a disaster simulation. I'll conclude my comments on GDIN with two observations. First, GDIN was founded to address a demonstrated gap in disaster information services, and from the start looked for a fit with existing international infrastructures, many of which are well developed. One of its early and critical tasks was to undertake a survey of existing infrastructures, and determine their strengths and remaining information needs. This is a good first principle for building tools for understanding and managing planetary emergencies. Duplication is avoided, mission clarified, and collaborators identified. However, to fulfill its financial plan, GDIN will have to demonstrate performance to governments and the private sector, to whom it plans to offer services and products on a for fee basis. To do this, GDIN will have to pass beyond the expert body/volunteer phase and create a salaried organization that will be judged on a commercial basis. It is too early to predict whether GDIN can achieve such a result.

TOWARD INTEGRATED DECISIONMAKING TOOLS

Our three examples describe more difficulties than solutions. But we can begin our summary with the following list of desirable criteria for integrated decision making:

- **Inclusiveness** of actors and perspectives
- Transparent and specific **qualifying methods**; not all actors and perspectives are equal, but the rules must be clear
- Recognition of **spatial** and **temporal** displacements created by the nature of information **or** decision making
- Recognition of **embedded values,** which may lead to misinterpretation of information
- **Robustness** of information as measured by its ability to resist deformation and loss of integrity as it passes from one information system to another; robust information will resist this reshaping more than less robust information
- **Centrality**, measurable to some degree by its fit with other variables, its consistency, and its ability to predict likely scenarios along a critical path. If abstracted, prediction is not possible.
- Ability to **deconflict** contradictory information and decisions from competing systems, and to identify gaps
- **Operability** through existing actors and organizations

If we ask ourselves what kind of **quantitative** tools might help us achieve better **qualitative** decision making along the lines indicated above, a different list of considerations emerges:

- we need a calibration tool or **protocol** to mediate the different time scales characteristic of political, economic, and scientific information systems; this may take the form of an accounting tool that includes financial/political responsibility for unevenly distributed large scale effects of decisions
- we need a "marker" that helps us track and **measure the integrity** of information as it passes from one system to another;
- we need to make **rigorous distinctions** between critical quantifiable and nonquantifiable factors used in decision making. The best example is the calculate of current and future equity in sustainable resource allocations. At this juncture, no agreement exists on how to calculate equity, and attempts to incorporate equity in quantitative models will reduce analyses of the model to a discussion of the assumptions about equity;
- we need a continuous feedback mechanism for global negotiations that helps us track small deviations in approach and quantify the impact on the likely end point, and to calculate the impact on the environmental conditions the agreement is intended to affect. Such a tool will help us avoid the deviation from original purpose that characterized the Kyoto negotiating process from 19998 through 2000.
- we need a tool that can more effectively valorize costs and benefits over longer periods of time that is acceptable to economists, and more dynamic and able to distribute costs and benefits to different geographic and political units of the planets and to different groups of people across time.
- we need to measure the density, linkages, and overlaps of our existing international intellectual infrastructure to identify gaps, conflicts, and redundancies . **Information network design** practice may help.
- we need to introduce **economic altruism** in a mathematical way to the handling of planetary emergencies, especially in application to intense local emergencies that collectively gain global significance because of their economic and social costs.

All these tools will be difficult to devise, but even more difficult, we need to have them all in a package for use as a group, so we can begin to describe the range and sensitivities of regional and subregional responses across the many dimensions of life on our planet. The goal is not a global simulator for climate change, which we certainly need, but a global simulator for sustainable development on earth. This has to be a scientific endeavor of highest quality and importance, built on existing knowledge and infrastructure but not limited to them. Unfortunately, we are today in our ability to analyze potential planetary emergencies about where we were in the late 1960s when environmental impact assessments began on economic development projects. We have some ways to go, clearly, and we will need to find a way to broaden the subjects included within the toolkit beyond natural and human caused catastrophes, and widescale planetary environmental changes, to include non-environmental causes, such as rapid changes in the global financial markets, and regulation of multinationals. There will be countervailing forces in operation. One the one hand, nations and groups of states like the EU will want to limit controls on their regulatory authorities and companies, while they want to gain new influences over the global activities of firms that operate within their political jurisdictions and economic boundaries. On the other, there will be a tendency to put certain extremely difficult issues beyond single national competence to decide effectively into the UN system, such as a recent French call for a negotiation on a convention on human cloning. Where there is no international infrastructure in place to manage issues such as ethical uses of biotechnology, the pathway is by no means clear, and the outcome less certain. It is not even clear that more limited issues of universal importance like the goals of stem cell research can be decided universally in a world still governed politically by the idea of national boundaries and national sovereignty. And this, I suppose is the challenge to international science and its practitioners, both to provide an current and continuing element of rigor to other information systems and modes of decision making, and beyond this, to look ahead to those emerging problems which will have the greatest need for scientific discipline and habits of mind, in order to prepare the analytical groundwork for those difficult discussions ahead. This seems to me to capture the two spirits of Erice: excitement and responsibility.

8. WATER — TRANSBOUNDARY WATER CONFLICTS

WATER, CONFLICT, AND COOPERATION

AARON T. WOLF
Department of Geosciences; 104 Wilkinson Hall; Oregon State University;
Corvallis, OR 97331-5506, USA; Tel: +1-541-737-2722; Fax: +1-541-737-1200;
email: wolfa@geo.orst.edu

SUMMARY

River basins and groundwater aquifers which cross international boundaries present
increased challenges to effective water management, where hydrologic needs are often
overwhelmed by political considerations. While the potential for paralyzing disputes are
especially high in these basins, the record of violence is actually greater *within* the
boundaries of a nation. Moreover, history is rich with examples of water acting as a
catalyst to dialog and cooperation, even among especially contentious riparians.

WATER AND CONFLICT—INTRANATIONAL WATERS

The scarcity of water for human and ecosystem uses leads to intense political pressures,
often referred to as "water stress." As a consequence, water resources have contributed
to tensions between competing uses around the globe, from neighboring irrigators to
neighboring nations; from towns versus agriculture to environmentalists versus high tech
manufacturers. While water *quantity* has been the major issue of the 20th century, water
quality has been neglected to the point of catastrophe. Water demands are increasing,
groundwater levels are dropping, surface-water supplies are increasingly contaminated,
and delivery and treatment infrastructure is aging (United Nations, 1997).

These tensions have spilled into violence on occasion, mostly at the intranational
level, generally between tribes, water-use sectors, or states/provinces. Examples of
internal water conflicts ranging from interstate violence and death along the Cauvery
River in India, to California farmers blowing up a pipeline meant for Los Angeles, to
much of the violent history in the Americas between indigenous peoples and European
settlers.

While these disputes can and do occur at the sub-national level, the human
security issue is more subtle and more pervasive than violent conflict. As water quality
degrades—or quantity diminishes—over time, tensions can spill across boundaries, and
the overall effect on the stability of a region can be unsettling.

BACKGROUND TO INTERNATIONAL WATERS

There are 261 watersheds, and countless aquifers, which cross the political boundaries of two or more countries. International basins cover 45.3% of the land surface of the earth, affect about 40% of the world's population, and account for approximately 80% of global river flow (Wolf et al. 1999).

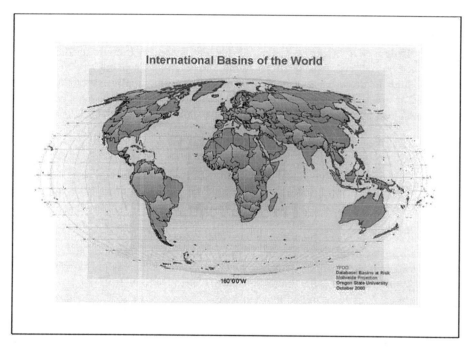

Fig. 1. International Basins of the World.

These basins have certain characteristics that make their management especially difficult, most notable of which is the tendency for regional politics to regularly exacerbate the already difficult task of understanding and managing complex natural systems.

Disparities between riparian nations—whether in economic development, infrastructural capacity, or political orientation—add further complications to international water resources management. As a consequence, development projects, treaties, and institutions are regularly seen as, at best, inefficient; often ineffective; and, occasionally, as a new source of tensions themselves.

TRADITIONAL CHRONOLOGY: DEVELOPMENT, CRISIS, CONFLICT RESOLUTION

A general pattern has emerged for international basins over time. Riparians of an international basin implement water development projects unilaterally first on water within their territory, in attempts to avoid the political intricacies of the shared resource. At some point, one of the riparians, generally the regional power[1] will implement a project which impacts at least one of its neighbors. This might be to continue to meet existing uses in the face of decreasing relative water availability, as for example Egypt's plans for a high dam on the Nile, or Indian diversions of the Ganges to protect the port of Calcutta, or to meet new needs reflecting new agricultural policy, such as Turkey's GAP project on the Euphrates. This project which impacts one's neighbors can, in the absence of relations or institutions conducive to conflict resolution, become a flashpoint, heightening tensions and regional instability, and requiring years or, more commonly, decades, to resolve.

It feels both counterintuitive and precarious that the global community can let water conflicts drag on to the extent they often do—the Indus treaty took ten years of negotiations, the Ganges thirty, and the Jordan forty—while all the while water quality and quantity degrades to where the health of dependent populations and ecosystems are damaged or destroyed. A re-read through the history of international waters suggests that the simple fact that humans suffer and die in the absence of agreement apparently offers little in the way of incentive to cooperate, even less so the health of aquatic ecosystems. This problem gets worse as the dispute gains in intensity; one rarely hears talk about the ecosystems of the lower Nile, the lower Jordan, or the tributaries of the Aral Sea—they have effectively been written off to the vagaries of human intractability.

GETTING AHEAD OF THE CURVE: PREVENTIVE DIPLOMACY AND INSTITUTIONAL CAPACITY BUILDING

There is some room for optimism, though, notably in the global community's record of resolving water-related disputes along international waterways. For example, the record of acute conflict over international water resources is overwhelmed by the record of cooperation. The last 50 years has seen only 37 acute disputes (those involving violence) and, during the same period, 157 treaties negotiated and signed. In fact, the last (and only) war fought specifically over water took place 4,500 years ago, between the city-states of Lagash and Umma along the Tigris River. Total number of events in the last 50 years are equally weighted towards cooperation: 507 conflict-related events, and 1,228 cooperative. Moreover, almost two-thirds of all events are only verbal and, of those, more than two-thirds are reported as having no official sanction at all. The most vehement enemies around the world either have negotiated water sharing agreements, or are in the process of doing so as of this writing. Violence over water seems neither strategically rational, hydrographically effective, nor economically viable. Shared interests along a waterway seem to consistently outweigh water's conflict-inducing

262

characteristics.

Number of Events by BAR Scale

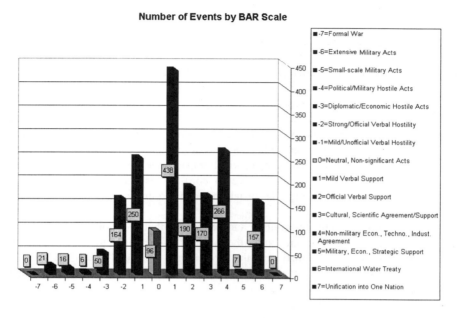

Fig. 2. Number of Events by BAR Scale.

Furthermore, once cooperative water regimes are established through treaty, they turn out to be impressively resilient over time, even between otherwise hostile riparians, and even as conflict is waged over other issues. For example, the Mekong Committee has functioned since 1957, exchanging data throughout the Vietnam War. Secret "picnic table" talks have been held between Israel and Jordan, since the unsuccessful Johnston negotiations of 1953-55, even as these riparian nations were in a legal state of war until recently. And, the Indus River Commission not only survived through two wars between India and Pakistan, but treaty-related payments continued unabated throughout the hostilities.

These patterns suggest that one valuable lesson of international waters is as a resource whose characteristics tend to induce cooperation, and incite violence only in the exception. The greatest threat of the global water crisis, then, comes from the fact that people and ecosystems around the globe "simply" lack access to sufficient quantities of water at sufficient quality for their well being.

LESSONS FOR THE INTERNATIONAL COMMUNITY

Despite their complexity, the historical record shows that water disputes *do* get resolved, and that the resulting water institutions can be tremendously resilient. The challenge for the international community is to get ahead of the "crisis curve," to help develop institutional capacity and a culture of cooperation in advance of costly, time-consuming crises, which in turn threaten lives, regional stability, and ecosystem health.

One productive approach to the development of transboundary waters has been to examine the benefits in a basin from a multi-resource perspective. This has regularly required the riparians to get past looking at the water as a commodity to be divided, and rather to develop an approach which equitably allocates not the water, but the benefits derived therefrom.

The most critical lessons learned from the global experience in international water resource issues are as follows:

1. Water crossing international boundaries can cause tensions between nations which share the basin. While the tension is not likely to lead to warfare, early coordination between riparian states can help ameliorate the issue.
2. Once international institutions are in place, they are tremendously resilient over time, even between otherwise hostile riparian nations, and even as conflict is waged over other issues.
3. More likely than violent conflict occurring is a gradual decreasing of water quantity or quality, or both, which over time can affect the internal stability of a nation or region, and act as an irritant between ethnic groups, water sectors, or states/provinces. The resulting instability may have effects in the international arena.
4. The greatest threat of the global water crisis to human security comes from the fact that millions of people lack access to sufficient quantities of water at sufficient quality for their well being.

REFERENCES

1. "Power" in regional hydropolitics can include riparian position, with an upstream riparian having more relative strength *vis a vis* the water resources than its downstream riparian, in addition to the more-conventional measures of military, political, and economic strength. Nevertheless, when a project is implemented which impacts one's neighbors, it is generally undertaken by the regional power, as defined by traditional terms, *regardless* of its riparian position.

BORDERS CROSSING BORDERS: EFFICIENCY AND EQUITY CONSIDERATIONS OF GROUNDWATER MARKETS IN THE CIUDAD JUÁREZ/EL PASO REGION ALONG THE MEXICO/UNITED STATES BORDER

DAVID S. BROOKSHIRE, JANIE CHERMAK, MARY EWERS
Department of Economics, University of New Mexico[1], USA

INTRODUCTION

Water is a scare commodity worldwide. The problems are most severe in regions that suffer from low rainfall and episodic droughts. Semi-arid regions can be found on all continents. Problems with water availability and allocation stem from climate events, population issues and population growth that are far beyond the carrying capacity of these regions at the current time. In addition, a critical confounding factor for the allocation of water both efficiently and equitable is the varying institutional structures that exist for the division of water rights and the mechanisms for the allocation of those rights, within states and countries. The problems and difficulties of efficiency and equity allocations are further exacerbated with international treaties and borders.

While flowing waters know no boundaries, human beings create geopolitical entities and sub entities with often conflicting and sometimes overlapping demands on the same resource. Each competes for an ever-scarcer resource and each depends on that resource for present and future economic growth.

This paper addresses issues along the border of Mexico and the United States. We focus on this semi-arid region to discuss issues of water scarcity. There are several shared water resources along this border; with some of the rivers flowing south and others north. For example, the San Pedro flows north from within Mexico to the state of Arizona in the United States. Alternatively, the Rio Bravo/Grande[2] flows south from the United States through the state of Colorado, to New Mexico, and Texas then becoming the international border of Mexico and United States. The waters of the Rio Bravo/Grande and the subterranean aquifers in the southwest are controlled by human beings through a complex framework of constitutional, regulations and law overlaying a complex cultural and economic setting. Our focus is upon the Rio Bravo/Grande. While we discuss surface water issues, our primary focus is on groundwater resources, in particular the sister cities of Ciudad Juárez and El Paso area in Mexico and United States. This area is heavily populated and the groundwater is the primary source and predicted to be depleted by 2030[3]. Figure 1 shows the location of our focus area.

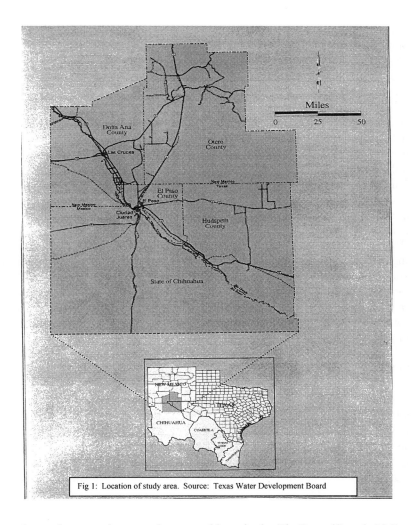

Fig 1: Location of study area. Source: Texas Water Development Board

In our focus on the groundwater problems in the Rio Bravo/Grande Valley in the areas of El Paso and Ciudad Juárez, we address the issues of common pool ground water resources. Common pool resources depletion occurs when individual property rights structures lead to overuse and degradation of the resource. In the case of the resources in the Ciudad Juárez/El Paso region, the property rights are shared by two nations that act as "individuals" that do not have an agreement as to the appropriate management of the resource. This has lead to a situation whereby the resource is not efficiently exploited, which will lead to depletion of the aquifer.

This problem is not unique to the Ciudad Juárez/El Paso regions but is, to varying degrees, a problem along the entire border. In fact, this is a problem world wide, especially in the Middle East, wherever you find a groundwater resource that crosses a border that is political in one sense, physical in another sense and where the agreements for sustainable use of the groundwater resource are not in place.

A tremendous amount of effort has gone into resolving these issues in the forms of treaties and international frameworks. While there have been some successes, much remains to be done. We note in considering the issue that substantive proposals often fall short of actual implementation procedures. That is, how will the resource be divided or how will the property rights be agreed upon so that there is an equitable division of the resource? This is essential for any resolution of the problem. This is beyond the scope of our paper. However, if the resource is equitably divided, how do the respective nations address the problem of allocation? In doing so, the institution that is chosen for the allocation and management process must in some fashion maintain the equitable division. That is, once the property rights are agreed upon what will be the mechanisms for efficient allocation? Thus, the questions of efficient allocation and equitable use of the resource are paramount.

Highly in vogue today are discussions regarding the role that markets can play in allocating resources. Generally markets are viewed by many as the resounding answer to all questions. Others retort that markets are destructive in that issues of equity and justice can be overridden or the markets simply do not operate as expected[4]. We consider these issues in this paper. If a market for allocating groundwater resources was designed, what would be the characteristics of that market and how would the twin goals of efficiency and equity be addressed, if at all?

THE RIO BRAVO/GRANDE SETTING

The Rio Grande, which translates to "Great River", is approximately 3,000 kilometers (km) in length, making it the fifth longest river in North America and the twenty-fourth longest in the world. From its headwaters in the Southern Rocky Mountains of Colorado to its mouth in the Gulf of Mexico, it drains an area about equal to 870,000 square km. In addition to the flowing river, the basin is the home to several groundwater aquifers including the Jornada Del Muerto, Hueco-Tularosa, Diablo Plateau and Mesilla aquifers as shown in Figure 2. The combination of water flows and aquifers supply water to two countries, the United States and Mexico, eighteen Native American Pueblos[5], and seven states. The states in the United States include Colorado, New Mexico, and Texas, while the Mexican states are Chihuahua, Coahuila, Nuevo Leon, Tamaulipas and Durango.

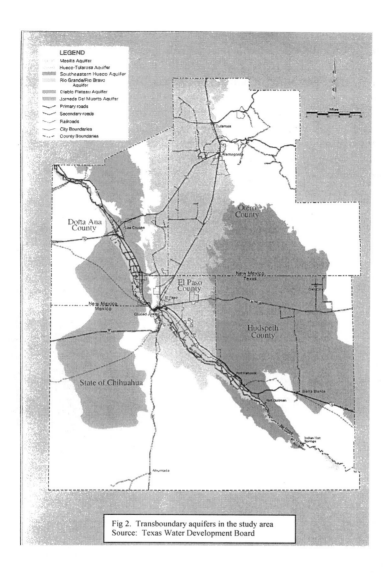

LEGEND
- Mesilla Aquifer
- Hueco-Tularosa Aquifer
- Southeastern Hueco Aquifer
- Rio Grande/Rio Bravo Aquifer
- Ciablo Plateau Aquifer
- Jornada Del Muerto Aquifer
- Primary roads
- Secondary roads
- Railroads
- City Boundaries
- County Boundaries

Fig 2. Transboundary aquifers in the study area
Source: Texas Water Development Board

While the basin is approximately 96% arid land, 1% wetlands and 2% protected areas, more than 800,000 hectares of the land are irrigated. Slightly more than half of this irrigated land is in Mexico. Irrigated crops vary, depending on the altitude. They include potatoes, alfalfa, cotton, citrus, and other vegetables. In addition to the agriculture, several cities are found on the banks of the river including Albuquerque, Las Cruces, El Paso, and Ciudad-Juárez.

The traditional agricultural use of water in the basin, which accounts for approximately 80% of the total water use, coupled with growing urban, recreational, and ecological demands has resulted in the demand for water outpacing the available supply of approximately 2.5 million acre feet annually. This is evident by the fact that the river, which once flowed continuously, is now, in actuality broken into an upper and a lower river. The upper river flows from Colorado to El Paso, Texas. At El Paso it is entirely diverted. The lower river resumes at the Rio Conchos, 402 kilometers downstream. During times of drought the river may not flow into the Gulf of Mexico, but may dry up short of its end[6]. The aquifers in the Rio Grande Basin are experiencing similar difficulties. Population growth has resulted in the mining of many of the aquifers, rather than sustainable, steady state use plans. Figure 3 shows the annual pumpage from the Hueco Bolson aquifer.

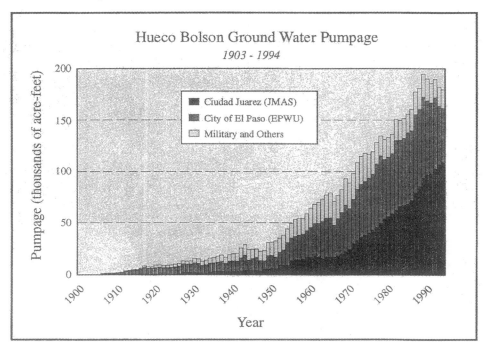

Fig 3. **Ground-water pumpage from the Hueco Bolson; 1903 - 1994 (source of data, City of El Paso Public Services Board).**

Focusing specifically on the El Paso - Ciudad-Juárez border area, El Paso's legal population is estimated to be 606,500[7] while Ciudad-Juárez's is estimated to be in excess of 1,300,000[8]. On the United States side of the border the growth rate is around two percent per year, while on the Mexican side the rate is estimated to be around four percent[9]. Per capita incomes are also significantly different. While the United States GDP per capita is $33,900, Mexico's is $8,500 (CIA, 2000). The economic differences between the populations are further exemplified by the decline of the peso to the United States dollar during the first half of the 1990s. In 1993 the exchange rate was 3 pesos/$1, but by 1994 the rate plummeted to 10 pesos/$1. It held fairly steady for the second half of the 1990s and in 2000 the exchange rate was 9.5 pesos/$1.

Water usage is also distinctly different on the two sides of the border with United States consumers having a substantially higher per capita use rate. While the average consumer on the United States side of the border uses between 640 (El Paso) and 830 (Laredo) liters of water per day, a Mexican border resident consumes around 380 liters of water per day. These large differences are hypothesized to be influenced by many factors, including the differences in income discussed above, as well as lifestyle differences. For example, while over 90% of the El Paso population has access to sewer service, only 66% of Ciudad- Juárez residents have access to sewer service.

The projections for the El Paso - Ciudad-Juárez area are for continued growth. Current growth forecasts estimated the combined population for the area would top six million people by 2025. However, the maximum number of people that can be sustained on the available water sources is about two-thirds of that, or four million people. Given the shear magnitude of people in the area, coupled with a water shortfall of 33%, the outlook for the border area is bleak. Without a plan to either increase the available water supply or to control population growth the consequences to this area may be dire.

BORDERS, INSTITUTIONS AND WATER RIGHTS

There is a mixture of discrete and overlapping borders, institutions and water rights on the Rio Bravo/Grande for surface water. A series of compacts between states, treaties and Federal and state laws define the allocation of surface water. In the United States there exists the United States Winters Doctrine (1906) and the Rio Grande Compact (1939). The Winters Doctrine established the notion of a reserved right for Native Americans based upon a priority date of when the Tribal entity became a sovereign nation. These rights supercede all other rights other than Native American aboriginal rights. This aspect is especially important in the Rio Bravo/Grande basin as there are over twenty Native American pueblos and tribes who claim aboriginal water rights in the region. These rights have not all been fully adjudicated and thus the overall Native American claim is not fully known within the United States.

The Rio Grande Compact between the States of Colorado, New Mexico, and Texas and the United States established amounts of water to be delivered to each state at a series of gauging stations along the river. The compact essentially created three rivers as a result of well-defined water delivery requirements both spatially and temporally.

The United States Endangered Species Act (ESA) of 1973 has become a dominant force for the allocation of water. In some situations, the ESA has trumped all other property rights. This has called into questions many of the accepted norms for allocating water in the southwest.

A series of treaties overtime have defined the border and water allocation between Mexico and United States. The Treaty of Guadalupe Hidalgo (1848) ended the Mexican

American War and established the Rio Bravo/Grande as the international boundary between Mexico. The Convention of May 21, 1906 established the allocation of water between Mexico and United States for the Rio Bravo/Grande. Further, The Utilization of Waters Treaty of 1944 between Mexico and United States established the distribution of waters for the international segment of the river from Fort Quitman, to the Gulf of Mexico and authorizes the two countries to construct and operate dams on the river.

International Boundary and Water Commission Minute 242 was signed by the United States and Mexico in 1973. The overall purpose of the agreement was to address the salinity problem of the Colorado Basin and address transboundary water resource issues in the Arizona Sonora border area. Further, each country agreed to consult with the other on future developments of either surface or groundwater in order to avoid adversely affecting the other. Finally, the North American Free Trade Agreement (1991) has exacerbated the water allocation problems in the region through spurred economic and population growth.

Turning to groundwater, which is a heavily used resource, there are no agreements, only intentions as expressed in Minute 242. While serious efforts have been made, the situation is one of negotiations. The Bellagio Draft Treaty provides a framework for international aquifers to be managed by mutual agreement in order that the common property problem is overcome. The draft treaty sets out various guidelines. The management of critical zones would be the focus, not a comprehensive framework for the entire international boundary. Enforcement would reside with each country with limited discretion to an overarching agency. In order to address the common property nature of the resource, mechanisms are addressed for planned depletions, drought events, water quality issues, and procedures for settling disputes[10]. However, no institutional framework has been agreed upon for the management of water which would address the need for continual reallocation, yet maintain equitable distribution.

COMMON POOL RESOURCES

When Hardin (1968) wrote of the "tragedy of the commons" he was describing the overuse that occurred on property where individual use was not defined by private property rights. The resulting problem is that the summation of the individual use choices, which may be efficient at the individual level, result in overuse and degradation of the property.

The groundwater aquifer that underlies El Paso and Ciudad Juárez is a classic example of a common property resource. Any groundwater aquifer, with more than one user can result in a tragedy of the commons, or the depletion of the common property. In the case when the aquifer straddles borders and the owners are from both countries, the problems associated with the use of the resource may be exacerbated. Not only is there the difficulty of a resource that doesn't have well-defined property rights until capture, but in addition there may be institutional, technological, economic, and even cultural differences that increase the potential of an efficient or an equitable solution.

The common property problem can be observed in many resources that are common property or common pool. Conrad (1999, p32) defines a common property resource as one that is not recognized as private property until it is captured. Examples include, for example, fisheries, oil or gas reservoirs, as well as ground water aquifers. Since the private property aspect is not recognized until capture, the rational user of the resource, in order to maximize his or her benefits from consuming the resource, will consume the resource to the point where the marginal benefit (MB) of the last unit consumed is exactly equal to the marginal cost (MC) of the acquiring that last unit. This is depicted in Figure 4.

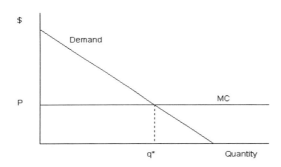

FIGURE 4: Optimal Individual Consumption Choice

The demand curve represents an individual consumer's value, or MB the consumer places on each additional unit of a resource consumed from a common pool. The horizontal MC curve represents the cost to the consumer of consuming or acquiring one more unit of the resource. The net benefits to the consumer are represented by the area under the demand curve, above the MC curve, and to the left of the intersection of the two curves. Given these specific demand and MC curves, the consumer's net benefits are greatest when the consumer chooses a consumption level of q*.

The problem that arises, from society's perspective, is that the individual consumer of a common property resource doesn't recognize, or doesn't care, about the impact that the collective individual consumption choices has on the resource. A characteristic of common property resources is that they are either exhaustible or depletable. In the case of an exhaustible resource, exploiting a single unit results in one unit less that can be exploited in the future. As such, there is an option value associated with the consumption of each unit in the current period. The option value is equal to the value of the extracted resource unit in the current period that is forgone in the future. This option value is commonly referred to as a user cost.

A depletable resource is one in which the units extracted can be regenerated. In order for this to happen, however, the extraction levels have to be such that a steady state is achieved. That is: the extraction level is exactly equal to the regeneration level. If the

272

extraction level exceeds the regeneration level, then the resource is being depleted and it will be exhausted. Therefore, depletable resources also have user costs or option values associated with them.

The difficulty with such resources is that the optimal production or exploitation choices must be considered inter-temporally. The optimal consumption level, from society's perspective in each time period is that quantity where the net benefits of the last unit consumed are exactly equal across all periods. This can best be illustrated with a two period graph, as in Figure 5.

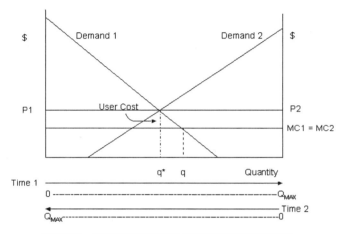

FIGURE 5:   Optimal Consumption of an Exhaustible Resource

Demand 1 is the demand curve for time period 1 (T1) and Demand 2 is the discounted demand curve for time period 2 (T2). If naïve consumption occurs in T1, that is, the finite stock of the resource, Q_{MAX}, is not recognized, the optimal consumption level would be consistent with q. This is the exactly the situation depicted in Figure 1. However, when the finite nature of the resource is recognized, the optimal consumption levels in the two periods will be those levels where the summation of the net benefits across the two periods is maximized. This occurs where the MB in T1 is equal to the MB in T2, or at the intersection of the demand curves. The offset in value above the MC to produce the last unit is equal to the user cost. It is the lack of recognition of the user cost in the private individual's consumption decision that results in over-consumption from a common property resource and leads to the depletion of the stock of the resource. The user cost is a burden on society, but does not impact the private consumer, and results in an externality, which is a market failure.

ECONOMIC SOLUTIONS

Several economic mechanisms have been suggested for common property market failures and some have been employed with varying degrees of success. We discuss a number of these in the following paragraphs, including the mechanism for implementation, as well as the pros and cons of each mechanism. In order to discuss the mechanisms, it is important to define the objective of the policy maker in order to gauge how effective each mechanism may be. We consider two objectives. The first is the classic economic objective of efficiency. That is, the maximization of net social welfare. The second objective considered is that of equity.

EFFICIENCY MECHANISMS

One alternative that is always available for any market failure is to maintain the status quo, or the "do nothing" strategy. In the case of the El Paso – Ciudad Juárez aquifer, if nothing is done, the aquifer will continued to be depleted and at some point new water sources will have to be found for the cities. Indeed, this appears to be the path of choice for the city of El Paso. The obvious difficulty with this solution is that the problem is not alleviated and the water will run out. This solution, however, may be optimal if the transaction costs of all other available solutions were such that the net benefits of continuing on the depletion path were greater than the net benefits of any alternative. In this case, the outcome is not optimal, but it the most efficient.

A second possibility is to let the market prevail without regulatory oversight. The difficulty with "assuming a market" in common property resources is that efficient competitive markets are predicated on the absence of market failures[11]. The externality created by the divergence of private and social costs is a market failure and, as such, an unregulated market would have an effect similar, or worse, than the "do nothing strategy". If an unfettered market were established between the United States and the Mexican for the water from the aquifer, not only would the depletion of the aquifer continue, but also there is a strong possibility that the majority of the water would end up on the United States side of the border. This is due to the differences in the United States and the Mexican economies. Given this, we might expect to see a market develop that is represented by Figure 6.

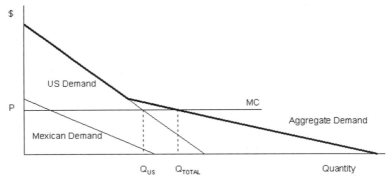

FIGURE 6:   Aggregate Market

The individual United States and Mexican demands are shown, as is the aggregate demand, which is the horizontal summation of the two. The intersection of aggregate demand and MC yields the market price and quantity for the market[12]. Given the market price, the total consumption of water would be Q_{TOTAL}. The quantity consumed by United States consumers is Q_{US}. The difference between Q_{TOTAL} and Q_{US} is the consumption quantity for Mexico. The expected results of an unregulated market for water between El Paso and Ciudad Juárez therefore are; 1) the continued depletion of the aquifer due to the divergence of private and social costs and 2) the consumption of the majority of the water by United States concerns, given the differences in the economic situations between the two countries.

Given that an unregulated market is not feasible for this problem, a potential solution is to develop a regulated market between the two cities. The most simply regulated market would be to implement a mechanism that would alleviate the divergence between the private and social costs. This could be done by instituting a tax on water production where the tax rate would be equal to the user cost associated with the marginal unit of water that is produced. This would reduce the total quantity of water consumed to the point where it was equal to the social optimal. This would be efficient, from the perspective of maximizing net social benefits. However, this could exacerbate the second problem we identified in the unregulated market. That is, the negative impact on the proportion of water consumed on the Mexican side of the border. Figure 7 illustrates the potential impact. The market price and quantity in this regulated market includes the user cost (as a tax) in order to bring the private and social costs into alignment. By internalizing the externality, the relevant supply curve is represented by the MC + User Cost function. This offsets price from MC by an amount equal to the user cost associated with the last unit of water consumed. This improves the efficiency of the market. However, the increase in price, in the example (which shows the extreme case) reduces the Mexican consumption to zero. While the actual reductions may not be as extreme, the introduction of a tax in the market to bring the private and social costs into alignment has the potential to place a larger burden on consumers with the relatively lower economic status.

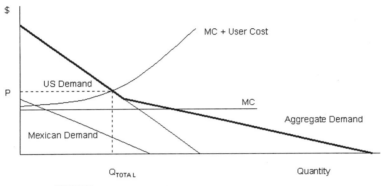

FIGURE 7: Regulated Market

A fourth consideration would be to unitize the aquifer and appoint a single operator for the entire aquifer. This is a solution that is used in many United States oil or gas fields that have a single reservoir owned by a number of individuals. The single operator produces the resource and the revenues and costs are split between all property owners, consistent with the percentage of the fled that each owns. The single operator brings the private and social perspectives more closely into alignment. A mechanism such as this should have a similar effect on the production of an aquifer. However, even though the efficiency may be improved, there are still the equity consideration that was pointed out in the previous two mechanisms. In addition, it has not been an easy task in the oil and gas industry for owners to decide on the operator or on an agreed objective. In addition, there are additional transaction costs associated with unitization due to the costs of monitoring the operator. It is not apparent that these difficulties would be any less cumbersome in an international aquifer, and it is easy to see where, in some cases, the difficulties may be more severe.

EQUITY MECHANISMS

While we've presented at least two mechanisms in the above paragraphs that would improve the efficiency in the production and consumption of the El Paso – Ciudad Juárez aquifer, neither mechanism results in what might be considered an equitable solution. That is, under the efficiency mechanisms, we can reduce the externalities associated with the common property resource, but the fact of the matter is that under these mechanisms the water will flow to where the money is; that is the water will flow to the United States side of the border. Equity consideration will require additional regulations, which will require additional cooperation between the governments on each side of the border.

Either the taxed market or the unitization mechanisms could be additionally modified to consider equity. For brevity's sake, we limit our discussion to the regulated market mechanism. In either case, a working definition of equity would have to be determined. This in itself is, most likely, not an easy task.

Questions abound as to how to achieve an equitable initial distribution. Naïve possibilities include distribution by historical use from some point in time, to a current per capita division using the population of each border city. All of these are problematic. What is needed is an equitable division that rises above historical events and/or current institutional settings. A "Rawlsian Blind"[13] is such a possibility. In this setting, actors from each interested country would meet to determine the equitable distribution. The underlying element that drives an impartial diversion is what happens after the agreement is reached. Each actor is returned permanently to a country, which may not necessarily be their home country. Thus, an actor from a wealthy country might well end up in a poor country. Thus, the incentive is to consider a range of solutions that are potentially not self-serving.

Returning to pragmatic considerations, let us assume that there was an agreement that an equitable distribution of water would be where each side of the border were guaranteed to receive a minimum per capita quantity of water. In order to achieve this objective a transfer payment would be required from United States consumers to Mexican consumers. This transfer would, in essence, be seen as a tax on the United States side, which would shift the United States demand curve to the left (as water became relatively more expensive). From the Mexican perspective, the transfer would be a subsidy to the consumer. This would shift the Mexican demand to the right (as water became relatively less expensive). Figure 8 depicts the potential outcome.

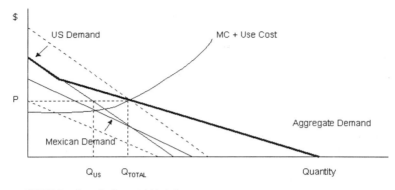

FIGURE 8:   Transfer Payment Market

The hatched lines show the original United States and Mexican demands, while the solid thin lines represent the demand curves for the two countries considering the transfers between the countries. The aggregate demand is the horizontal summation of the two. The supply curve is equal to MC + User Cost. Mexican consumption is the difference between Q_{TOTAL} and United States consumption, Q_{US}. This solution is efficient because it brings the private and social costs into alignment. If the transfer payments are designed appropriately, that is if the transfer from the United States consumers to the Mexican consumers is such that minimum per capita consumption levels in the two countries are met (the working definition of equity assumed in this example), then the outcome is also efficient.

The difficulties associated with this mechanism, however, are many. First, policymakers on both sides of the border must to committed to reaching an agreement on the working definition of efficiency. Second, the transfer payments from the United States consumers to the Mexican consumers will have to be accepted by the consumer groups. This is a potentially difficult problem for elected officials because the cross-border transfer payment agreement is a solution to a long-term problem. The time horizon until the next election may be substantially shorter and thus, the re-election objective of elected officials may be time incompatible with the objective of sustainable water use. This brings about not an economic problem, but rather a political problem. Third, in order for this mechanism to work, there is a requirement of truth, or accurate information. Without a mechanism to instill truthful revelation from both sides, the incentive is given to advantageously report the facts to either reduce the transfer (the United States) or increase the transfer (Mexico) payment level. Without providing the design detail, the transfer payment mechanism would have to be augmented with an incentive compatibility constraint, which would result in honesty being the best policy for both sides. In general, the incentive mechanism results in some loss of efficiency, as players are rewarded for truthfulness. Fourth, the monitoring that would be required for the transfer method may result in high transaction costs. Fifth, there are added difficulties concerning the sanctity of contract. Given the dynamic characteristic of democratic governments, there is the added risk of contract renegotiations under future elected governments.

CONCLUSIONS

We can design an economic mechanism that is efficient (the regulated market) or we can design a mechanism that is equitable (the transfer payment with incentive compatibility). But we cannot design a mechanism that is both efficient and equitable. This, however, may be of secondary importance to other factors. In order for either the efficient or the equitable solution to work, there is a requirement for policymakers from both sides of the border to work together to define the rules of the game. The cultural, economic, and social differences make this a difficult task. Furthermore, the incentives of elected officials on both sides of the border must be consistent with the incentive of sustainable resource use. If the players involved cannot agree, it is unlikely that any economic mechanism will be effective.

REFERENCES

1.   Central Intelligence Agency, 2000, The World Fact Book 2000, NTIS: Washington, DC.
2.   Conrad, J.M., 1999, Resource Economics, Cambridge University Press: New York.
3.   Hardin, Garrett, 1968, "The Tragedy of the Commons," Science, Volume 162:1243-48.
4.   Hayton, Robert D. and Albert E. Utton, 1992, "Transboundary Groundwaters: The Bellagio Draft Treaty," Natural Resources Journal, Volume 32:665-689.
5.   Rawls, John, 1971, "A Theory of Justice", The Belknap Press of Harvard University Press, Cambridge Massachusetts.
6.   Wolak, F.A. and R.H. Patrick, 2001, "The Impact of Market Rules and Market Structure on the Price Determination Process in the England and Wales Electricity Market," NBER Working Paper Series, Working Paper 8248, National Bureau of Economic Research: Cambridge, MA.

NOTES

1.   Respectively, Professor of Economics, Associate Professor of Economics and Research Associate at the University of New Mexico, USA. Contact David S. Brookshire brookshi@unm.edu. This material is based on work supported in part by SAHRA (Sustainability of semi-Arid Hydrology and Riparian Areas) under the STC Program of the National Science Foundation, Agreement No. 9876800.

Support is acknowledged from Ettore Majorana Foundation and Centre for Scientific Culture.

2. The Rio Bravo, as it is named in Mexico and the Rio Grande as it is named in the United States are the same river along the international border.

3. We fully recognize the conjunctive use in the region. However, for tractability and focus we concentrate on the efficient and equitable allocation of groundwater between the two nations.

4. The recent events in the deregulation California market in the United States is a prime example of a market generating large wealth transfers in society. Regarding water markets, there is literature suggesting that the use of market forces in Chile has not necessarily had desirable outcomes from the standpoint of society. Again, it appears wealth transfers are disproportionate.

5. The Pueblos, which include Taos, Picuris, San Juan, Santa Clara, Nambe, San Ildefonso, Pojoaque, Tesuque, Cochiti, Jemez, Santo Domingo, Zia, Santa Ana, San Felipe, Sandia, Isleta, Laguna, and Acoma, are all located in the State of New Mexico on the United States side of the border.

6. For example, the river dried up about 275 meters short of the Gulf of Mexico in February 2001.

7. It is estimated that there is an illegal immigrant population of approximately 200,000 people in El Paso. (From www.tcada.state.tx.us).

8. Statistical information is from *Frontera NorteSur* archives (1997-2000) and various editions of the *El Paso Times*.

9. Given the growth forecasts in the El Paso and Ciudad-Juárez it is estimated the combined population might top six million people by 2025. The maximum number of people that can be sustained on the available water sources is about two-thirds of that, or four million people.

10. "The overriding goal of the draft treaty is to achieve joint, optimum utilization of the available waters, facilitated by procedures for avoidance or resolution of differences over shared groundwaters in the face of the ever increasing pressures on this priceless resource" (Hayton and Utton, p.665).

11. An example of the difficulties that can occur when policies are based on the assumption of a competitive market for an industry that doesn't possess the characteristics of a competitive market is the England and Wales and the California electricity markets (see, for example, Wolak and Patrick 2001). We should note that difficulties in the electricity markets were due to the failure to recognize potential market power by the generators, rather than by common property.

12. For ease of presentation, we assume constant MC of production. The graphical results will hold under several cost considerations.

13. So named after John Rawls, see Rawls, "A Theory of Justice."

9. CLIMATIC CHANGES — GLOBAL MONITORING OF THE PLANET

THE INCREASING CONCENTRATION OF ATMOSPHERIC CO_2: HOW MUCH, WHEN, AND WHY?

GREGG MARLAND, TOM BODEN
Environmental Sciences Division, Oak Ridge National Laboratory
Oak Ridge, Tennessee 37831-6335, USA, Phone 865-241-4850, Fax 865-574-2232, e-mail gum@ornl.gov

INTRODUCTION

There is now a sense that the world community has achieved a broad consensus that:

1. the atmospheric concentration of carbon dioxide (CO_2) is increasing,
2. this increase is due largely to the combustion of fossil fuels, and
3. this increase is likely to lead to changes in the global climate.

This consensus is sufficiently strong that virtually all countries are involved in trying to achieve a functioning agreement on how to confront, and mitigate, these changes in climate. This paper reviews the first two of these components in a quantitative way. We look at the data on the atmospheric concentration of carbon dioxide and on the magnitude of fossil-fuel combustion, and we examine the trends in both. We review the extent to which cause and effect can be demonstrated between the trends in fossil-fuel burning and the trends in atmospheric CO_2 concentration. Finally, we look at scenarios for the future use of fossil fuels and what these portend for the future of atmospheric chemistry. Along the way we examine how and where fossil fuels are used on the Earth and some of the issues that are raised by any effort to reduce fossil-fuel use.

ATMOSPHERIC CO_2

Systematic monitoring of the atmospheric concentration of CO_2 dates back to 1958 and the creation of the monitoring program at the Mauna Loa Observatory in Hawaii. This virtually continuous record now shows an increase from an average of 316.0 ppm (parts per million, by volume) in 1959 to 369.4 ppm in 2000 (Keeling and Whorf, 2001) (Fig. 1). This increase represents an additional 114×10^9 metric tons of carbon in the atmosphere as CO_2. Analysis of proxy records, such as gas bubbles trapped in glacial ice, at a variety of places throughout the world suggests that the atmospheric concentration of CO_2 has varied considerably over geologic time but generally remained in the range 280

+/- 10 ppm for several thousand years prior to the onset of the industrial era. Over the 420,000 years preceding the industrial era the concentration appears not to have exceeded 300 ppm (Barnola et al., 1999).

Atmospheric CO_2 Mixing Ratios: 1958-2000

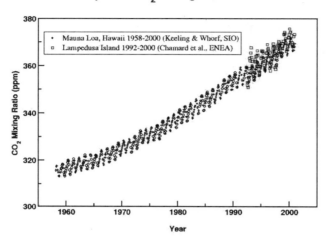

Fig. 1: The atmospheric concentration of CO_2 at Mauna Loa Observatory, Hawaii, and at Lampedusa Island, Italy.

The atmospheric increase in CO_2 recorded at Mauna Loa is not unique to Mauna Loa but has been documented at many monitoring sites throughout the globe. An Italian monitoring site on Lampedusa Island, just south of Sicily, shows an increase from 360.8 ppm in 1993 to 371.3 ppm in 2000 (Chamard et al., 2001) (Fig. 1), and the CDIAC data files (http://cdiac.esd.ornl.gov) contain records from over 50 different sampling sites. It seems clear that the atmospheric concentration of CO_2 is increasing and that it is now in a range that has not been experienced for perhaps 20 million years (IPCC, 2001).

FOSSIL FUEL COMBUSTION

Fossil fuels are burned in order to derive useful heat. Fossil-fuel burning involves the oxidation of hydrocarbon compounds and CO_2 and H_2O are necessary and essential products. We can use the combustion of methane (CH_4) to illustrate the reaction that takes place, and the essential difference for the other fossil fuels is their H to C ratio.

$$CH_4 + 2\,O_2 = CO_2 + H_2O$$

One consequence of the varying ratio of H to C is that different fuels have different rates of CO_2 emissions per unit of useful energy (Table 1). For the basic fuel types there is a strong correlation between the C content and the energy content, so we can derive these coefficients that allow us to estimate CO_2 emissions when fuel consumption is expressed in energy units.

Table 1: CO_2 Emission Rates for Fossil-Fuel Combustion (kg C / 10^9 joules gross heating value).

FUEL	CO_2 Emission Rate
Natural Gas	13.78
Petroleum	19.94
Hard Coal	24.15
Lignite and Brown Coal	25.22

CO_2 EMISSIONS FROM FOSSIL-FUEL COMBUSTION

Total CO_2 emissions from fossil-fuel combustion for 1998 have been estimated at 6.4 x 10^9 metric tons of carbon, with another 0.2 x 10^9 tons C released during the calcining of limestone to make cement (Marland et al., 2001). This large and growing anthropogenic release of carbon to the atmosphere is a relatively recent phenomenon, having risen from a sum of 1.6 x 10^9 tons C in 1950. At the start of the 20th century it was 0.5 x 10^9 tons C (see Fig. 2). The fossil-fuel release occurs largely from energy consuming activities in the developed countries. North America and Western Europe contributed 38% of the total in 1998. That emissions growth is faster in the developing parts of the world is reflected in the fact that the contribution from North America and Western Europe was 71% of the total in 1950. Figures 3 and 4 show the history of emissions from major geographic regions. We show Germany as a separate region in order to preserve the discrete, historical representations of Eastern and Western Europe.

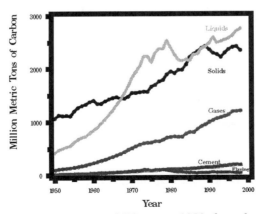

Fig. 2: Global total emissions of CO_2, since 1950, from the combustion of fossil fuels and manufacture of cement.

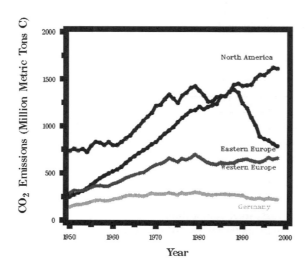

Fig. 3: Emissions of CO_2, since 1950, from fossil-fuel combustion and cement manufacture in North America and Europe.

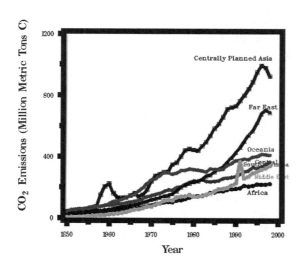

Fig. 4: Emissions of CO_2, since 1950, from fossil-fuel combustion and cement manufacture in geographic regions other than North America and Europe.

Given the apparently very large differences among emissions from various parts of the world, it is interesting to look at per capita emissions for some of the major countries. Figure 5 shows per capita emissions of CO_2 from fossil-fuel burning and cement manufacture from the 20 countries with the largest total emissions in 1998. Per capita emissions differ by greater than a factor of 10 from highly energy-intensive countries like the USA (5.4 t C/capita) to less energy-intensive countries like India (0.3 t C/capita) (Fig. 5). At the global average, CO_2 emissions amount to 1.13 tons C per capita.

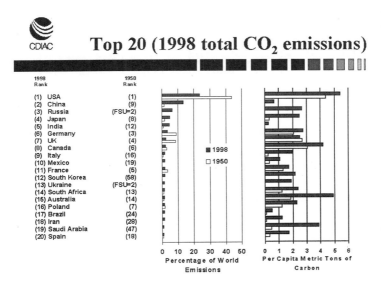

Fig. 5: *Total and per capita emissions of CO_2, for 1950 and 1958, for the 20 countries that had the largest total emissions in 1998.*

In Tables 2 and 3 we contrast emissions data from 4 countries (USA, Italy, Norway, and China) to illustrate how the different countries and their varying economies and access to resources are reflected in their CO_2 emissions. Table 2 shows that per capita emissions from the USA are nearly 3 times those of the 2 European countries and 9 times those of China. Part of the differences among countries is seen in row 3 of the table, which reveals that China is the most dependent on coal while Norway derives a large fraction of its total energy supply from non-fossil sources. Row 4 of Table 2 gives insight into the structures and efficiencies of the economies. Countries with larger contributions from the primary and heavy, and hence energy-intensive, industries will have high emissions per unit of GDP while those in which the information and service industries play a larger role will have lower emissions per unit of GDP. Similarly an economy characterized by low efficiency will have higher C emissions per unit of GDP. Table 3 quickly reveals, for example, that electric power in Norway is largely non-fossil

288

(e.g. hydro) and that the fossil-fuel based transportation system in China is much less developed than in the other countries.

Table 2: CO_2 Emissions From Fossil-Fuel Burning (1998) (data from International Energy Agency, 2000).

	Italy	USA	Norway	China
Total emissions (10^6 metric tons CO_2)	426	5410	34	2853
CO_2/capita (t CO_2/capita)	7.5	20.1	7.8	2.3
CO_2/total primary energy supply (t CO_2/Tj)	61	59	32	66
CO_2/GDP (kg CO_2/1999 US$)	0.36	0.77	0.22	3.54

Table 3: CO_2 Emissions by Sector (1998) (in percent, data from International Energy Agency, 2000).

	Italy	USA	Norway	China
Electric power (includes public heat)	31	44	1	40
Other energy industries	5	5	38	5
Manufacturing/Construction	19	10	21	33
Transport	26	30	36	8
Residential	16	6	3	7
Other	3	4	1	7

If we look at greenhouse gas emissions in the context of the Kyoto Protocol it is interesting to note that the Protocol makes no explicit acknowledgement of the many differences between countries that are reflected in the foregoing figures and tables. The Protocol does recognize the desire of developing countries to enjoy economic growth, and hence these countries have no commitments to limit emissions, and the negotiated targets do express minor differences in national commitments, as negotiated by the Parties largely on political grounds. Table 4 emphasizes that the Kyoto Protocol covers the full array of greenhouse gases (although CO_2 from the energy sector is the dominant greenhouse gas in all of the developed countries except New Zealand) and uses the USA as an example to show how the commitments are calculated. CO_2 is used as the reference and all gases are converted to "carbon equivalents" according to their integrated net impact on the global radiation balance over 100 years.

Table 4: USA Emissions Commitment From Kyoto (emissions in 10^6 tons carbon equivalent).

Greenhouse gas	Emissions in 1990 base period	Computation for 2008 to 2012 commitment period
CO_2	1372	
CH_4	179	
N_2O	38	
HFCs/PFCs	19	
Total, 1990	1608	
Target for 2008 to 2012	7457	1608 x (1.00-0.07) x 5 = 1495 x 5

Collectively the Kyoto Protocol would, if it were to enter into effect, require that the 38 countries listed in its Annex B (developed countries and countries with economies in transition) reduce emissions for the period 2008 to 2012 to an average of 5.3% less than comparable emissions in 1990. Figure 6 shows how emissions from those countries listed in Annex B (and those countries not listed in Annex B) have progressed over the first 8 years following the reference year. Emissions from the Annex B have declined by 4% over the 8-year period while emissions from non-Annex B countries have increased by 30%. Severe economic problems in Eastern Europe have had a very large impact on the collective growth rate of emissions from the Annex B countries.

Kyoto-Related Fossil-Fuel CO_2 Emission Totals

	Annex B Countries		Non Annex B Countries	
	Fossil-Fuel CO_2 Emissions (million metric tonnes C)	Bunkers (million metric tonnes C)	Fossil-Fuel CO_2 Emissions (million metric tonnes C)	Bunkers (million metric tonnes C)
1990	3851	78	2126	41
1991	3751	88	2306	41
1992	3663	92	2291	43
1993	3610	92	2341	48
1994	3607	92	2487	50
1995	3624	95	2607	52
1996	3674	95	2704	58
1997	3696	97	2775	61
1998	3690	100	2756	62

Source: Gregg Marland and Tom Boden (CDIAC, Oak Ridge National Laboratory).

Table details provided as links.

Updated: 08/03/2001

Fig. 6: CO_2 emissions, beginning in the 1990 base year, for those groups of countries that do (Annex B) and do not (non Annex B) have emissions limitations under the Kyoto Protocol. Emissions from fuels used in international commerce (bunker fuels) are shown with the country where the final fuel loading occurred, but these emissions would not be limited under the Protocol.

290

Before looking at the relationship between the atmospheric concentration of CO_2 and emissions from fossil-fuel combustion, we note that CO_2 has also been added to the atmosphere as a result of changes in land use and the destruction of terrestrial vegetation. Until the beginning of the 20th century emissions from land clearing were greater than those from fossil-fuel burning, but the latter now dominate by a factor of about 3 (Fig. 7). Total emissions from 1850 to 1990 from land-use change amounted to about 124 x 10^9 t C.

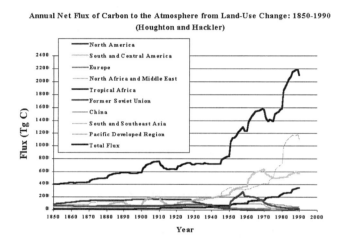

Fig. 7: Emissions, by region, of CO_2 from land-use change and the destruction of terrestrial vegetation. (Data are from Houghton and Hackler, 2001.)

ESTABLISHING CAUSE AND EFFECT

Returning to the atmosphere, we find that the cumulative total of CO_2 emissions from fossil fuels (and cement manufacture) since the beginning of the industrial era have amounted to 270 x 10^9 t C (Fig. 8). This anthropogenic emission to the atmosphere is about 45% of the pre-industrial atmosphere stock of CO_2, and the increase in atmospheric CO_2 over this period represents about 42% of the fossil-fuel emissions. Over the decade of the 1990s the annual increase in the atmospheric concentration has ranged from 0.6 to 2.6 ppm, seemingly independent of the rate of releases from fossil fuels, and we have to ask whether the increase in atmospheric CO_2 can be unambiguously attributed to releases from fossil fuels.

The global cycling of carbon involves complex interactions among the atmosphere, the oceans, the marine and terrestrial biospheres, volcanic eruptions, and rock weathering. Figure 9 provides a rough illustration of the major reservoirs and exchanges and shows that while anthropogenic, fossil-fuel emissions represent a sizeable perturbation on the system, some of the other stocks and flows are quite large by comparison.

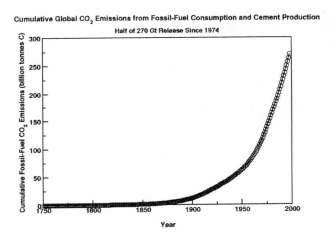

Fig. 8: Cumulative emissions of CO₂ from fossil-fuel combustion and cement manufacture since the beginning of the fossil-fuel era.

292

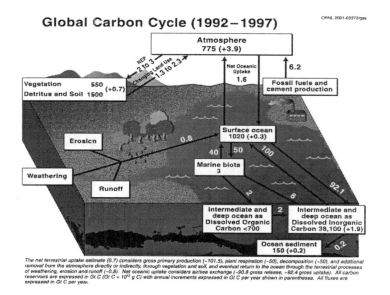

Global Carbon Cycle (1992–1997)

The net terrestrial uptake estimate (0.7) considers gross primary production (~101.5), plant respiration (~50), decomposition (~50), and additional removal from the atmosphere directly or indirectly, through vegetation and soil, and eventual return to the ocean through the terrestrial processes of weathering, erosion and runoff (~0.8). Net oceanic uptake considers air/sea exchange (~90.8 gross release, ~92.4 gross uptake). All carbon reservoirs are expressed in Gt C (Gt C = 10^15 g C) with annual increments expressed in Gt C per year shown in parentheses. All fluxes are expressed in Gt C per year.

Fig. 9: The major stocks and flows of the global carbon cycle.

Nonetheless, there are at least 4 lines of evidence that argue persuasively that it is the human perturbation, particularly the combustion of fossil-fuels that is driving the observed increase in atmospheric CO_2. First, if we plot the cumulative increases in both atmospheric CO_2 and fossil-fuel emissions, we find that the two curves are tightly linked for the 40+ years of the Mauna Loa record. These cumulative curves smooth over the year-to-year variability in the functioning of the biosphere and the circulation of the ocean. Second, if we examine the latitudinal gradient of the concentration of atmospheric CO_2, as revealed in data from many monitoring sites, we observe that the concentration is growing more rapidly in the Northern Hemisphere, as expected with the fossil-fuel source some 95% in the Northern Hemisphere. Although CO_2 mixes throughout the atmosphere, there is a lag in the mixing between hemispheres that can be correlated with the latitudinal mix of sources and sinks. Third, the changing $^{13}C/^{12}C$ ratio of atmospheric CO_2 tells something of the source of the excess C. Plants, and thus fossil fuels, preferentially concentrate the lighter ^{12}C. An increase in atmospheric CO_2 derived from burning fossil fuels or terrestrial vegetation should be accompanied by a decline in the $^{13}C/^{12}C$ ratio whereas excess releases of CO_2 from the ocean or volcanoes, for example, would leave the atmospheric $^{13}C/^{12}C$ ratio relatively unchanged. The observed changes in atmospheric $^{13}C/^{12}C$ have been consistent with a CO_2 source from biologic materials. And, fourth, the declining concentration of atmospheric O_2 provides a compelling connection with increasing CO_2. Only recently has it been possible to provide a sufficiently accurate record of atmospheric O_2 to reveal that the increase in CO_2 has been accompanied by the O_2 decline that would be expected with a combustion source of CO_2 (Keeling et al., 1996; Bender et al., 1998; Battle et al., 2000).

The evidence that increasing atmospheric CO_2 is largely a consequence of fossil fuel burning seems very strong.

WHERE NOW?

Given the connection between atmospheric CO_2 and fossil-fuel burning to date, it should be possible to model the future relationship. This does require a mathematical model that captures changes in ocean uptake, biologic responses, and concomitant changes in all portions of the global carbon cycle.

In 1992 the Intergovernmental Panel on Climate Change constructed and analyzed 6 scenarios, i.e. 6 possible paths for future emissions of greenhouse gases (see IPCC, 1995). Over the 6 scenarios CO_2 emissions in the year 2100 ranged from 4.6×10^9 to 35.8×10^9 t C. The IPCC "business-a-usual" scenario (IS92a) suggested 2100 emissions at 20.3×10^9 t C. Using these emissions scenarios in several different models of the global carbon cycle, for the business-as-usual scenario the models suggested an atmospheric concentration approaching 700 ppm CO_2 in 2100 (almost 2.5 times the pre-industrial concentration). Even for the lowest emitting IPCC scenario, in which emissions peak at 8.8×10^9 t C in 2025 before declining to 4.6×10^9 t C in 2100, the atmospheric concentration of CO_2 continued to climb throughout the century, reaching a value near 480 ppm by 2100.

To see what it would take to get atmospheric CO_2 to stabilize, Enting et al., 1994, conducted an interlaboratory model comparison exercise in which they 1.) described paths for the atmospheric concentration of CO_2 stabilizing at various levels (at 350, 450, 550, 650, and 750 ppm), and 2.) asked scientists with mathematical models of the global carbon cycle to calculate what it would require in fossil-fuel emissions to get the atmospheric concentration to move along the prescribed paths to stable levels of CO_2. The results are shown in Figure 10. The conclusion is that if the atmospheric concentration of CO_2 is to stabilize at a level even as high as 750 ppm, the path of emissions is probably going to have to diverge significantly from the business-as-usual path by the early decades of this century.

294

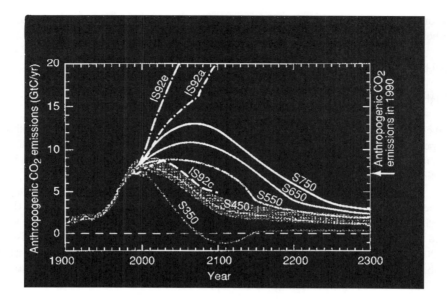

Fig. 10: Computer model representations of the global carbon cycle. Curves show the emissions of CO_2 from fossil-fuel burning that are consistent with prescribed paths for the eventual stabilization of the atmospheric concentration of CO_2. (from IPCC, 1995.) The numbers on the curves give the atmospheric concentration (in ppm) at which stabilization will eventually occur. Also shown are the paths of atmospheric CO_2 for the high, low, and business-as-usual scenarios from the 1992 IPCC analyses (see IPCC, 1995).

REFERENCES

1. Barnola, J.M., D. Raynaud, and C. Lorius, 1999. Historical CO_2 Record from the Vostok Ice Core, *in* Trends Online: A Compendium of Data on Global Change, Carbon Dioxide Information Analysis Center, Oak Ridge National Laboratory, at http://cdiac.esc.ornl.gov.
2. Battle, M., M. L. Bender, P. P. Tans, J. W. C. White, J. T. Ellis, T. Conway, and R. J. Francey, 2000. Global Carbon Sinks and Their Variability Inferred from Atmospheric O_2 and del ^{13}C, Science 287: 2467-2470.
3. Bender, M. L., M. Battle, and R. F. Keeling, 1998. The O_2 Balance of the Atmosphere: A Tool for Studying the Fate of Fossil-Fuel CO_2, Annual Reviews of Energy and Environment 23:207-223.

4.   Chamard, P., L. Ciattaglia, L. di Sarra, F. Monteleone, 2001. Atmospheric Carbon Dioxide Record from Flask Measurements at Lampedusa Island, *in* Trends Online: A Compendium of Data on Global Change, Carbon Dioxide Information Analysis Center, Oak Ridge National Laboratory, at http://cdiac.esd.ornl.gov.

5.   Enting I. G., T. M. L. Wigley, and M. Heimann, 1994. Future Emissions and Concentrations of Carbon Dioxide: Key Ocean/Atmosphere/Land Analyses, Technical Paper No. 31, CSIRO Australlia, Canberra, 118 pp.

6.   Houghton, R. A., and J. L. Hackler, 2001. Carbon Flux to the Atmosphere from Land-Use Changes: 1850-1990, *in* Trends Online: A Compendium of Data on Global Change, Carbon Dioxide Information Analysis Center, Oak Ridge National Laboratory, at http://cdiac.esd.ornl.gov.

7.   IPCC, 1995. Climate Change 1994: Radiative Forcing of Climate Change and an Evaluation of the IPCC IS92 Emission Scenarios, Intergovernmental Panel on Climate Change, Cambridge University Press, Cambridge UK.

8.   IPCC, 2001. Climate Change 2001: the Scientific Basis, Summary for Policymakers and Technical Summary of the Working Group I Report, Intergovernmental Panel on Climate Change, Geneva, 98pp.

9.   International Energy Agency, 2000. CO_2 Emissions from Fuel Combustion 1971-1998, OECD/IEA, Paris.

10.   Keeling, R. F., S. C. Piper, and M. Heimann, 1996. Global and Hemispheric CO_2 Sinks Deduced from Changes in Atmospheric O_2 Concentration, Nature 381: 218-221.

11.   Keeling, C. D., and T.P. Whorf, 2001. Atmospheric Carbon Dioxide Record from Mauna Loa , *in* Trends Online: A Compendium of Data on Global Change, Carbon Dioxide Information Analysis Center, Oak Ridge National Laboratory, at http://cdiac.esd.ornl.gov.

12.   Marland, G., T. Boden, and R. J. Andres, 2001. Global, Regional, and National Annual CO_2 Emissions from Fossil-Fuel Burning, Cement Production, and Gas Flaring: 1751-1998, *in* Trends Online: A Compendium of Data on Global Change, Carbon Dioxide Information Analysis Center, Oak Ridge National Laboratory, at http://cdiac.esd.ornl.gov.

WOULD GLOBAL WARMING RESULT IN A NEW ICE AGE?

JAY OREAR

Floyd R. Newman Laboratory, Cornell University, Ithaca, New York, USA.

INTRODUCTION

There is evidence that the north polar ice cap is shrinking and may disappear in a few decades. (Ref. 1 and 2 report that the average thickness of arctic ice has decreased by about 40% in the last two decades.) Air passing over an ice-free Arctic Ocean would contain an order of magnitude more moisture than it does now and the precipitation on neighboring lands might be an order of magnitude higher than now. The annual accumulation of snow could exceed the annual melting and regions that are presently arctic tundra would quickly transform into ever thickening glaciers. These glaciers could in a short time subtend a larger solid angle as seen from the sun than the present polar ice cap with the result of an increased albedo and a decrease of solar heating. This positive feedback loop would increase the rate of glacial growth and increase the intensity and extent of the resulting ice age . If the melting of the north polar ice cap is due to past human activity, then there would be no way to prevent the resulting ice age.

DOCUMENTATION

Reference 1 concludes that "New studies indicate that the Arctic Ocean's ice cover is about 40% thinner than it was 20 to 40 years ago, and the area of its perennial ice could be shrinking at a rate of about 7% per decade." If the ice keeps melting at the same rate the most of the polar ice would disappear within 40 years. The present ice thickness at the north pole is now only 2.2 m in September. It is so thin and broken up in summer that the tourists who go by ship (a converted ice breaker) each summer to the north pole are not permitted to disembark.

I have made measurements to confirm that open surface water at temperature of 4 deg C evaporates at about 1 mm per day. And also that ice at 0 deg F sublimates at about 1 mm per month. I repeated the experiment with the same results for salt water except that salt crystals started growing on top of the ice. So we can conclude that air moving across the water would pick up an order of magnitude more moisture than the air moving across the ice. Then precipitation rates over the arctic tundra should correspondingly increase an order of magnitude and the average rate of snow accumulation could very well exceed the average rate of melting. (Ref. 3 states: "geological data support the idea that greenhouse warming, which is expected to be most pronounced in the Arctic and in the winter months, coupled with decreasing summer insolation may lead to more snow deposition than melting at high northern latitudes and thus to ice-sheet growth.") Up until now the arctic tundra has been a desert-type of climate with a small depth of snow in the winter which disappears in the summer and exposes edible plants for the mass migration of elk.

Figure 3 of Ref. 3 shows the equilibrium curve for mean summer temperature as a function of annual snowfall in the tundra regions. Under the present conditions the temperature and snowfall are represented by a point to the left of the curve where the annual melting would exceed the annual snowfall. An order of magnitude increase in annual snowfall is represented by a point to the right of the curve which is the region where there is annual accumulation of snow. Thus in just a few years after melting of the ice cap the tundra regions could change from green in the summer to white all year. And as viewed from the sun there would be more white cross sectional area than at the present time when the earth has only a white polar ice cap. This net increase in average albedo would be a positive feedback loop resulting in a rapid onset of the new ice age. According to References 3 and 4 ice cores from Greenland and Antarctica show that the beginning and ending of ice ages can occur very quickly in a matter of decades. So it might be that during the lifetimes of our children and grandchildren there could be a sizeable global cooling in areas where under present conditions crops are able to grow.

CONCLUSIONS

Combining the shrinkage in area with the shrinkage in thickness gives a shrinkage in volume of 48% for the polar ice cap since 1980. This is a gross effect in a very short time and could result in a gross change in climate. If the melting rate continues at this rate and is uniform throughout the ice cap, the Arctic Ocean could completely open up by the year 2022. The local surface effect would be a warming, but adjoining land surfaces should experience an increase in snowfall. And if the average precipitation rate is greater than the average melting rate the ice thickness would increase with time. This effect has already been seen in Greenland which is mainly surrounded by open water. According to Ref. 5 the Greenland ice sheet is thickening at a rate of 23 cm per year. Note that melting of floating ice will not raise sea level, but formation of new ice on land will instead lower the sea level.

If the cause of rapid melting of the polar ice cap is due to natural climatic fluctuations, then we are at the mercy of nature; but if it is due to man-made gases in the atmosphere it is too late to prevent the inevitable cooling of northern temperate zones. The agricultural output of regions cooled by the new ice age would be adversely affected. The human race should work together to improve the agricultural output of the Southern Hemisphere. Also we should put much more effort in study of the above effects. For example, the ice cap thickness measurements were mainly done using upward pointing sonar from nuclear submarines. The data from 1958 to 1976 and from 1993 to the late 1990s have been analyzed. Apparently data exist in the 1976 to 1993 gap, but they are

298

still classified by the Navy. I see no reason why these data should not be declassified immediately and promptly analysed. Also it is important to take daily snow depth, temperature and wind measurements over all the northern latitude regions in both hemispheres. Let us find out just when the peak annual snow levels start increasing and when they start accumulating.

REFERENCES

1. Search and Discovery: The Decreasing Arctic Ice Cover, PHYSICS TODAY, Jan. 2000, pp. 19-20.
2. The BIG Meltdown, TIME, Sept. 4, 2000, pp. 52-55.
3. G. H. Miller and A. de Vernal, Will greenhouse warming lead to Northern Hemisphere ice-sheet growth? , NATURE 355, 244-246, (1992).
4. Severinghaus and Brook "Abrupt Climate Change at the End of the Last Glacial Period Inferred from Trapped air in Polar Ice", SCIENCE, 29 Oct. 1999, pp. 930-934.
5. H. J. Swally, "Growth of Greenland Ice Sheet: Interpretation", SCIENCE 246, 1589 (1989).

THE POLICY IMPLICATIONS OF INTERNATIONAL TREATMENT OF METEOROLOGICAL DATA AND ITS IMPACT ON FORECASTS, WARNINGS, AND COMMERCIAL WEATHER INDUSTRY DEVELOPMENT

BARRY LEE MYERS
Executive Vice President and General Counsel
AccuWeather, Inc., State College, Pennsylvania, U.S.A.

INTRODUCTION

In 1995, the World Meteorological Organization, a United Nations organ, adopted Resolution 40 which was designed to allow countries to restrict and thereby withhold, charge for, and limit certain meteorological data and information. The effort, led mostly by European countries, was to allow for government meteorological agencies (Met Services) to control weather information and to limit or eliminate the development of a commercial weather industry in those nations.

In contrast, the United States, which has some of the most diverse and severe weather in the world, is the best prepared for such events because its system, laws and attitude have allowed free and unrestricted availability of meteorological information and this approach, coupled with a limited mission for the U.S. National Weather Service, has fostered the development of a commercial weather industry. That industry not only moves government warnings to the public quickly and explains their meaning, but also provides a vast quantum of weather forecasts, warnings, displays and information through all media: radio, television, newspaper and Internet. The U.S. National Weather Service has, for many years, taken the position that the commercial weather industry is a national benefit and has encouraged, rather than discouraged, its development. That is not the case in most other nations in the world. The United States, as a result, has what are believed to be overall, the best weather information, warning and distribution mechanisms of any country.

Although almost all of the weather data in the U.S. originates from the government, over 85% of the weather forecasts and warnings reaching the general public are issued by or through commercial weather companies in the private sector, not the government; and, almost all of the weather presentations and displays are created by the private sector. Household names, such as AccuWeather and The Weather Channel plus scores of other companies, account for much of this distribution, along with various news media like CNN and The Associated Press, which utilize the services of such companies.

If nations were to open up weather data for free and unrestricted access, thereby increasing opportunities for commercial weather companies, they could find the quality and volume of weather information to their populations would increase many-fold. As it stands now, however, many nations have a negative attitude towards allowing commercial weather companies to develop in, or enter, their countries. And, many of those countries that do allow them, have erected economic barriers or imposed restrictions on certain weather data that originated within that country and might be used there by anyone other than the national Met Service.

Despite expectations of these governments, little money is actually raised from these activities and some countries have given up or are considering giving up such practices. Yet, many nations still cling to such policies and disadvantage their own populous as a result.

BACKGROUND

Each sovereign nation of the world can make decisions affecting the gathering of weather information and its release to its own citizens and to other nations and peoples of the world.

Each sovereign nation has been making these decisions, either on its own or in concert with one another, and no doubt will continue to do so.

Viewed broadly, weather is a worldwide resource. Its vagaries respect no political boundaries. Its energy potential and its destructive forces pay heed to no government.

Accordingly, weather is a resource similar to the world's oceans and to the global ozone layer. But it is a global resource more immediate and intense in its affects on the world's people than others are. While we may be concerned about the oceans and ozone over decades of use and abuse, the weather gives little time for study. It is a daily epidemic, with no regard for national boundaries, able to spread death and destruction in unrelated areas of the world with little warning. A violent storm may cause flooding and wind damage within minutes. It is the advance knowledge of impending disaster, which can allow actions by individuals, groups, businesses and governments to mediate the impact on property and people. Also, knowledge of non-severe weather, allows for planning of personal, business and government activities to maximize the benefit of those activities.

Therefore, in gathering weather data, time is of the essence.

In analyzing it, understanding its implications and in distributing information on the results of scientific analysis of weather observations, time is critical.

And in getting this scientific analysis into the hands of those who need it to protect life and property, and to make the most economic use of it, not only is time critical, but broad distribution to all in need of it is paramount.

THE U.S. EXPERIENCE AND APPROACHES BY OTHER COUNTRIES

The recent book, Isaac's Storm[1], provided a well-researched story of the hurricane that struck Galveston, Texas on the Gulf of Mexico just over 100 years ago. The storm killed an estimated 6,000 people, making it the worst natural disaster in U.S. history. Some quotes from Erik Larson's well-documented book that are of particular interest show the following:

"The [U.S. Weather] bureau's men in Cuba said that the storm was nothing to worry about; Cuba's own weather observers, who had pioneered hurricane detection, disagreed. Conflict between both groups had grown increasingly intense, an effect of the unending campaign of Louis Moore, Chief of the U.S. Weather Bureau, to exert ever more centralized control over forecasting and issuance of storm warnings." (p.9)

"In August, Moore moved to hobble the competition, once and for all. The War Department was then still in charge of Cuba. Moore persuaded the War Department to ban from Cuba's government-owed telegraph lines all cables about the weather, no matter how innocent, except those from officials of the U.S. Weather Bureau – this at the peak of hurricane season." (p.102)

"... the Weather Bureau under Willis Moore wanted hurricanes all to itself." (p.103)

"The bureau even sought the help of Western Union. On August 28, [1900], Willis Moore, then serving as acting secretary of agriculture, wrote to Gen. Thomas T. Eckert, president of Western Union. "The United States Weather Bureau in Cuba has been greatly annoyed by independent observatories securing a few scattered reports and then attempting to make weather predictions and issue hurricane warnings to the detriment of commerce and the embarrassments of the [U.S.] Government [weather] service." (p.105)

" To [William B.] Stockman [Weather Bureau manager], the tropical storm then making its way over Cuba... did not add up to much. On Saturday... he released the Bureau's evaluation of the storm... 'A storm of moderate intensity (not a hurricane)'..." (p.107)

"In Havana, Wednesday, Julio Jover sent an 8:00 A.M. dispatch – by mail to La Lucha: 'We are today near the center of the low pressure area of the hurricane.' " (p.111)

"Again, that dreadful word. When William Stockman read Jover's report, he surely laughed... Stockman saw Jover's report as further justification for the telegraph ban--it was another example of alarmist forecasting by the Cubans, who seemed to care more about drama and passion than science. Stockman did not consider the storm worthy of much further attention." (p.112)

302

"A week later, with Galveston in ruins, Cuba's Julio Jover paid a visit to Colonel Dunwoody. Emboldened by disaster, Jover sought to confront Dunwoody on the telegraph ban, but the conversation expanded to include the efficiency of hurricane prediction. As the interview gained heat, Dunwoody grew frustrated. He told Jover. '...a cyclone has just occurred in Galveston which no meteorologist predicted.' " (p.114)

"Jover, incredulous, paused a moment. He said, slowly, as one might address an inmate of an asylum: 'That cyclone is the same one which passed over Cuba..' 'No sir,' Dunwoody snapped. 'It cannot be; no cyclone ever can move from Florida to Galveston.'" (p.114)

"Six days after the storm, the War Department, apparently fed up with Stockman and Colonel Dunwoody, revoked the ban on Cuban weather cables. Moore was furious. In a letter to the secretary of agriculture, he fumed, 'I know that there have been many secret influences at work to embarrass the Weather Bureau. I regret that the restriction that heretofore has been placed on the transmission of private observations and forecasts over the Government lines has been removed.'" (p.253-4)

While the U.S. has learned a lesson from these events of over a century ago, many other nations have not. In the field of meteorology we are faced today with a problem similar to 1900 in Galveston, but on a global scale: governments restricting meteorological data through encryption, fees, long release periods and other delays. These are efforts premised in part on the belief that government meteorologists have the right answers when predicting severe and routine weather but that their counterparts in private industry do not and as such they should be restricted or monopolized by government agencies.

Gathering of weather data and making it available costs money for each country which undertakes to gather and distribute it.

These data–gathering activities have been government functions for over a hundred years, both in the United States and elsewhere around the world. They may involve surface weather observations, radar observations, orbiting satellite and other data–gathering platforms.

In the United States, the National Weather Service, a federal government agency, has a specific role to play and private commercial weather service companies have a specific role to play. Together they make up what has been referred to by General Jack Kelly, head of the U.S. National Weather Service, as the American Weather System.

This system of public and private cooperation[2] helps to save countless lives, prevents property damage, and effects economic efficiencies, all estimated in the tens of billions of dollars per year.

The United States government, and commercial weather companies operating in the United States, are dedicated to the proposition that weather information is a highly time-sensitive and perishable, scientific public good, which, if utilized quickly and communicated to people who are in a position to act, effects real economic efficiencies.

They believe that the beneficial impact on the U.S. economy far outweighs the cost of operating the data collection and dissemination network.

Another belief is that all scientists should be free to access scientific data so that they may render timely viewpoints and opinions on what the future weather may be. That is the essence of a weather warning or forecast. And, given equal access to information and analysis tools, meteorologists of equal skill, whether in a government agency or a commercial weather company, will develop forecasts and warnings of equal quality.

The freedom of access to scientific data and its free use for the benefit of society can stand alone as a guiding principle. In the United States, this freedom of access extends to all areas of society and is founded upon principles having also to do with free speech and freedom of information, which have been enacted into law[3].

Other countries of the world have constructed their relationship with regard to weather information differently. Some governmental agencies make weather information available free of charge; others impose various costs and restrictions on certain information; and, still others actively engage in commercial weather activities of their own[4].

Among some Met Services, there has been interest in developing various restrictions and costs for weather data which moves beyond the hands of the government. The basis for this concept is found to be incorporated into the data classification scheme of WMO Resolution 40, which creates division between so called "essential" (or "basic") data and so called "additional" data[5].

The reason for this interest is to make money for a government or national Met Service, to acquire funds which can offset the cost of running the government weather observation, gathering and distributing network and to control competition between private enterprise and the government weather service.

The UK Met Office in restricting and charging fees for weather data, apparently found, that if revenue from intergovernmental transfers was excluded, the cost of administering the program of data control and price assessment equaled or exceeded revenue, so there was no financial benefit to the government with this approach. This is part of the reason why the UK Met Office is abandoning its process of restricting and then charging for additional data.

In addition to the structure of WMO Resolution 40, which creates essential data and additional data, WMO Resolution 40, in and of itself, was devised and created as a mechanism to discourage commercial weather activities in many countries of the world. It has been reported that since the passage of WMO Resolution 40, commercial weather activities in Europe by U.S. firms has been significantly reduced or eliminated[6].

Additionally, since the enactment of Resolution 40, there has been little or no identifiable increase in the amount of data provided for exchange by participating WMO countries, either as part of the essential data set or as additional data. This expected increase was a major reason given by supporters of Resolution 40 before its enactment.

Even the research and academic community, which has had access to both the essential data sets and additional data, find that procedures to access the data have been complicated in some instances.

In spite of all this, there has been little or no significant new added revenue to European Met Services reported as a result of the monopoly position that they hold[7]. Rather, these restrictions have resulted in more narrow dissemination and less value from the data that is collected by these Met Services. And, technical assistance to developing Met Services is decreasing and cooperation among Met Services worldwide is decreasing. Both are negative implications for the scientific community.

SOME IMPACTS OF WEATHER DATA RESTRICTIONS

WMO Resolution 40 has perpetuated the perception among government decision-makers in developing countries and some developed countries that Met Services can be partially or totally funded through their own commercial endeavors[8]. Yet many of the Met Services of the world are not capable of providing the quality of forecasts and warnings that are currently produced by the commercial weather industry and they may not be for possibly decades to come. The concept that commercial weather companies in the United States and some other countries like Japan and the Netherlands share, is that restrictions by a monopolizing government agency are damaging to the very countries which are attempting to exclude development of a commercial weather industry.

The cornerstones of fundamental research are freedom of inquiry, full and open availability of scientific data on an international basis and the open publication of results[9]. This is also true with regard to scientific data in the field of meteorology, which is needed for the real-time protection of life and property.

Withholding, restricting or charging for such data is little different than the refusal of a government to share information on the spread of disease under epidemic situations.

In addition to WMO Resolution 40 data, some other weather data, such as from orbiting satellites, is restricted by certain government entities. The purpose is to help fund the cost of the data creation, in this case putting satellites in orbit and operating them. The largest availability of satellite based weather data and information comes from countries such as the U.S., which does not restrict the data. But, other critical satellite information such as that from EUMETSAT is encoded, with the decoding keys requiring revenue payments if any commercial companies wish to make use of the data.

The point that is consistently missed here, and in WMO Resolution 40, is that commercial weather companies will not pay much money to receive these kinds of data because the commercial weather industry is a relatively small industry world-wide, and the risk/reward analysis is questionable. But, all the data sets taken together, can lead to significantly enhanced weather forecasts and warnings, and lead to the generation of income that could be taxable like any other corporate profit. Of even more economic importance is the benefit to businesses, commerce and the people from the enhanced protection of life and property that more widespread data dissemination would bring about.

The adverse effects on scientific and technical progress, on public weather warnings and disaster forecasting and related life and property decisions, are consistently undervalued in the scenarios that many of the Met Services of the world have constructed. The general economic and social costs inherent in restricting and discouraging the downstream application and transformative uses of such data and information is also consistently undervalued[10].

Open and unrestricted international exchange of meteorological and climatological data was the norm for the better part of a hundred years. Beginning in the 1800s typical meteorological data exchange included observations of temperature, wind, pressure, precipitation, and other factors. Cooperative government arrangements were based upon the realization that information on weather occurring upstream proved useful in anticipating weather downstream, that one nation could not generate all the data required, and an understanding that the benefits to science and society in general are maximized by data sharing. Even through the Cold War period, major antagonists like the U.S., USSR, and China freely exchanged weather data in real time, with the exception of certain crisis periods. Excellent discussion of the formative years of this process between 1850 and 1950 can be found in "The Road to Resolution 40 and Beyond," an unpublished paper by Robert C. Landis, and the acceleration of global meteorological science and services that occurred from 1950 to 1980 is well documented there.

The start of significant conflict between the commercialization of meteorological services and public funding which began in the 1980s and is continuing to this day is also well-documented and discussed by Landis. It is correct to say that the ultimate adoption of WMO Resolution 40 in 1995 was really based not on key principles of data sharing, but rather on the idea that the private commercial sector in the meteorological community was "disloyal competition." WMO Resolution 40 was adopted to enact a protocol within the United Nations World Meteorological Organization that would attempt to delay and stifle competition from private companies interested in providing weather services.

The United States and certain other nations have declared all meteorological data "essential" under the terms of WMO Resolution 40 and therefore open and unrestricted. The U.S. does not utilize the "additional" data classification, or what has sometimes been referred to as "restricted data." The U.S. makes all of its data, including its satellite data, available to all users with no restrictions placed upon its use or redistribution. U.S. data is made available to all scientists and researchers regardless of national origin, and all of the U.S. data is made available to international data archives. The U.S. continues to foster development of private commercial weather service companies, thereby increasing the amount and diversity of weather information and services available to business, industry, government and the public, although there is debate even in the U.S. as to the proper mission of the government weather service and monitoring is ongoing with respect to actual or potential competition by the government with the commercial weather industry even though such competition is prohibited by law.

The concept of the WMO at its founding, which is acknowledged even by the staunchest critics of "free and unrestricted" from France, was of free and unrestricted weather data. This was WMO's mission for fifty years[11]. The focus of the WMO at this point seems to be on supporting the individual Met Services within member countries to control data and competition, rather than with assuring access to data and information.

This is not simply an issue within the mindset of these national Met Services. Some of this is generated from external pressure by their respective governments to generate revenues from all government sources; and, therefore, some Met Services search to find ways of doing this. And, some is generated from a belief that only the government should be providing weather forecasts. But, such a methodology is counterproductive, both from a revenue standpoint and in terms of the life and property implications for each nation.

There have even been suggestions that all of Met Services' data be supplied to and through the WMO, which would essentially constitute the WMO as an operational data supplier and put one international organization at the control of scientific data from all of its members. This could be a disastrous approach. In fact, it would be entrusting the sheep, not to the shepherd, but to the wolf.

It should be noted that the World Federation of Scientists has gone on record opposing polices similar to WMO Resolution 40 with regard to the copyrighting of databases which would lead to the restriction of their availability and use, both within government and within the private sector[12].

The Coalition for the Open Exchange of Global Data, in Washington D.C., in January of 1997 opposed both WMO Resolution 40 and the European Economic Interest Group (ECOMET) in their efforts to restrict data in the field of meteorology[13].

DISCUSSION

The creation and collection of data, of course, is not an end in itself but is rather a means to an end, the first step in the development of new information, knowledge, and understanding[14]. In the field of meteorology, the end is the achievement of accurate forecasts for the protection of life and property and planning daily events.

The process itself, in which the original data are continually refined and recombined to create new insights, adds value at each level of processing--ultimately synthesizing a new product, allowing for the interpretation of the original data in new ways[15], and leading to better weather forecasts and warnings.

Processing of data leads to what has been called an enhancement paradigm. The original unprocessed data, or data with minimal processing, may be the most difficult to understand or use except by a primary scientist, but with each successive level of processing, greater understandability is introduced and the value to the non-expert user increases. As the processing and formatting for easier use continues, the data becomes much more commercially valuable.

But the issue really is, where does this value lie? Many countries of the world, through their Met Services, have attempted to place value on the raw data itself or on data at the model output level. The real commercial value, however, comes by making the end product easier to understand and more readily available to the public, to industries, businesses, end users, and the public. It is at that level that the marketplace decides what the value really is.

In many nations various industries and businesses will pay for specialized weather services and products. Even government agencies such as emergency managers, police departments, highway departments and fire departments will determine from their budgets what value such a product or service has.

The public has indicated it will generally not pay for this information, and yet 85% of the weather information reaching the general public in the United States comes free to the public from private businesses in the commercial weather industry through radio, television, newspapers and the Internet. Availability of this information, which occurs in near real time, coveys forecasts and warnings to the public, carried by local and national media, paid for by advertiser support or by the radio stations, television stations, cable networks and providers, newspapers, and Internet web sites themselves, who believe that such information is of value to their listeners, viewers, readers or users.

The government of the United States does not attempt to impose some arbitrary value to the raw data. For economic, constitutional and legal reasons this would not be practical in the United States. Nevertheless, income taxes are levied against the ultimate net value for which these services are sold by commercial weather companies to the media or on the advertising support that is paid to make these available to the public. Also, the media, business, and industry benefit from enhanced efficiency and profitability from the weather information is taxed, generating a multiplier effect.

While there are a variety of studies that suggest the gross revenue of all the private weather companies in the United States is only in the 500 million dollar range annually, the estimates of the benefits to the public, business, industry, agriculture and to government of having this weather information available has been estimated to range into the tens of billions of dollars.

It can be argued that governments charging for weather data that can help to save lives and protect property is, in such cases, holding the public hostage and that it should be provided for free by the government Met Services. These Met Services often charge even their own publics, business, industry, and government agencies for this information, and by restricting competition, ensure that this cost is far higher than it would be in an unrestricted environment. In essence they are treating this data as a government owned commodity to be commercialized by the government. Even when the data is sold by the government to such customers it is often further controlled by requiring users to sign restrictive legal agreements on how, where and for what purpose the information may be used.

Open and unrestricted access to data and open competition in using the data is healthy because it enhances the quality of the products and information available, it increases the diversity and density of products, and it leads to new uses for weather information and weather products.

Government based charges operate in such a way as to decrease the availability of timely, critical weather information in each nation. Life and property are endangered more than would be the case if the information was freely and, thereby, widely available and flowing without control.

In nations with well-established market economies, governmental agencies should have nothing to fear from free and open availability of data and the operation of commercial weather companies. The public benefits through the dissemination of low-cost, easy-to-access, understandable weather forecasts, warnings and data.

For countries with developing market economies or with fewer resources which are part of the global observational network and are expending funds to maintain those observations, but may feel that they are not receiving benefits back in kind, there is the understandable desire to recover costs.

But these nations, as well, could benefit from the development or importation of weather products and services including forecasts and weather depictions and weather presentation and communication systems. This is beneficial for their governmental agencies, for their farmers, for their truck drivers, for their road builders, for their airports, for rail transport and for virtually all other weather related activities. These benefits will more than outweigh the maintenance costs for observational sites.

If maximum economic development for the nation is the objective, policies should encourage commercial weather companies to operate in these countries in free and open competition and in working relationships with government agencies, rather than policies which restrict data and markets and exclude commercial weather activities.

If a country wants to gather weather information and share it globally, but keep it from its own citizens or restrict their access to it, I would respectfully suggest the nation is doing its citizens a disservice.

Threats of data restrictions through taxing real-time scientific information will, in some countries, such as the United States, cause private companies and government to question whether or not restrictions should be imposed on the delivery of its data and model output to countries which have taken a hostile approach towards the international market for commercial weather activities.

The possibility or prospect of a trade war with regard to weather information is an event that countries of the world have worked very hard to avoid[16]. The United States, its government and its commercial weather companies have in the past supported, and continue to support policies of free and unrestricted exchange of weather information worldwide, in the belief that ultimately the citizens of all countries and the economies of all countries will benefit. The concept is similar to the policy adopted by the World Federation of Scientists in Erice in 1999.

While WMO Resolution 40 speaks to this in principle, it allows restrictions on some data and also fosters a hostile environment toward development of a global commercial weather industry.

Nations which assert their data is an economic good which belongs to them should more strongly consider that none of the countries of the world, through their Met Services or otherwise, can produce quality weather products for broad distribution without the mutual exchange of data and without a robust commercial weather industry.

RECOMMENDATIONS

1. All meteorological data that is gathered or developed by governmental agencies should be made available freely and unrestricted.
2. Government Met services should be focused on core missions and assisted in securing funding for core missions. Core missions include development of data and information, research to enhance basic understandings of the atmosphere, development and operation of computer modeling that can lead to enhanced forecast capability and the issuance of weather warnings.
3. Governments should encourage the development of the commercial weather industry and its dissemination of as much weather information as possible.
4. Policies that have a chilling effect on the development of weather information and its access by the research community and the commercial weather industry should be eliminated.
5. Governments should not compete with the commercial weather industry through Met Services or other agencies with respect to products and services that are provided by or can be provided by the commercial weather industry.

REFERENCES

1. Larson, Erik, 1999, *Isaac's Storm,* (New York: Crown Publishers).
2. See, for example, "Policy Statement on the Role of the Private Weather Industry and the National Weather Service." 56 Fed.Reg. 1984 (1991). See also Glenn E. Tallia, "Policy Issues in the Dissemination and Use of Meteorological Data and Related Information: Summary of Presentation," International Seminar on Nuclear War and Planetary Emergencies, 25th Session, Majorana Centre for Scientific Culture, Erice, Italy, 19-24 August 2000.
3. See, for example, OMB Circular A-130, 61 Fed.Reg. 6428 (1996), and Paperwork Reduction Act of 1995, Pub. L. 104-13, 44 U.S.C. Chapter 35.
4. Wiess, Peter N. and Peter Brickland, 1996 "International Information Policy in Conflict: Open and Unrestricted Access versus Government Commercialization. Will Inconsistent Government Policies Inhibit Development of a Global Information Infrastructure?" Harvard Information Infrastructure Project.
5. WMO Resolution 40 (Cg – XII), Geneva, 26 October 1995.
6. Landis, Robert C., 2001, "The Road to Resolution 40 and Beyond (Evolution of International Atmospheric Data Exchange)," Unpublished Paper.
7. Landis, *loc.cit.*
8. Landis, *loc.cit.*
9. Landis, *loc.cit.*
10. Uhlir, Paul F., 1999, "The Trend Toward Increasing Legal Protection of Digital Databases: Potential Impacts on Basic Research," International Seminar on Nuclear War and Planetary Emergencies, 24th Session, Majorana Centre for Scientific Culture, Erice, Italy, 19 - 24, August 1999.

11. Landis, *loc.cit.*
12. "Ensuring Access to Data for Science in the Face of Increasing Protectionism", World Federation of Scientists, Erice, Italy, 23 August 1999.
13. Coalition for the Open Exchange of Global Data, Position Paper on International Data Exchange, January 1997, Alexandria, Virginia.
14. Uhlir, *loc.cit.*
15. Uhlir, *loc.cit.*
16. Landis, *loc.cit.*

LONG-RANGE WEATHER PREDICTION III: MINIATURIZED DISTRIBUTED SENSORS FOR GLOBAL ATMOSPHERIC MEASUREMENTS[*]

EDWARD TELLER
Hoover Institution, Stanford University, Stanford CA, USA.
Lawrence Livermore National Laboratory, Livermore, CA, USA

CECIL LEITH
Lawrence Livermore National Laboratory, Livermore, CA, USA

GREGORY CANAVAN
University of California, Los Alamos

LOWELL WOOD
Hoover Institution, Stanford University, Stanford CA, USA.
Lawrence Livermore National Laboratory, Livermore, CA, USA

ABSTRACT

We continue consideration of ways-and-means for creating, in an evolutionary, ever-more-powerful manner, a continually-updated data-base of salient atmospheric properties sufficient for finite differenced integration-based, high-fidelity weather prediction over intervals of 2-3 weeks, leveraging the 10^{14} FLOPS digital computing systems now coming into existence.

A constellation comprised of 10^6-10^9 small atmospheric sampling systems – high-tech superpressure balloons carrying early 21st century semiconductor devices, drifting with the local winds over the meteorological spectrum of pressure-altitudes – that assays all portions of the troposphere and lower stratosphere remains the central feature of the proposed system. We suggest that these devices should be active-signaling, rather than passive-transponding, as we had previously proposed only for the ground- and aquatic-situated sensors of this system.

---

[*] Prepared for presentation at the 26th Symposium on Planetary Emergencies, Majorana Centre for Scientific Culture, Erice, Italy, 20-23 August 2001. Work performed in part under the auspices of the U.S. Department of Energy, in the course of Contracts W-7405-eng-36 and -48 with the University of California.
Opinions expressed are those of the authors only.

Instead of periodic interrogation of the intra-atmospheric transponder population by a constellation of sophisticated small satellites in low Earth orbit, we now propose to retrieve information from the instrumented balloon constellation by existing satellite telephony systems, acting as cellular tower-nodes in a global cellular telephony system whose 'user-set' is the atmospheric-sampling and surface-level monitoring constellations. We thereby leverage the huge investment in cellular (satellite) telephony and GPS technologies, with large technical and economic gains.

This proposal minimizes sponsor forward commitment along its entire programmatic trajectory, and moreover may return data of weather-predictive value soon after field activities commence. We emphasize its high near-term value for making better mesoscale, relatively short-term weather predictions with computing-intensive means, and its great long-term utility in enhancing the meteorological basis for global change predictive studies.

We again note that adverse impacts of weather involve continuing costs of the order of 1% of GDP, a large fraction of which could be retrieved if high-fidelity predictions of two weeks' forward applicability were available. These $\sim\$10^2$ B annual savings dwarf the $<\$1$ B costs of operating a rational, long-range weather prediction system of the type proposed.

INTRODUCTION AND SUMMARY

Weather significantly impacts a very wide variety of human undertakings, including all outdoors recreational, economic and defense activities. We have previously estimated its adverse impacts to have aggregate costs of the order of one percent of GNP.[1] Thus, improved ability to forecast weather – over longer distances in time, in higher geographic resolution and with greater fidelity-of-forecasting – and thereby to mitigate its disruptions may be expected to lead to aggregate economic benefits of several tens of billions of dollars per year in the U.S. alone, and several times this world-wide.

Computational Capability Gains Emphasize Geophysical Data Shortages
Improved global measurements of the state of the atmosphere, correspondingly enhanced physical models, and advanced computational capabilities all would be required to support such rational, enhanced predictive capabilities. It appears likely that the requisite computational capabilities – arithmetic engines with $\geq 10^{14}$ Floating-Point Operations per Second, or 0.1 petaFLOPS capabilities – will come into existence during the next few years for a wide variety of other digitally-based physical simulation purposes, ranging from nuclear explosions (e.g., USDoE ASCI, which has just taken delivery on a 1.2×10^{13} FLOPS system) to folding proteins (e.g., IBM 'Blue Gene', which is projected to attain 10^{15} FLOPS capability).

---

[1] E. Teller et al., "Long-Range Weather Prediction Enabled by Probing of the Atmosphere at High Space-Time Resolution," Univ. of Calif. Lawrence Livermore Nat'l. Lab. Report UCRL-JC-131601 (August 5, 1998), presented at the 23rd Symposium in 1998.

The progress rate-limiting consideration thus appears to be improved knowledge of the time-dependent status of the atmosphere (and, to lesser extents, that of the upper oceans and the land-surface) and the enhanced physical models that may be expected to naturally arise swiftly from such improved knowledge. In this paper, therefore, we continue to focus on swift attainment of large improvements in this geophysical knowledge base.

Improving Global Change Forecasting

We note that 'new' atmospheric science learned in the course of greatly improving measurements of atmospheric conditions could also remove impediments that have reduced the utility of weather models for climate studies and of climate analyses for weather forecasting. Such amelioration of this fundamental logical disconnect would improve the utility of both types of modeling, as well as enhance the overall level of confidence in the corresponding predictions regarding long-term effects of anthropogenic forcing of the atmosphere, land and oceans.

The research program needed to evaluate the usefulness of a large population of rather uniformly distributed sensors for global fluid-geophysical measurements could be based on a modest extension of current technology. It would also provide a more quantitative assessment of the eventual impact of such measurements on improved weather and climate forecasting.

CENTRAL CONCEPT

We continue to suggest that global atmospheric measurements are most aptly performed by small balloons designed to float at various altitudes of meteorological significance and which carry high-technology means for conducting *in situ* measurements of local position and wind velocity, as well as ambient temperature, water-vapor density and other such other physical variables as may be deemed essential for the specification of the Earth's atmosphere and calculation of its time-evolution. These balloons act as Lagrangian trace particles, with their motions providing a direct, accurate measurement of local wind velocity, which is arguably the most important meteorological variable for accurate weather predictions, but which is difficult to measure or infer accurately remotely. Carried-along sensors attached to the balloons can measure local atmospheric pressure and temperature, insolation, water-vapor and -droplet densities, and trace constituents.

A modest constellation of such balloons, more or less uniformly distributed and numbering about 3×10^4 (and perhaps having a cost-to-create-at-the-margin of a few million dollars) could provide measurements at the horizontal resolution D – 200-300 km – characteristic of most contemporary general circulation models. In general, the number of balloons required increases as $1/\{D^2 \ln D)$, where D is the mean horizontal spacing between the balloon micro-airships; the logarithmic term accounts approximately for the greater vertical resolution required as the horizontal sampling-length shrinks. As we will discuss in more detail below, the resulting weather-prediction times only increase logarithmically with the horizontal resolution D. However, a comparatively modest number of balloons ($\sim 10^5$) could double the time-intervals over which predictions are accurate, relative to the best present-day ones, and data returned from an eminently practical number of balloons (10^8-10^9) could support atmospheric measurements at the theoretical limit of the weather's predictability.

We note that detailed results of our collaborators, which we presented at this meeting two years ago,[2] offered a first-order demonstration-in-simulation that such Lagrangian tracers actually remain quite uniformly distributed in GCM model-studies over multi-month intervals. We also invite recollection of the Global Atmospheric Sampling Program (GASP) results of several decades ago, in which the mean lifetime of superpressure weather balloons in the troposphere was demonstrated to be of the order of six months, with icing-out in tropical storms being the dominant loss mechanism.

We find it remarkable that a constellation comprised of 10^9 such atmospheric monitoring devices – which would apparently have a marginal creation cost of the order of $1/unit – could sample the atmosphere at the currently-perceived spatial resolution limits of its intrinsic, chaos-limited predictability, which apparently is between 2 and 3 weeks in future-time, and that an exponentially more modest-sized constellation of 10^7 such devices likely could provide atmospheric sampling data sufficient for predictions of 2 weeks' validity in future-time. We now discuss such scaling considerations in a little detail.

Constellation sizes can be estimated geometrically, as the balloons are intended to give roughly uniform coverage over the whole Earth. If the average distance between balloons is D, each balloon 'covers' an area $\pi(D/2)^2$. The area of the Earth's surface is $4\pi R_e^2$, R_e the Earth's radius. Thus the number of balloons required to give n vertical measurements with a mean horizontal separation of D is

$$N^* \approx nN = n\{4\pi R_e^2/\pi(D/2)^2\} = n\{(4R_e/D)^2\} \qquad (1)$$

For a single layer of sampling, e.g., at the 500 mb pressure-altitude, uniform horizontal resolution with D = 1,000 km would require a constellation population N of 650 balloons; while a D of 300 km would involve 7,200, and one of 10 km would take 6.5 million. Adding vertical measurements that increase logarithmically with the horizontal spatial-sampling frequency – e.g., from 2 vertical-sampling layers at 3000 km horizontal resolution to 10 such layers at 3 km horizontal resolution – would increase the constellation-size for 1,000, 100 and 10 km mean horizontal spacing to about 2,100, 390,000, and 56 million, respectively. These data are among those presented in Figure 1. As the balloons are estimated to have average lifetimes on the order of a year, maintaining any of these constellations would require launching roughly these numbers of balloons annually.

[2] E. Teller et al., "Long-Range Weather Prediction And Prevention Of Climate Catastrophes: A Status Report," Univ. of Calif. Lawrence Livermore Nat'l. Lab. Report UCRL-JC-135414 (August 18, 1999), presented at the 24rd Planetary Emergencies Symposium in 1999.

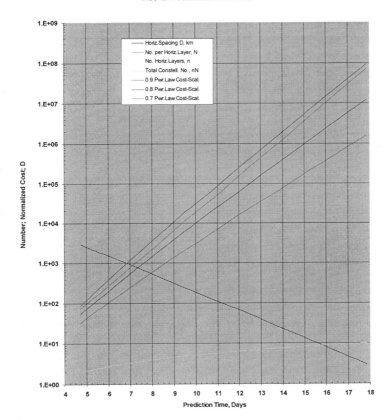

Fig. 1. *Constellation Parameters*

Prediction time-intervals vary with the resolution of meteorological measurements, particularly the air's vector velocity, at the initial time for the prediction. The Earth's large-scale flows are essentially two-dimensional turbulence; hence, they are only partially predictable. There are no fundamental theories for their evolution; current numerical predictions are simply brute-force integration of approximate equations of motion for the largest scales with some prescription for the effects of sub-grid scales. Currently, imperfect knowledge of initial conditions on smaller distance-scales degrades the accuracy of such integrations at about 4-5 days into future-time. However, turbulence models have advanced to the point where they can make semi-quantitative estimates of the time required for uncertainty about the time it takes for uncertainty about the smallest scales to propagate to larger scales and contaminate their prediction. One such model, the

Lagrangian History Direct Interaction Approximation, can estimate the rate of error propagation and is thought to be accurate to within numerical parameters on the order of unity.

Such models confirm the intuitive notion that the time-scale for contamination of the information at a given scale is roughly the circulation time for turbulent eddies of that size, i.e., the time for an eddy of size D to rotate about its axis. The result is that each factor-of-two improvement in resolution should add about one day to the future-time over which the large scales that drive mesoscale weather are predictable. Given resolution of a few kilometers, the ultimate predictability of the atmosphere is estimated to be from 14 to 21 days, depending on the strength of the two-dimensional turbulence in the scales affected. A reasonable representation of the predictability estimates, in days of future-time T, for moderate levels of turbulence is

$$T \approx 3.3 + 1.9 \ln(R_e/D) \text{ days}, \tag{2}$$

which gives predictability times of ~4.7 days at D = 3000 km, 9.1 days at 300 km, 13.5 days at 30 km, and 18 days at 3 km.[3] Equation (1) can be solved for D and substituted into Equation (2) to produce

$$T \approx 3.3 + 1.9 \ln(\sqrt{(N/n)}/16) \text{ days} \tag{3}$$

Figure 1 plots total constellation size N* (i.e., nN), number of vertical layers n, number per horizontal layer N, mean horizontal spacing D, and normalized costs for constellation creation/maintenance for learning-curve power-coefficients of 0.9, 0.8 and 0.7, which is believed to span the range of likely interest, all as functions of future-time of prediction validity T. It indicates that the current prediction time of 4-5 days could be supported by about 200 balloons, 9 days by 30,000 balloons, and 14 days by 10^7 balloons. These numbers quickly become large with increasing future-time of prediction validity, although they're certainly not excessive in overall cost-benefit terms. Fundamentally, it appears that the number of balloons in the atmospheric-sampling constellation might aptly be increased progressively, in synchrony with capabilities for data return and for computational analysis, with the total size of the constellation deployed being determined by incremental costs versus incremental societal benefits realized for longer-duration, higher geographical accuracy weather predictions.

The atmospheric-sampling interval required is roughly the Lagrangian turnover time for eddies of the size of the separation between balloons. Smaller sizes are noise, which must be averaged out. Larger sizes produce the mean motion that only advects the eddies. In fixed-position sensors, this produces rapid time variation. Advection of a frozen field of eddies of spatial frequency $k = 2\pi/D$ by a large scale random velocity U

---

[3] G. Canavan, "Value of Global Weather Sensors," Los Alamos National Laboratory Report LA-UR-99-0018 (1999).

produces a time-varying signal e^{ikUt}. Its correlation function is $e^{-(kUt)^2}$, so measurements decorrelate on an Eulerian time T_E

$$T_E \approx 1/kU \tag{4}$$

For $D = 100$ km, $k \approx 6\times10^{-5}$/m, so for $U = 30$ m/s, $T_E \approx 1/(6\times10^{-5}$/m x 30 m/s) ≈ 500 s, so that sensors distributed with this mean horizontal spacing would have to make measurements in a time short compared to this.

In a three-dimensional Kolmogorovian inertial cascade, the turnover time for an eddy of size D is $t \sim D/u$, where $u \sim (\varepsilon D)^{1/3}$ is the eddy's Lagrangian velocity, and $\varepsilon \sim U^3/L$ is the rate of energy cascade from the largest eddies of velocity U and scale L through smaller ones. However, in the large-scale, two-dimensional portion of the spectrum which characterizes the actual atmosphere, the semi-conserved quantity is enstrophy, or specific angular momentum, which cascades at a rate $\sigma \sim (U/L)^2/(L/U) \sim (U/L)^3$ to larger scales. σ is not known with precision, but it has the dimensions of $1/\text{time}^3$, so inertial range eddy velocities must scale as $u \sim (\sigma D^3)^{1/3} \sim D\sigma^{1/3}$. Eddies of size D are 'contaminated' on their turn-over time-scale, which is their Lagrangian time

$$T_L \sim D/u \sim D/\sigma^{1/3}D = 1/\sigma^{1/3}, \tag{5}$$

which is independent of eddy size. Hence, the total predictability time is the number of scales times the time-scale to corrupt each, which is $1/\sigma^{1/3}$. For resolution $D = 3$ km, the ratio of scales is $\approx 6,371/3 \approx 2000$, so the number of scale-doublings is $\approx \log_2 2000 = 11$. Thus, the maximum predictability time of $T \approx 18$ days results from 11 successive contaminations, each of which $1/\sigma^{1/3}$, so $T \approx 11/\sigma^{1/3}$, and $\sigma^{1/3} \approx 11/T \approx 0.55$/day. The local velocity for $D = 1,000$ km is $\approx D\sigma^{1/3} \approx 10^6$ m x 5×10^{-6}/s ≈ 5 m/s, and that for 100 km ≈ 0.5 m/s, but both must be sampled on time-scales short compared to $1/\sigma^{1/3} \approx 2$ days in order to properly capture local dynamics. This time is longer than the Eulerian measurement time of Eq. (4) by a factor

$$t_L/t_E \sim (D/u)/(D/U) \sim U/u \sim \sigma^{1/3}L/\sigma^{1/3}D \approx L/D, \tag{6}$$

which for 100 km eddies is a factor of $\approx 6371/100 \approx 63$.

We therefore take the required atmospheric sampling time-interval to be $\sim 10^{-2}$ x $1/\sigma^{1/3}$, or $\sim 10^3$ seconds for the smallest distance-scales. A data-frame of ~ 100 bytes, supplemented by headers, error-detection/-correction syndromes, etc., would involve data-packets of $\sim 10^3$ bits in total length, so that each atmospheric sampling platform would return a time-averaged data-rate of 1 bit/second. A constellation of $\sim 10^6$ platforms thus would have a total data return-rate of 10^6 bits/second. Over the IRIDIUM satellite telephony constellation, for instance, this would amount to $\sim 10^4$ bits/second-satellite, which is of the order of 1% of available channel capacity. It's therefore clear that atmospheric-sampling constellations of the scales presently contemplated would utilize only tiny fractions of the capabilities of even extant satellite telephony systems. The

318

constellations of ocean- and land-surface-sampling platforms would have comparable data-return rates as their atmospheric cousins but would be substantially less populous, thereby adding only incrementally to the overall loading of the telecommunications sub-system just estimated.

Atmospheric Measurement Technologies
In our initial study of this subject, presented at this Symposium three years ago, we proposed the use of entirely passive atmospheric transponders, maximally simple devices that merely changed the frequency- or polarization-dependence of their reflectivities as the local temperature and relative humidity changed. The time-dependent positions of these Lagrangian particles were sensed by probe laser-beams emitted by a constellation of small satellites in low Earth orbit, with pairs of position-sensings made at the ends of a small, known time interval providing the vector velocities of movement. The local temperature and humidity of course were 'read out' during each such probing event, both to moderate (but necessarily not really high) accuracy.

While we continue to believe that such a system is eminently feasible from a technical standpoint, its relatively large initial cost – associated with creation of the LEO constellation of transponder-interrogating smallsats – is now recognized as a significant programmatic impediment. We therefore have considered alternative approaches to the same basic atmospheric sensing capability with substantially lower 'up front' costs. Our thoughts have naturally turned toward leveraging commercial off-the-shelf capabilities, ones that can be purchased, rather than developed *ab initio*. In this spirit, we're naturally attracted by 'free' capabilities, ones that may be borrowed or shared with others, e.g., Global Positioning System (GPS)-based sensing of local position and velocity.

Sensing of Atmospheric Properties
We presently believe that the cost-efficiency optimum of any system architecture directed to present ends centers on somewhat larger-than-previous superpressure balloons deployed in the 'meteorological' portions of the atmosphere – ones of 30-100 cm diameter having individual volumes of a few dozen to several hundred liters deployed over the first ~15 km in altitude above sea-level. Each of these micro-airships would carry as a minimal, 'asymptotic' payload a modern top-end cellular telephony chip complete with embedded GPS receiving and control-processor circuitry which supervises various micro-miniaturized environmental sensors, along with a small battery and a photovoltaic power-supply, with a total payload-system mass of the order of 10 grams. The notional layout of a monolithic-silicon implementation of this system is depicted in Figure 2. When at their various operational altitudes – i.e., when fully-inflated – such superpressure balloons could very readily and mass-efficiently deploy a sub-micron thickness of GaAs solar photovoltaic array and the set of L-band antennae needed for multi-channel GPS reception, as well as the cellular telephony antenna, all on the same thin plastic balloon-wall ($_r\sim10^{-2}$ cm). All of these electromagnetic structures could be implemented as either thin semiconductor sheets or traces of metal having very modest mass budgets, and yet could source of the order of a single time-averaged watt of

electrical power into the secondary battery and balloon-payload, as well as serve as quarter-wave dipole antennae for cellular telephony transceiving and GPS signal-reception.

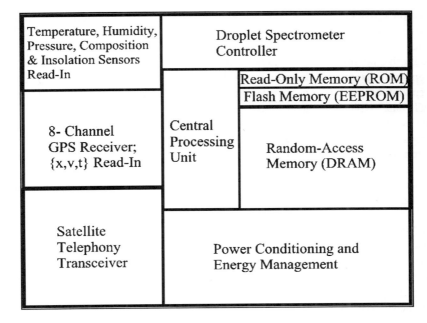

Figure 2. Schematic Layout of Micro-Airship's Single-Chip 'Silicon Payload'

Maximally-Simplified Data-Return

While in the atmosphere over continental territory which has been sufficiently highly developed to have cellular telephony towers, such devices could simply "call home" periodically, in order to download their data to the central data-processing facility through the global public switched telephony network (PSTN), thereby providing wind velocity, temperature, humidity, pressure, insolation, etc. data-streams along the space-time trajectory defined by the motion of the atmospheric packet in which each of them is embedded as a Lagrangian particle. Indeed, this "maximally commercialized" mode of data-return would be eminently suitable for initial proof-of-principle studies, as it would return highly useful data-streams at absolutely minimal investment. Quite importantly, its bi-directional character would permit the central data-processing facility to re-program the balloons' embedded digital controllers in a serial number- or location-dependent manner, by uplinking new software-loads or, even more readily, by simply changing key parameters being referenced by the on-board control processor. Each balloon could thus serve as a truly interactive mobile weather-station, dynamically modifying its data-collection and -return behavior in order to optimize the global performance metrics of the main Weather-Prediction Engine, as that Engine's environmental data requirements might vary in time.

However, it appears that this data-return scheme would be useful only over economically highly-developed portions of the Earth's land-mass, and couldn't possibly function satisfactorily when the balloons were carried by air-motion over mountains, deserts, jungles, lakes and oceans, where cellular telephony towers are yet to be found – which is likely to be at least 95% of the Earth's total surface, recalling that only 15% of the Earth's landmass is truly arable, and most of that is yet to be 'blessed' with cellular telephony. Data-streams returned by balloons just coming into range of a cellular telephony tower, e.g., on a western continental margin, thus would be cluttered with "stale data" typically collected dozens to hundreds of hours previously – and thus would have relatively little content of then-current value for weather-prediction purposes. Of comparable importance, ground tower-based cellular telephony is intrinsically incapable of returning geophysical data from most all members of the ocean- and land-based constellations-of-environmental-monitors, data that are essential as boundary-condition inputs to the numerical integration procedures which creates the long-range weather forecasts.

Satellite Telephony Data-Return

However, this scarcity-of-ground-based-towers picture is quite likely much too bleak. There are already multiple satellite telephony services in commercial operation, and any one of these having truly global 'reach' could potentially provide world-wide real-time connectivity between any one of our atmospheric-monitoring micro-airships and the main Weather-Prediction Engine. One of these satellite telephony systems, the IRIDIUM one discussed at a Symposium session several years ago, following an interval of financial and organizational turbulence, currently has the U.S. Defense Department as its 'anchor tenant' and presently provides unlimited service to 20,000 DoD-owned user hand-sets around the world. With only a small fraction of its data-handling capacity, IRIDIUM could service the data-traffic of literally a million our of balloon-borne atmospheric monitors, each of which might be expected to return a data-packet of 1 kilobytes length – e.g., ~32 data-frames of ~32 bytes each, taken regularly at tenth-hour intervals – in a few-second-duration, burst-type 'call home' every few hours, comprising a steady-state loading on the IRIDIUM system of 'only' 300 calls, or a half-dozen calls per satellite.[4]

---

[4] Some concern might arise as to whether the substantial mass and volume of contemporary satellite telephony handsets – the smallest of which have ~300 gm masses – are compatible with carriage by micro-airships with total payloads of the order of 10 grams. Our response to this is that these handsets are built to support continuous communication over intervals of more than one hour, and have power supply and heat-dissipation means that are very conservatively rated for such service, as customer tolerance of any malfunctions in such premium means of communication is minimal. In acute contrast, we contemplate few-second-duration, burst-type data-communication every few hours, for which the heat capacity of the transmitter chip is adequate heat-sink and the pulse capacity of the tiny albeit high-energy-density (e.g., Li ion) on-board battery is sufficient power-source. In addition, the balloon's size supports deployment of physically optimal antennae, in marked contrast to those of hand-held satellite telephony user-sets. We therefore conclude that gram-scale, silicon-intensive communications systems will be quite

Interestingly, data return could be scheduled to take place normally only outside of urban areas, so that de-confliction relative to the geographical peaks of commercial traffic-loadings could be attained quite readily; the incremental loading of the system by the constellation of micro-airships thus would be virtually invisible to IRIDIUM system operators (or to other system customers).

Since the U.S. Defense Department is a not-unlikely sponsor or co-sponsor of any American component of any international long-range weather-prediction program, use of the IRIDIUM system for both initial and steady-state data-return purposes is an especially interesting prospect from both technological and economic perspectives, one which could greatly facilitate commencement of operation of the capability which we propose. Of course, there are other commercial satellite telephony systems that offer alternate paths to this initial operational capability, so that this particular programmatic option should be regarded as a quite feasible, low-risk one.[5]

The overall layout and data connectivity of the global environmental monitoring system that we contemplate as a baseline one is indicated schematically in Figure 3. We emphasize that all of the in-space features of this system already exist – and have been in routine operation for years. Only the balloon-borne and surface-emplaced constellations-of-sensors need to be created – and the two fluid-envelope constellations even self-deploy!

LIFETIMES AND COSTS

As already noted, well-designed superpressure balloons have been GASP-demonstrated to have lifetimes in the troposphere of the order of six months – long before modern

---

capable of sourcing the highly-intermittent, short-duration meteorological data-traffic from the atmospherically-mobile balloon-borne weather-stations that we propose.

[5] Despite the attractiveness of generic satellite telephony, we consider deployment of a second constellation of balloons in the upper stratosphere as an alternate architecture for the data-return portion of the atmospheric monitoring system of present interest. The purpose of these much larger, much longer-lived superpressure balloons is to serve as the cellular towers of this global telephony system or, from the above perspective, to stand-in for the wireless telephony satellites in orbit.

Employing mesospheric balloon technology which is currently semi-commercially sourced, we contemplate mesospheric data-relay systems with a payload of the order of a few kilograms – about 10^{-3} of the largest payloads currently lofted in such systems – which receive data-calls from the lower-lying atmospheric sampling balloons, out to their local Earth-horizon at a few hundred km distance. The greater size of these mesospheric balloon systems would permit the generation of kilowatt-scales of electrical power, the storage of megajoule quantities of energy, and the deployment of the larger antennae and high-Q filters appropriate for receiving calls from multi-hundred kilometer-distant transmitters having effective radiated powers of a single watt. Each of these systems would then uplink suitably-conditioned data-packets into the global comsat system for return to the main Weather-Prediction Engine.

Of the order of one thousand of these mesospheric "satellites in the atmosphere" would provide world-wide connectivity between the much smaller, much lower-power atmospheric monitoring balloons beneath and the central Weather-Prediction Engine. Each would service between 10^3 and 10^6 monitoring-balloons, at a call-density of 0.1-100 at any given moment, i.e., $10^{-2.5 \pm 1.5}$ of the typical service-level of a single terrestrially-deployed cellular tower.

322

materials became available. Their lifetimes appears to be limited by catastrophic icing in tropical storms; they circulate about the planet of the order of a dozen times until they are caught in such a storm, take on a relatively heavy load of ice, and fall into the underlying ocean before the ice-burden melts. Our baseline balloon system architecture therefore contemplates a mean lifetime of 0.5 years. [To be sure, we plan to exploit the superpressure feature of our balloons by providing them with an 'emergency lift' capability which may be invoked under control of their embedded processors to expand their displaced volume abruptly by as much as 100% with no increase in mass, thereby giving them the ability to rapidly ascend out of icing regions of storm-clouds and thus escape death-by-icing. Use-as-needed of this feature may extend the average life of our micro-airships by several-fold.]

We thus contemplate a requirement for creating and launching at least 10^6 and no more than 10^9 third- to half-meter-diameter balloons each year – roughly 0.03-30 per second, continually. As we noted three years ago, there is nothing particularly challenging about this requirement, technically, operationally or economically.

Fig. 3.

Cost Determinants

The experience of the modern chip business suggest that the 'cost-driver' of such balloon-borne weather-station creation likely will be the attachment of the silicon chip+mini-battery module onto the metal bonding-pads on the balloon-wall and the associated functionality-testing; balloon-creation and -inflation to its appointed pressure-altitude seemingly will be next most costly, with procurement of the chip+battery module being third most expensive.

Balloon-associated steps likely will be the more expensive ones, simply because they won't be already-mass-production ones when the program gets seriously underway. On the other hand, disposable cellular phones are just now coming onto the market, with imputed costs of the order of $5 for the phone that's discarded after a total use-interval of the order of one hour; the associated silicon cost can't be much in excess of $1 – and this is for unit quantities of an item sold at retail to a rather price-insensitive customer-community. Underscoring this swiftly-emerging trend toward commodity telephony silicon, Motorola, the largest U.S. cellular telephony manufacturer, recently announced that it's commencing to sell its entire cellular telephone component line – including its vaunted "single-chip solutions' – as turn-key packages and kits to any-and-all comers, signaling to industry observers that it's departing a line-of-business which it pioneered in favor of selling mass-market commodity parts, like it's long sold linear semiconductors. Already, cellular phones are going on-sale with built-in GPS receivers and single-button-push means of transmitting the contents of its coordinate registers. We're therefore reasonably confident that total balloon-creation costs-at-the-margin will be $2-5/unit, with unit marginal silicon costs being of the order of $1, when micro-airships are created by the millions per year.

System Operating Costs

Our estimate for system operations costs thus is less $10^{7.5\pm1.5}$/year and are dominated by balloon-replacement expenses, assuming that cellular telephony costs are kept reasonably in-line. If the system's cellular telephony costs are made to be negligible by such gambits as IRIDIUM usage, if balloons can evade most fatal icing incidents by means such as active pressure-altitude control and thus attain year-scale mean operational lifetimes, and if mass-production cost-scaling can attain a learning-curve power-law index of 0.8, these operating costs for even a maximum-scale system might of the order of 10^8/year, as is indicated in Figure 1. This represents an order-of-magnitude improvement over our cost-estimates of 3 years ago – and even those far-higher costs were still two orders-of-magnitude less than the gain in economic benefits from long-term, high-fidelity weather predictions. Much more importantly, however, the buy-in costs to start up the system that we presently propose are the comparatively negligible ones of prototyping powered-silicon-bearing balloons, as compared to those of creating a constellation of smallsats in LEO together with transponder-balloon prototyping.

What Price Distant-Future Knowledge?

Finally, we note that the last few days of durability in future-time of weather-prediction accounts for essentially all of the cost of the type of atmospheric monitoring, data-return and computation system that we contemplate. We have estimated above that extension of forecasts of a given level of fidelity by 1 day requires nearly a doubling of resolution in the spatial frequency of atmospheric sampling, i.e., a 3-fold increase in the constellation-size of micro-airship weather stations, with a nearly linear increase in cost. Only the enormous benefit-to-cost characteristics of this forecasting system could possibly justify driving the atmospheric-monitoring constellation size up to the 10^9-balloon level and on out to the ~18-days in future-time theoretical limit of forecast validity estimated above; indeed, much more detailed analyses of cost-benefit-at-the-margin likely will be required to justify 'buying' the last 2-3 days of future-time forecast validity with 90% of the maximum-scale system's cost.

It is a curious, but potentially quite important, point that the large majority of the maximum-scale system's cost buys knowledge of future weather that seems destined to have the lowest unit value. [Most – but assuredly not all, farmers being an obvious exception – of us care much more about the weather in the next 3-7 days than we do about it on days 14, 16 and 18 from now.]

SOME SYSTEM-LEVEL CONSIDERATIONS

We now address a few system-level considerations, including system operation in the real world, safety, esthetics and benefits.

Operations In The Real World

The balloon lifetime-limiting feature in the lowest 3-5 km of the atmosphere isn't icing, but rather the pilot's bane of CFIT: controlled flight into terrain. Higher-altitude terrain in general and mountains in particular pose as much danger to lower-altitude balloons as they do to powered aircraft, and avoiding crash-landings is the primal motivation for the already-mentioned engineered-in capability employed to evade death-by-icing in major storm systems. After all, every circum-Earth cycle, or roughly every two weeks, all balloons with pressure-altitudes of ≤3-5 km – a third to a half of the constellation – will have to ascend significantly in order to clear the Andes in the Southern Hemisphere or the Himalayas and/or the Rockies in the Northern one. While flying over land, the lowest layers of them will also have to slowly but continuously adjust their pressure-altitudes so as to track reasonably close to the local underlying solid-Earth surface without flying into it.

We contemplate providing such altitude-control capability via tiny, e.g., piezoelectric or rotary-drum, traction-motors attached to filamentary cables within each balloon whose other ends are anchored in the very thin plastic walls. When the balloon's embedded digital controller wishes to reduce the balloon's (pressure-) altitude, the micro-cables are synchronously reeled in by micro-motor action slowly doing the requisite PdV work – about 1 kilojoule's worth, for a 2-fold volume-change in a 20 liter full-volume balloon operating with a base 1 bar pressure-altitude (i.e., a worst-case) – on the balloon's pressurizing-gas and thereby decreasing the balloon's total volume. Conversely, these tensioned filaments may be released (swiftly) under program control when (rapid) ascent is called for. [Other, quite different means of effecting this variable-lift function seem obvious; resistively heating a tiny, insulated-and-valved reservoir containing a thermally unstable material such as LiH in order to reversibly input ~1 gram of hydrogen gas into the balloon's main buoyancy-generating volume is one such alternative.] Obviously, all of these altitude-control means will be GPS-informed, so that their operation will be entirely automatically implemented by the balloon-resident digital microcontroller, which will carry an all-planet, km-resolution-scale elevation map in a gigabit ROM.

Safety

These balloons that we contemplate are much smaller in size and bear payloads of far smaller mass than do ordinary weather balloons and associated radiosondes. This is the *prima facie* argument for their safety relative to manned aircraft. However, their far greater numbers – which is the source of their socially-valuable weather-predictive power – might give grounds for concern with respect to rare collisions between one of them and a manned aircraft. We note that their contemplated mean density is 0.1-0.001/km^3, so that an airplane with a typical cross-section of perhaps 5 m^2 and a speed of the order of 200 meters/second will have a mean time-to-collision of $10^{10\pm1}$ seconds, or 3000-30 years (for constellations of 10^7-10^9 balloons, respectively). Even when such a collision occurs, it will have far less adverse consequences than that with a small bird – for few birds that are 'seen' by planes have masses as small as 10 grams. We therefore conclude that there are no significant human safety issues associated with deployment of even large quantities of these gossamer-like micro-airships.

Esthetics

We point out that the micro-airships which we contemplate will be quite invisible from the Earth's surface. The lowest-altitude ones, floating at about 1 km above the local surface, will have an angular subtense smaller than the angular resolving capability of all but the best human eyes. We expect that their underside skins will be optimally mottled in order to further obscure them from Earth-surface view, so as to entirely obviate visual-esthetic concerns.

We are conscious of the need to dispose of fallen balloons in an environmentally-sound manner. The battery-chip module has a mass of the order of 1 gram and the physical scale of a small pebble; covering one-trillionth of the Earth's surface with such readily weather-degraded synthetic pebbles each year seems a very minor environmental impact indeed. When a balloon grounds out—or hits the ocean surface—we contemplate employing the battery's entire energy-store to cut its ultra-thin plastic skin into cm-scale flakes by electrically-ignited means and to similarly open the protective shell of its battery-chip module to weathering, if the micro-airship descends on land; the latter will swiftly descend into very long-term abyssal storage, if the final descent is into the ocean.

326

While the assessed environmental impacts are not completely negligible, we suggest that they are offset by many orders-of-magnitude by the benefits of long-range, high-fidelity weather prediction, of which enhanced management of large-scale floods is merely a single example.

Benefits
The likely benefits from the proposed globally distributed atmospheric measurement capability fall into three broad categories: science, weather, and climate.

Science Benefits Atmospheric science is now dominated by uncertainties ranging from predictability to cloud physics, i.e., what will the weather be in a few days (and will we ever be able to improve such predictions) to why does it rain? Remote measurements from the ground, aircraft, and satellites have been too difficult and too expensive to really resolve such basic issues. Detailed knowledge of winds, mixing on all scales, and *in situ* measurements, e.g., of cloud droplet size-density distribution, microparticle density and ion concentration, could provide fundamental insights leading to resolution of such uncertainties.

Weather-Prediction Benefits
Such insight could make it possible to improve weather forecasts from the 4-5 days of the best current techniques, employed under the most favorable prevailing conditions, to one of 2-3 weeks during potentially supportable by more accurate measurements made with much higher space-time resolutions. In the process, the R&D program needed to develop that capability could provide direct information on the atmospheric quasi-turbulent energy spectrum and mesoscale phenomenology needed to refine theoretical estimates of predictability. We again note that the benefits of as much as 90% of ultimate-duration weather predictions may be available for as little as 10% of the costs of the 'ultimate' system which we've discussed in the foregoing, so (exponentially) demanding are the technical requirements for forecasts of the longest-possible duration.

Climate Change Benefits
Fundamental advances in the understanding of atmospheric physics and weather prediction could also be expected to provide the calibration now lacking for general circulation models of the Earth's response to altered radiative forcing. Those models are now used to predict the impact of anthropogenic effects on time-scales of decades to centuries, despite the fact that current models do not necessarily converge to the long-term circulation patterns actually observed on the Earth when run to very long times. That is, a better understanding of atmospheric physical processes, of real-world mass, momentum and energy flows, and of cloud formation and evolution vis-à-vis both atmospheric science and weather forecasting could also lead to a real scientific basis for long-term climate impact forecasts, thereby permitting discarding of the present-day *ad hoc* approximations.

We emphasize the unique utility of continuously and densely measuring *in situ* the vertical profiles of tropospheric temperature simultaneously with ground- and ocean-surface temperature measurements made by the associated surface-based set of boundary-condition monitoring stations, which we discussed in a little detail three years ago and which we now contemplate linking to the main Weather-Prediction Engine via the same satellite telephony means as is proposed for use by the micro-airship constellation. Such measurements can be expected to swiftly resolve the present-day discrepancy between long-term temperature trends in the lower troposphere versus those on the ground. We also invite particular attention to the peculiarly great ability of the continuously-operating constellation of atmospheric monitoring stations to diagnose cloud physics in a statistically compelling manner, e.g., by space-time-dense *in situ* measurements of the dependence of water droplet size and concentration on position and state variables, throughout large storm systems; most all currently controversial issues, e.g., long-range advective transport of droplets in tropical storm systems, should be objectively resolved by the first few months of reasonably full-scale constellation operation.

The knowledge gains deriving from such advances in space-time-dense, *in situ* measurement could be worth most of one trillion dollars/year of world-wide avoided costs, if such improved understanding led to substantially more cost-efficient means of minimizing anthropogenic aspects of global change than those now contemplated.

A REPRESENTATIVE PROGRAMMATIC PATH

We now sketch what we consider to be a representative path for a program that culminates in the system which we've just outlined. Obviously, many alternate routes to the same goal are feasible, and others may be much preferred, for reasons not obvious to us now.

'Breadboard' Prototyping
A proof-of-principle could be attained quickly by attaching to a standard superpressure balloon of a few meters' diameter – i.e., a few kg of payload-mass – a suitably integrated control processor, temperature, pressure, humidity and insolation sensors, GPS receiver, battery-pack, photovoltaic array and satellite telephone. If inflated for a pressure-altitude of 10 km, this proof-of-principle system will never risk loss via CFIT. Due to the 0.3 bar pressure-altitude selected, it may be expected to avoid an ice-out fate, and thus can be employed to provide arbitrarily finely time-spaced meteorological data for a time-interval of at least a year. Suitably instrumented and programmed, it will even report on its mode-of-demise, whenever this may occur.

'Brassboard' Prototyping
While a small number—perhaps a dozen—of these 'breadboard' prototypes are being so exercised and incidentally are returning quite novel scientific datastreams, a chip-set suitable for a 'brassboard' prototype can be assembled, tested and produced in limited quantity. This chip-set would feature a much higher level of integration, with single

chips for power management, for full 8-channel GPS reception, for system control and data-management and for satellite telephony. The solar photovoltaic array, GPS antennae and telephony antenna all would be lightweighted and integrated into the balloon structure, and the sensor-module would be ruggedized and miniaturized. Roughly one hundred of these brassboard systems might be deployed for operational evaluation, at various pressure-altitudes between 5 and 15 km so as to endow them with reasonable *a priori* operational lifetimes.

'Steelboard' Prototyping

While brassboard field evaluation was taking place—and the central data-collection facility was being more extensively exercised—the main development of the 'steelboard'-level prototyping components would be conducted. These components would be comprised of a single chip implementation of all powered silicon functions, antennae implemented as metal traces and the photovoltaic array as a semiconducting film-structure on the wall of a lightweighted superpressure balloon outfitted with electrically-controlled buoyancy control. The sensor module of the steelboard prototypes would include a microminiaturized water droplet size-and-density spectrometer, and such other environmental sensors as the program sponsor would consider appropriate. At least one thousand of these steelboard prototypes likely would be deployed for operational evaluation, which naturally would even more vigorously exercise the data-collecting facility. [If as many as 10,000 were to be deployed, the resulting data-set quality would support computational weather forecasting to attain as high quality over the entire Earth as that presently available anywhere, as we noted above.]

The steelboard prototyping components would go into mass production as the field evaluation interval concluded, during which time high-rate serial production facilities would be implemented and operated at low initial rates.

Computational Program

Concurrent with the three stages of prototyping of the atmospheric monitoring balloon-stations, one or more pre-existing weather prediction codes would be modified to accept actual atmospheric data in the formats and at the rates being made available at the various prototyping levels. These codes then would be ready to perform on commercially-available computing systems – commercial variants of the ASCI or 'Blue Gene' systems—of appropriate high aggregate processing-rate, as the full-scale system ramped into existence. [As previously noted, as few as 10^4 balloons deployed—possibly even within the 'steelboard' prototyping level – should support the generation of world-wide weather predictions of the 3-5 day duration and reasonable fidelity now available for only the most-advantaged locations. Computing systems of the corresponding scale are already rather widely available from multiple commercial sources.]

Program Time And Dollar Scales

We expect that roughly one year should suffice for each of the three prototyping phases, so that the full-scale long-range, high-fidelity weather-prediction system could be in operation roughly a half-decade after the program commenced. The cost of the program would rise roughly exponentially in time from a base of ~\$5 M/year to ~\$100 M/year over this half-decade interval.

Risk Management
It is important to note that each of these prototyping and evaluation phases of the proposed program has clearly defined objectives, objective decision criteria for continuation and graceful exit strategies. Sponsor technical and programmatic risk-profiles are stringently minimized in each phase, and the sponsor's forward commitments are for at most a single year's duration. We anticipate that this set of features will be especially congenial to the program-sponsoring community in the U.S.

INTERNATIONAL COOPERATION

We have emphasized from the commencement of our consideration of the prospects for long-range weather-forecasting, e.g., in our presentation to this Symposium three years ago, the great desirability of the fullest-possible international cooperation in the design, development and operation of the type of system that we propose.

We continue to welcome comments and suggestions for improvement in all respects, and we look forward to active collaborations with colleagues from all nations, as this program gets underway.

Long-term weather is intrinsically a truly global phenomenon, and long-term weather-prediction, by the nature of physical law as it's expressed in terrestrial meteorology, *necessarily* develops forecasts of world-wide validity, just as it *demands* near-real-time knowledge of the state of the atmosphere from all over our planet. For instance, the day-to-day meteorological phenomena in the western U.S. become Europe's weather, less than a week later. Thus, there are few present-day areas of technical endeavor in which international collaboration is more necessary – or in which the large-and-obvious benefits may be more naturally, uniformly and immediately shared among all mankind.

CONCLUSIONS

We have presented a top-level discussion of a proposed system for generating high-fidelity, high geographical resolution weather forecasts – necessarily, of global extent – over future time-durations of 2-3 weeks. We believe that it's feasible to realize such a system in a risk-minimized manner on a half-decade time-scale, at a total program cost of perhaps a third-billion dollars. Continuing operations thereafter, for a full-scale system, would involve comparable annual expenditures. Order-of-magnitude smaller continuing expenditures – of the order of \$30 M/year – could 'buy' the real-time geophysical data to support global weather forecasts of 80-90% of the future-time duration of the full-scale system. The benefits associated with such expenditures, as we've pointed out previously, would be several orders-of-magnitude larger: as large as $100 B/year to the U.S. alone, and 3-4 times that level, world-wide.

Independently, the data that can be gathered uniquely by a distributed, dense collection of airborne and Earth-surface sensors will quickly put global change research on a far more sound scientific basis, as well as resolve some of the outstanding issues currently in play between the several schools-of-thought as to the anthropogenic component of global change.

Thus, there appears to be excellent scientific and societal justifications for near-term commencement of the proof-of-principle phase of the proposed program, naturally with maximal international cooperation.

10. INFORMATION SECURITY

INTERNATIONAL INFORMATION SECURITY CHALLENGES FOR MANKIND IN THE XXI CENTURY

ANDREI KROUTSKIKH
Head Directorate of the Department on Security and Disarmament of the Ministry of Foreign Affairs of Russia, Moscow, Russia.

In 1998, after a discussion in the United Nations on international information security, Russian Foreign Minister, Mr. I. Ivanov, noted in his letter to the UN Secretary General that mankind is witnessing the formation of a truly global information society in which information is acquiring a new, revolutionary quality, significance, and influence both nationally and universally.

A single technological line is formed by computers, telephone, radio/TV, and space-based communication systems. Society today greatly depends on its smooth functioning. In fact, it is experiencing a land-slide with the worldwide introduction of hi-tech telecommunication, and cybernetic means. Local and global networks have created a new quality of transborder information exchange. All this directly affects politics, economy, culture, international relations, national, regional, and international security.

A single worldwide information space is emerging as a global development factor and, as such, is determining the main trends of social progress. Information is becoming the major strategic resource of states.

The global information-technological revolution we are living through today has brought obvious boons and is promising more. At the same time it has created fundamentally new threats. The scientific and technological achievements can be abused to reach aims that have nothing in common with international peace, stability and security, rejection of the use of force, non-interference in domestic affairs of other states, and respect for human rights. It takes no wisdom to predict the information-technological threats evolving into serious challenges to twenty-first century international security.

New types of extremely destructive weapons can be developed—the information kind. The pace of their development and growing interdependency of national and international information infrastructures leaves, as it seems, no chance to any state to be immune to possible hostile transborder actions, either with the use of information technologies or against critical information resources.

So far, the term "information weapon" has not yet received an exact definition. It was first used by the American military in 1991 after the Gulf War. It is hard to define because of the bulk of the information technologies are of dual or non-military application. But, whatever the terms are, the huge potential of the information-computer technologies can be used to ensure military-political domination, the use of force and blackmail and cannot exclude a possibility that in the foreseeable future, punitive expeditions against international outcasts will use information weapons rather than cruise missiles and bombs. This will turn conflicts into information warfare.

Sophisticated scenarios of information warfare took our breath away in horror films of the past: computer viruses secretly introduced into electronic systems of a state and its economic administration, and military command coming to life and paralyzing them: acting at long distances, scoundrels used electronic devices to remove money from the adversary's bank accounts and to cripple industry, communication, power production, transport, municipal services, ecological monitoring, atomic power stations, airports, strategic forces' command points; powerful generators of electromagnetic impulses destroyed software and deleted vitally important data bases of protected computer systems. All this created panic among civilian population and deprived the state leaders of correct information.

Little by little alike TV horrors reached real life. Back in the mid-eighties, the United States embargoed Iranian bank deposits during the American-Iranian hostage crisis by using a computer program. Military experts called the Desert Storm punitive operation that efficiently employed radioelectric means of warfare the first "information Hiroshima."

Deliberate information impact on the enemy has a history that is as old as the world itself. Today, thanks to the latest technologies, it is developing from scattered acts of information sabotage and disinformation into a fully-fledged method of international policy applied on a mass scale.

Information weapons are used: to achieve information superiority, damage information, information resources, processes, and systems; to improve traditional and create new types of armaments and military technology aimed at a further direct armed impact on the enemy; to put out of commission civilian objects and life-support systems; to disorganize state administration, to cause economic chaos and sabotage; to damage national financial systems based on information-computer networks; psychological brainwashing of the population to achieve social disorganization. Any of the above technologies when used by one state against another can be called information warfare or war.

The greatest damage is done when information weapons are applied against military and civilian objects that should function uninterruptedly and online (early warning systems; anti-air and anti-missile defense systems; power production complexes, especially atomic power stations; industry). The results may be catastrophic and comparable with those produced by mass destruction weapons.

Information weapons are qualitatively universal, highly efficient, and easily accessible. They offer a wide choice of time and place of use; they do not require large armies which makes information warfare relatively cheap. Their application can easily pass for routine action, at the same time it is hard to pin in on any particular state. Information weapons are indifferent to long distances and state frontiers.

The weapons can be used without a declaration of war; they do not need large-scale and obvious preparations. Sometimes a victim remains unaware that information impact is applied to it. It is much harder to respond to information aggression because there are no systems and methods to assess the threat of attack and warn about it.

The information weapon has produced a revolution in warfare. Many traditional military concepts such as "defense" or "assault" have been transformed. In local clashes, there is no longer need to seize territories or take POWs; it has become possible to reduce loss of life of one's own army and to entrust initiative in combat assignments to pilotless means.

Sham humanitarian nature of information weapons should be also emphasized. Many methods of information warfare such as crippling telecommunication systems, virus programs, jamming, blocking communications systems, etc., while dealing a heavy blow at economy do not cause directly bloodshed, loss of life and visible destruction common in conventional warfare. As a result, no one is deprived of food, dwelling, etc., needed to maintain life. There will probably be no refugee problem. This may lower the moral threshold of political decision-making. All talk about the humanitarian nature of the information-cybernetic means and methods of military-political impact may produce dangerous light-heartedness and tolerance where their use is concerned. There may be a tendency to excuse their use as unilateral sanctions on the ground that no blood was spilt.

Man-in-the-street may approve such means and methods because they do not require building-up of the armed forces; they even lead to their shrinking. Development of military-information potential camouflages itself as part of technological progress. Budget allocations for military purposes can easily be passed for a realized as spendings for large-scale peaceful programs. The miracle weapon looks tempting. Indeed, according to information from a U.S. financial control department, about 120 countries are engaged in or have completed elaboration of possible information-computer impact on a potential enemy's information resources.

Further development of the information weapons and progress in the use of civilian information-cybernetic networks and means for military purposes may let out a "technological genie" of a new generation to supplant the nuclear one.

There are practically no international laws to regulate the use of information weapons, to limit them as this is done under treaties with other weapon types and military activities. This cannot but aggravate the situation.

The emergence and proliferation of this weapon and militarization of peaceful information technologies are a powerful destabilizing factor in international relations. The present military strategic balance, the local and global balance of forces, and greater risks of an attack or blackmail are the payment for the new technological experiment. The entire system of international agreements on maintaining strategic stability, curbing arms race, at the regional level as well, will be put to a serious test.

It never rains but pours: information-cybernetic technologies can be used by criminals and terrorists. On the one hand, the technologies are easy to use, and access to the means of communication and data transmission means is cheap, global information networks are cosmopolitan. On the other, the world-wide information resources and infrastructure are vulnerable.

Individuals and groups acting towards unsanctioned penetration into information-cybernetic systems irrespective of their affiliation, breaking protection systems, stealing or destroying information for mercenary reasons or out of hooliganism are criminals. Computer thieves, or hackers, are also criminals. There are hundreds of thousands of registered computer-associated crimes all over the world and they double every year. According to Pentagon figures, in 1995 alone, hackers penetrated the U.S. Ministry of Defense computers through the Internet over 165 thousand times. Criminologists believe that there is a new type of organized crime in the world specialising in overcoming computer protection of military departments or objects, credit and finance organizations and in stealing secrets and money.

Information terrorism differs from information warfare and crime not so much by its method as by its aims typical of terrorism, and the tactics employed. Relying on the same technological foundation, the three types of menaces make the problem of information security as topical as other global problems: nonproliferation and liquidation of nuclear missile, and chemical weapons, anti-terrorist and anti-drug struggle, etc. They are more or less the same in scope and are part and parcel of international relations. One or several countries united in a bloc cannot deal with them. It is equally impossible to cope with these global problems on the "every man for himself" principle. The world information space cannot be divided while the information systems are interconnected.

The world community should not and cannot afford to permit itself to be involved in a new—information technology this time—area of confrontation, to face a possible escalation of the arms race in this field and an endless chase of countermeasures for offensive inventions as it was typical for the nuclear age.

There is an objective need to legally regulate the world-wide processes in civilian and military informatization and to create a concerted international platform of information security. The UN members have to formulate their own assessments of such threats and provide related basic definitions, including those of "unsanctioned interference" and "illegal use of information systems and resources."

There is no doubt—the complexity and the extent of security problems in the realm of global information community are enormous. Legal, economic, ethic, political, military, technological and other aspects are involved, making the issue sound at both national and international levels. But at the same time, this complexity makes it even more convincing—it is indeed necessary to address the global problem at the global level. Therefore, for general consideration or guidance, at least at this first initial stage, it needs an appropriate universal approach facilitated by the forum of the UN General Assembly. It is also essential that such a consideration be carried out on a maximum wide, joint basis to identify all national existing approaches, positions, views or concerns and than accumulating them in a sort of international concept.

When those trends and approaches are identified, they can be initially summarized in a "general principles" document to form the basis for a future internationally adopted regime or code of conduct for States which would be aimed at confining the emerging threats and to strengthen overall international information security. Practically, those principles can be incorporated in a multi-lateral declaration or further enhanced in an appropriate international or an "umbrella" type legal instrument.

One can try to visualize a possible subject of international negotiations on the information-related aspects even if it is impossible to predict how far the world community is ready to go to assume specific obligations to limit current threats to international security in international legal anti-military spheres.

A ban on or voluntary rejection of elaboration, production and application of especially dangerous information-cybernetic technologies and methods of their use can be high on the list of priorities. The same can be said about possible exclusion of the information means and methods of destructive impact on man and human organism. It is equally important to introduce a non-proliferation regime of military informational technologies, means and programs. There is a need to coordinate, on a world scale, anti-terrorist and anti-crime efforts; it is advisable to create international norms and legal acts and, probably, general technical measures to protect national information resources, regulate transborder information exchange, minimize the negative information impact on mass consciousness, ban the use of information-cybernetic technologies for aggressive aims against certain objects.

This work may draw on the rich experience international diplomacy has accumulated when elaborating maritime law, the legal regime of the use of space, wide-scale international conventions and agreements.

These efforts can go side by side with coordinating national legislations on information activity through parliaments. To extend the efforts to maintain information security it would be wise to set up permanent international monitoring of information threats; centers of information technological aid to the countries that fall victim to information aggression or any other illegal use of information means; international groups of experts to promptly react to all cases of information terrorist threats.

As for which international bodies could start practical work in this direction, it could be a UN group of governmental experts reporting to the Geneva Disarmament Conference or the UN Secretary General and the General Assembly. These efforts can also be supported by other UN and multilateral entities dealing with related but more specific issues like the International Telecommunication Union, UNIDIR or International Institute for System Analysis.

To summarize:

- It is obvious that the matters of information security cause concern for the whole international community;
- The issue is quite versatile and at the same time involves interrelated problems;
- So far there are no necessary political or legal means to regulate the problems manifested;
- The UN General Assembly can give general guidance for further consideration of the issue;

- To deal with the matter in an applied, systematic way, a proper international forum is to be identified;
- There is a need for a thorough expert review of all the issues involved;
- This review should form a basis for a future international legal regime on information security.

A RUSSIAN CONCEPT FOR ENSURING INFORMATION SECURITY

D.S. CHERESHKIN
(in collaboration with Dr. Alexander A. Kononov, senior research fellow in the Systems Analysis Institute.) Institute of Systems Analysis, 9, 60-Letiya Oktyabria Street, Moscow, 117312, Russia, E-mail cheresh@isa.ru Tel./Fax 135-50-43

The ever-growing role of informational environments is characteristic of the current stage of social development and is turning into an important factor of public life. Today, the level of informational development determines the success and pace of social, political and economic reforms in Russia. The reasons for that are as follows:

- Given the constitutional rights of citizens to free economic, informational and intellectual activities, the growing need of a socially-active fragment of society for a wider information exchange both within national boundaries and with the outside world;
- Today, the efficiency of public and state performance is determined by the level of development of telecoms infrastructure, integration of the nation with the global informational space, informatisation of virtually every facet of public life, state activity and administration;
- Informatisation and telecoms are the most dynamic and profitable industry determining the high-tech intensity and competitiveness of industrial products;
- Given an ever-greater significance of global information space as a political factor, ensuring information security becomes an important task of national security in Russia;
- Individual, group and mass consciousness are increasingly dependent upon the performance of mass media and mass communications.

These new political circumstances can be treated as a basis for defining Russia's national interests in the information area. Russia's information security Doctrine deals with four provisions viewed as the key national interests of Russia in the information area:

1. Safeguarding constitutional rights and freedom of the individual and citizen with respect to the access to and use of information, citizen's access to the

public state information resources, preservation and consolidation of the moral values of society, the cultural and scientific potential of this country.

2. Informational support of the national policy of Russia as regards provision of true information to the national and international public on the official policy of Russia towards socially significant events.

3. Development of up-to-date telecom technologies and hardware capable of ensuring security and efficient use of national information resources.

4. Protection of information resources from unauthorized access, ensuring safety of the existing and future telecommunications systems in Russia.

It would be safe to say that the extent of the security of Russia's national interests in the information area from internal and external threats is an indicator of the informational security of this country. The Doctrine explored the current state of information security of Russia, which was found to be inadequate for the needs of both state and society:

1. More intensive contradictions between the public needs for an ever wider and easier information exchange and the necessity for the state to continue with the imposition of restrictions on its dissemination. This situation is determined by the current political and socio-economic tension in national development.

2. The inadequate and inconsistent legal framework for public relations in the information area which results in serious negative implications. This is true, in particular, relative to fixing the limits to mass media freedom and the impact thereof on the balance of personal, public and state interests.

3. The failure to meet citizens' right of access to socially significant information, in particular to that of the performance on different levels of government, undermines the socio-political stability of society.

4. There is practically no adequate legal, institutional and technical support of constitutional rights of citizens to personal immunity, personal and family secrets and secrecy of correspondence.

5. The lack of a clear government information policy results in a deformed structure of international information exchange and internal information market.

6. The inadequate government support of Russian information agencies as regards promotion of their products in foreign information markets.

7. Aggravated situation as regards the safety of state secrets.

8. The intensive introduction of foreign telecoms technologies and hardware, the development of open information systems and integration with them of different types of national information systems essentially increase the threat of criminal, terrorist, military and political influence of the information environment of Russia with a view to accomplishing unlawful ends.

The above specifics place a demand on state measures adequate to defend against potential threats of impacting the information environment of Russia. The Doctrine outlines general approaches to ensuring information security of Russia, classified into legal, institutional, technical and economic areas. It is worth noting that the Doctrine sets forth the key principles of Russia's national policy with respect to information security. The most significant of these are:

- Adherence to Russia's Constitution and laws, generally acceptable principles and rules of international law in exercising activities towards information security of Russia.
- Transparent execution of functions by federal government, regional governments and public associations. This involves public awareness of their activities with regard to the restrictions imposed by legislation of the Russian Federation.
- Legal equality of all parties to the process of informational interaction irrespective of their political, social and economic status based on the constitutional right of citizens to freely search, acquire, transmit, generate and disseminate information in any legitimate manner.
- Prioritize the development of sophisticated domestic information and telecoms technologies, manufacture of hard- and software capable of ensuring improvements in national telecommunications networks, their connection to global information networks with regard to the vital interests of the Russian Federation.

The priority policy measures in the area of information security of the Russian Federation are as follows:

- Development and introduction of practicable laws and rules regulating relations in the informational area, drafting a concept of legal support of informational security of the Russian Federation;
- Development and introduction of policies in view of more efficient guidance of public mass media, pursuant to government information policy;
- Drawing up and implementation of federal programs involving:
- Higher legal awareness and computer literacy of the population;
- Building public information resource archives at all levels of government;
- Improvements in the infrastructure of an integral informational space in Russia;
- Comprehensive counteraction of the threats of information warfare;
- Development of safe information technologies for the systems used in implementing vital functions of society and state;
- Fighting computer crime, development of a special-purpose information and telecoms system to be used by federal and regional authorities;

- Ensuring technological autonomy of this country as regards development and maintenance of defense-oriented information and telecoms systems;
- Harmonization of national standards relative to informatisation and information security of computer-based management systems, general and special purpose information and telecoms systems.

The Doctrine also lays down the key organizational functions and components of the information security system of the Russian Federation. The Doctrine is the first of this kind of documents, reviewed and approved by the President of the Russian Federation. It is also the first document, I believe, dealing with the problems of the informational development of Russia to be endorsed at this level. We believe that the flaws in the Doctrine, which include a number of statements which were too radical, will be adjusted in due course. This will turn the Doctrine into a powerful tool for promoting informational interaction between citizens both within the country and with the rest of the world.

In the second part I would like to consider my own opinion about some statements in the Doctrine.

PART 2

1. This is the first Russian document in which:

   - Russian interests in the informational area are formulated precisely;
   - The interests of citizens, society and state are considered as equally important in ensuring informational security.
   - Within the area of informational security, belong also the problems of social conscience and the impact of mass media upon it.

2. In my opinion, the contribution of state bureaucracy at the final stage of the Doctrine's preparation has made it too politicized, reflecting the reality of political fights in Russia. As a result, the Doctrine contains some statements aimed at the limitation of the freedom of mass media and prioritizing the interests of various military and intelligence structures.

3. However, after approving the Doctrine, the President made some remarks that soften some aspects of it. For example, during a press conference Mr. Putin mentioned the unjustified harshness of some of the Doctrine's statements, especially those concerning the freedom of the press. The coordination of implementation of the Doctrine has been entrusted to the Ministry of Posts and Communications rather than to one of the former KGB agencies. To some extent this ensures the softening of some of the Doctrine's statements.

4. The Ministry of Posts and Communications is now preparing the Federal Program of Realization of the priority policy measures for Information Security. In the Program

there is some provision for guaranteeing citizen's informational rights and for protecting the national informational infrastructure.

5. It should be noted that the Doctrine is the first document of this kind in the world. All the unavoidable errors will be corrected with the passage of time, accumulated experience and the interaction with other States in the solution of Informational Security problems for the whole world.

INFORMATION SECURITY THINKING: A COMPARISON OF U.S., RUSSIAN, AND CHINESE CONCEPTS

MR. TIMOTHY L. THOMAS

Foreign Military Studies Office, Fort Leavenworth, Kansas 66048, USA

INTRODUCTION

The advent of the information age forced many nations to reexamine their security procedures. As a result of the new and overwhelming dependence of critical infrastructure on information systems, information security is a priority concern to governments everywhere. However, not every nation has interpreted this concern in the same fashion, nor at the same pace.

Russia, for example, is ahead of everyone. It has produced an Information Security Doctrine that includes the security of the state, society and the individual in its evaluation of information security threats. This paradigm allows Russian security managers to consider information threats to citizens (their cultural, spiritual and psychological well-being) as well as to critical technologies and resources. Perhaps for this reason Russia's military doctrine lists external threats as information-psychological and information-technical issues. China has also been affected by the information age. It is concerned with influential technologies that might be used as a form of virtual deception or influence. The Internet is the focus of most of China's concern, most likely because it offers information that the Chinese government cannot control completely at this time. At the same time, China has an information technology security plan known as Plan 863, and has established a Ministry of Information. China has not produced an information security doctrine or related document of which the author is aware. The U.S., while considering psychological operations as a key sub-element of information operations, relegates less attention to the information threat to the minds of its citizens. Rather, U.S. legislation and rights/acts (human rights, freedom of the press, speech, etc.) provide cover for its citizens in this field. The U.S. focuses on critical infrastructure protection instead of technologies and resources as the Russians. The U.S. seldom uses the term "information security" when discussing cyber-based threats in official documents.

A brief look at how these three nations define information security supports this broad overview. The U.S. published two Presidential Decision Directives (PDDs) to counter information age threats to U.S. systems and its population, PDD-63 and PDD-68. PDD-63 focuses on critical infrastructure protection. The 22 May 1998 PDD-63 White

Paper describes the growing vulnerability to U.S. cyber-based systems, and establishes a series of steps to counter this vulnerability. These steps are to be in place by 2003, and include the analysis of foreign cyber/information warfare threats. The directive mentions the term information security only twice, and then only in regard to public outreach programs[1]. PDD-68 coordinates U.S. efforts to promulgate its policies and counteract bad press abroad. This directive created the International Public Information (IPI) group to coordinate the identification of hostile foreign propaganda and deception techniques that target the U.S., according to the group's charter[2]. The focus is hostile information programs that might not be truthful. The directive does not characterize its actions as information security related. To find a U.S. definition of information security, one must turn to the military. Joint Publication 3-13, *Joint Doctrine for Information Operations*, defines information security as "the protection and defense of information and information systems against unauthorized access or modification of information, whether in storage, processing, or transit, and against denial of service to authorized users. Information security includes those measures necessary to detect, document, and counter such threats. Information security is composed of computer security and communications security. Also called INFOSEC"[3].

Russia has a number of definitions for information security, perhaps because they have thought more about this subject than other nations due to their loss of ideology when the Soviet Union dissolved in 1991. Russia's September 2000 Information Security Doctrine defines information security as "the state of protection of its national interests in the information sphere defined by the totality of balanced interests of the individual, society, and the state." Just a few months earlier, in a May 2000 United Nations resolution, Russia defined information security somewhat differently as the "protection of the basic interests of the individual, society and the state in the information sphere, including the information and telecommunications infrastructure and information per se with respect to its characteristics, such as integrity, objectivity, availability, and confidentiality." The Russian Academy of Natural Sciences defined the term as "the protection of the information medium of the individual, society and the state from deliberate and accidental threats and effects"[4]. Information security, yet another source adds, is connected with information and its material carriers: the mind of a person and other carriers of information (books, disks, and other forms of "memory")[5]. Thus, differently than the U.S., Russia views both the mind and information systems as integral parts of its concept of information security.

Chinese academician Shen Changxiang, of the Chinese Academy of Engineering, defined information security in the People's Liberation Army's (PLA) newspaper of the General Political Department (the *PLA Daily*): "[information security] refers to the prevention of any leakage of information when it is generated, transmitted, used, and stored so that its usefulness, secrecy, integrity, and authenticity can be preserved; and so that the reliability and controllability of the information system can be ensured"[6]. State Council member Shen Weiguang notes that information warfare (IW) is brain warfare. According to official Chinese sources, the Internet has the capability to manipulate information, the truth, and the moral-psychological state of Chinese citizens. China realizes that the Internet cannot be controlled, and is thus a concern to the government. Thus China appears more like Russia than the U.S. in its understanding of information security, with its emphasis on the mental aspect of information security and its extended use of the term itself.

This paper briefly examines the information security policies of the three countries noted. There are both narrow and broad approaches to the subject, and situational context, culture, and history are the real forces that differentiate national approaches. An understanding of each country's concerns and their paradigm for interpreting information security can potentially alleviate misunderstanding in future international discussions among nations over this concept.

U.S. VIEWS ON INFORMATION SECURITY

With regard to information technology and critical infrastructure protection, PDD-63 demonstrated that the U.S. is very interested in establishing an "infrastructure to protect the infrastructure." The document builds on the recommendations from the October 1997 President's Commission on Critical Infrastructure Protection. PDD-63 opens by describing the growing potential vulnerability to America's cyber infrastructure, and then states the President's intent, the national goal, and the public-private partnership to reduce these vulnerabilities. Next general guidelines are issued, and then a structure and organization to meet the challenge are introduced (which included lead agencies for sector liaison, lead agencies for special functions, interagency coordination, and the appointment of a National Infrastructure Assurance Council). Tasks included:

- Vulnerability analysis
- Remedial plan
- Warning
- Response
- Reconstitution
- Education and awareness
- Research and development
- Intelligence
- International cooperation
- Legislative and budgetary requirements[7]

A National Coordinator for Security, Infrastructure Protection, and Counter-Terrorism is responsible for coordinating the implementation of the directive. This person will also chair the Critical Infrastructure Coordination Group (CICG). Further, the President authorized the FBI to create a National Infrastructure Protection Center (NIPC). This organization shall serve as a national critical infrastructure threat assessment, warning, vulnerability, and law enforcement investigation and response entity. Finally, the national coordinator will encourage the creation of a private sector information sharing and analysis center (ISAC). The center would be the mechanism for gathering, analyzing, appropriately sanitizing and disseminating private sector information to both industry and the NIPC, and could transmit NIPC information the other way[8]. However, throughout the PDD-63 White Paper, only twice (and not in the main section) is the term "information security" used.

U.S. civilian agencies seldom use the term "information security." For example, in a computer search of Congressional committees and bills produced by the 107th Congress, there are four hits for the term "information security": the Computer Security Enhancement Act of 2001; the FBI Reform Commission Act of 2001; the E-Government Act of 2001; and the bill "To revise, codify, and enact without substantive change certain general and permanent laws, related to public buildings, property, and works, as title 40, United States Code, 'Public...'". None of these address psychological security or manipulating the emotions or logic of individuals via information. Searches for the terms information-psychological security, and psychological security in the 50 bills of the 107th Congress resulted in zero hits.

Again, one has to turn to the military to find the use of the term "information security." The U.S. General Accounting Office report of March 2001 (GAO-01-341) is titled "Information Security: Challenges to Improving DOD's Incident Response Capabilities." The contents of this document demonstrate that, to the authors of this report anyway, information security is a computer and information system related problem. No reference is made to the individual, just to technical systems. The definition from the Joint Publication noted above emphasized the computer security and communications security aspects of information security.

PDD-68, mentioned above, discusses the requirement to identify hostile propaganda and deception targeted against the U.S.. However, PDD-68 apparently does not address information security. The actual document has still not been released. Even though the U.S. does not mention information security in terms of mental activity, other western nations do. Sweden, for example, has a psychological defense department that aims to uphold for every Swedish citizen a constitutional right to correct, clear and complete information on what is happening in society and in the world at large, especially in times of stress or crisis, or in times of particular importance[9].

Most Americans realize that the issue of the influence of information on the minds of Americans is covered by different terminology than "information security." Advertising agencies are forbidden to persuade citizens using certain technology such as the infamous "25th frame effect" demonstrated in the 1950s at one theater. The mind reportedly can process 24 frames of film a second, and a 25th frame becomes a subliminal message. The movie theater in question sold popcorn and drinks by inserting this extra frame. This was ruled unconstitutional and other restrictions on media advertising are written into U.S. law to protect human rights and privacy. Thus American citizens are protected from "mind attacks" by different segments of legal code using different terminology.

The U.S. has not had to consider the human aspect of cyber-based operations nor the loss of an ideology as some nations have faced. In the affected countries consideration was given to the "information-psychological security" factor as a result to a much greater degree than in the U.S., which did not lose its democratic ideology. Russia lost an ideology and is working hard to ensure that Russian culture, identity, and spirit do not also disappear along with it.

RUSSIAN VIEWS ON INFORMATION SECURITY

The approved Russian Federation's information security document represents the purposes, objectives, principles, and basic directions of Russia's information security thought. The document is divided into eleven sections. They are:

- National interests of the Russian Federation in the information sphere (observance of constitutional and individual liberties, information backing for Russian Federation policy, development of information technology and industry, and protecting information resources from unsanctioned access)
- Types of threats to Russia's information security (to constitutional rights in spiritual life, to information backing for state policy, to the development of the information industry, and to the security of information—the same as national interests!!)
- Sources of threats to Russia's information security (external and internal)
- The state of information security in the Russian Federation and objectives supporting it (tension between the need for the free exchange of information and the need to retain individual regulated restrictions on its dissemination)
- General methods of information security of the Russian Federation (legal, organizational-technical, economic)
- Features of information security (economics, domestic policy, foreign policy, science and technology, spiritual life, information and telecommunication systems, defense, law enforcement, and emergency situations)
- International cooperation in the field of information security (banning information weapons, supporting information exchanges, coordinating law enforcement activities, preventing unsanctioned access to confidential information)
- Provisions of state policy of information security (guidelines for federal institutions of state power and institutions, based on a balance of interests of the individual, society, and the state in the information sphere)
- Priority measures in implementing information security (mechanisms to implement the rule of law, an increase in the efficiency of state leadership, programs providing access to archives of information resources, a system of training, and harmonizing standards in the field of computerization and information security)
- Functions of the system of information security (summary of above points, 17 in all)
- Elements of the organizational basis of Russia's information security system (President, Federation Council of the Federal Assembly, the State Duma of

the Federal Assembly, the government of the Russian Federation, the Security Council, and other federal executive authorities, presidential commissions, judiciary institutions, public associations, and citizens).

The security doctrine's sections are further subdivided into three or four recognizable areas. First, there are some general principles associated with the section. Then the external and internal threats to that section's principles are listed, and this is followed by a list of measures to be taken to offset these threats. A look at the section on "subjects of the information security of the Russian Federation in the sphere of defense" follows to examine this methodology.

There are four general subject areas listed under this section. These are the information infrastructure of the central elements (the branches of the armed forces and scientific research institutions of the Ministry of Defense) of military command and control; the information resources of defense complex enterprises and research institutions; software and hardware of automated command and control systems; and information resources, communication systems, and the information infrastructure of other forces and military components and elements.

External threats to the Defense Ministry make up the next section. They include the intelligence activities of foreign states; information and technical pressure (electronic warfare, computer network penetration, etc.) by probable enemies; sabotage and subversive activities by the security services of foreign states using information and psychological pressure; and the activity of foreign political, economic, and military entities directed against the defense interests of the Russian Federation. Internal threats include the violation of established procedures for collecting, processing, storing, and transmitting information within MoD; premeditated actions and personal mistakes with special information and telecommunications systems, or with the latter's unreliable operation; information and propaganda activities that undermine armed forces prestige; unresolved questions about protecting intellectual property of enterprises; and unresolved questions regarding social protection of servicemen and their families.

The final section of the defense discussion outlines the main ways to improve the system of information security for the armed forces including the systematic detection of threats and their sources; structuring the goals and objectives of information security; certifying general and special software and information-protection facilities in automated military control and communications systems; the improvement of facilities and software designed to protect information; improvement in the "structure of functional arms" in the system of information security; training of specialists in the field of information security; and, most important in light of the Russian military doctrine's views on information security, the refinement of the modes and methods of strategic and operational concealment, reconnaissance, and electronic warfare, and the methods and means of active countermeasures against the information and propaganda and psychological operations of a probably enemy.

Interestingly, this discussion of the information security dimension of the defense sphere varies from the military doctrine of the Russian Federation. The latter states quite specifically that information-psychological and information-technical matters are the two external threats to Russia, and disruptive plans or technologies are the greatest internal threats. That these two important terms are not used in the information security doctrine is probably due to the fact that military people advised but did not write it. However, the two sections on the spiritual and cultural sphere, and the section on the scientific research sphere cover the military's information-psychological and information-technical concerns from the military doctrine. While not citing information-psychological activities directly, several sections imply this is a concern. For example, under constitutional rights, it notes that a threat is the unlawful use of special techniques influencing the individual, group, and social consciousness, as well as the disorganization of the system of cultural values. Under foreign policy concerns, an internal threat is the propaganda activities of political forces, public associations, the news media, and individuals who distort the strategy and tactics of the foreign policy of the Russian Federation. An important spiritual concern is the prevention of unlawful information and psychological influences on the mass consciousness of society and the uncontrolled commercialization of culture and sciences. Thus, information security in Russian is focused on both the mind and on technical systems. The impression is that internal threats are more real and consequential than external threats.

The information security doctrine uses the term "information-protection" quite often, and this term is the one item that the state could put on a par with the military's information-psychological and information-technical issues. Another area of obvious emphasis is the formation of a legal base for information security. The laws on "State Secrecy," "Information, Computerization, and Information Protection," "Participation in International Information exchange," and "Essentials of Legislation of the Russian Federation on the Archive Collection of the Russian Federation and Archives" are specifically mentioned. Legal issues are one of three methods of information security mentioned in the doctrine, with organizational-technical and economic the other two. The threat of information weapons to Russia's information infrastructure and the threat of foreign governments using information warfare techniques against Russia were also mentioned. Special concern focused on the inadequate development of telecommunication systems, the integrity of information resources, space-based reconnaissance, and electronic-warfare facilities.

On an international level, the doctrine is designed to do several things. These include:

- Banning the development, proliferation, and use of information weapons
- Supporting the security of international information exchanges, including the integrity of information during its transmission by national telecommunication channels
- Coordinating the activities of these law enforcement agencies of countries that are part of the world community to prevent computer crime
- Preventing unsanctioned access to confidential information in international banking telecommunication networks and world trade information-support

systems as well as information of international law-enforcement organizations fighting transnational organized crime, international terrorism, illicit trade in arms and fissionable materials, and in the distribution of narcotics and mind-altering substances, and the trade of human beings.

Near its end the doctrine notes that the "implementation of the guarantees of the constitutional rights and liberties of man and citizen concerning activity in the information sphere is the most important objective of the state in the field of information security." While this sounds pleasant enough, Russian citizens are concerned over the interpretation of this notion. Some citizens fear the fact the government was concerned about two issues: conveying RELIABLE information to the Russian and international community, and NOT ALLOWING propaganda that promoted the incitement of social, racial, national, or religious hatred and animosity. Measured against the government's handling of the Kursk incident and the war in Chechnya, the idea of reliable government information is questionable at best. The government considers the "information war" conducted by the press for public opinion as a very important aspect of keeping the emotions and loyalties of its people in check during crises. All governments do this to a certain degree, but the Russian government appears to have swung the pendulum far past reason in these two cases. The doctrine states near its conclusion that a basic function of the system of information security of the Russian Federation is "the determination and maintenance of a balance between the needs of citizens, society, and the state for the free exchange of information, and the necessary restrictions on the dissemination of information."

The last paragraph of the document states that:

"The implementation of the priority measures in support of the information security of the Russian Federation enumerated in this doctrine presupposes the drafting of the corresponding federal program. Certain provisions of this doctrine may be made more specific with reference to particular spheres of the activity of society and the state in the appropriate documents approved by the President of the Russian Federation."

Over the next few years more attention should be directed to these more recent documents. The first meeting to start the process was the 23 October conference on information security. Anatoliy Streltsov, deputy head of the Information Security Department (which has six members) of the staff of the Russian Security Council, said that the doctrine might promote a dialogue between the authorities and the press. The conference hoped to create a data bank for shaping state policy in the sphere of the mass media and the formation of the most effective basis for cooperation between the press services of ministries and agencies, on the one hand, and the mass media on the other[10].

CHINESE VIEWS ON INFORMATION SECURITY

The issue of information security appeared to become an important subject in China in early 2000. A series of articles on the subject appeared in the press throughout the year. Even though not directly addressed in the definition above of information security, the content of the articles showed that Chinese analysts consider the information security aspect of the mind more so than does the U.S. but less so than the Russians. For example, some Chinese analysts feel that the U.S. wants to use the Internet to shape the world's values in accord with those of the U.S. in order to maintain political, economic, military, and information benefits[11].

It was also of interest that many Chinese information security articles are published by military journals. The military cites problems due to a neglect of safeguards, backward technology, and a lack of comprehensive rules and regulations[12]. The first Chinese all-army forum held in late October 2000 set forth management rules and accelerated building an effective protection system for information security technology. There are now more than 1,000 computer networks of various kinds for the PLA to manage[13].

The military identified the "three shifts" required to ensure the security of network information. These shifts are from maintaining secrets to maintaining secrets and building firewalls, from merely emphasizing conventional security work to emphasizing the security of information, and from stressing administrative management to stressing both administrative management and prevention technology. Educators are trying to help officers renew their concept of security and establish a "firewall" in everyone's head. The armed forces are trying to train a group of "network guards" with high-tech expertise. This process has been accelerated through the establishment of an all-army center for information security examination, appraisal, and recognition[14].

Chinese information theorists believe that the modern fight is centered on the "right of controlling information," making information security one of the "commanding heights" for winning a war. "The right of controlling information" is synonymous with having an advantage in warfare. This makes the information confrontation an important operational component of modern war with information security and confidentiality the main battleground. Obtaining and counter-obtaining, control and anti-control, and destroying and counter-destroying will become an important operational mode in future warfare, as will developing the strategies and tactics of information security work.

PLA deputies to the National People's Congress called for legislation on information security, and pointed out a few hidden dangers—technological traps and weak preventative measures. Deputies called for a national defense information security law, and concentration of effort to speed up the development of information security technology in China[15]. On 23 October 2000, the Chinese State council recommended that the 18th meeting of the 9th National People's Congress consider a final draft of related regulations on maintaining network and information security. It appears that the issue is receiving top-level consideration, and that a policy may be forthcoming in the coming year[16].

Shen Changxiang wrote the most authoritative article available to this author on the subject of information security. He noted that the main threats to China's cyber-information system included (1) its being tampered with, altered, stolen or sabotaged by viruses, or by hackers taking advantage of the Internet to intrude into the system to collect and transmit sensitive information (2) using computers' CPUs, or pre-installed information collection and sabotage programs in the computer's operating system, data managing system, and application system (3) and monitoring and intercepting information, using the Computer's electromagnetic leaks and its peripheral equipment. Whoever excels in preserving information security controls information and wins the information war in future conflicts. In IW, the difference between a developed country and an undeveloped country will be far smaller than if it were a conventional war. This requires that China draw up an information security strategy and information defense system[17].

In the civilian sector, information security is as important a topic as it is in the military sector. In June of 2001 the weekly general affairs journal of China's official news agency Xinhua, *Liaowang*, stated that information security issues are now of the utmost importance. For example, the Internet is linked to vital national infrastructures as well as to the education, health, and minds of citizens. This makes the control of the Internet's information security aspect twice as important as it was in the past. In addition, it is estimated that over 33% of the U.S. economic growth is from the information industry. The U.S. has control over the Internet and online resources. *Liaowang* estimated that America has 70% of the world's web sites, 94 of the top 100 most visited web sites, accounts for nearly one half of the world's total number of cybercitizens, and conducts nearly 90% of the world's Internet business. It also controls 80% of the world's computer systems and software markets. The Internet has also produced new information security issues such as simply controlling the amount of information available[18].

The most visible expression of a Chinese information security "plan" was offered in the June 2000 issue of the Internet website Guojia Gao Jishu Yanjiu Fazhan Jihua (HTRDP) 863. This website is sponsored by the China Office for the High Technology Research and Development Plan, an institution under the State commission for Science and Technology. The web site offered a guide to help China focus on making technical breakthroughs regarding typical information security issues and tasks in politics, the economy, and culture. The guide is based on decisions from the "Emergency Information Security Technology Plan" formulated by a specialist group researching information security strategy for National Plan 863. The goals of the plan are: effectively alleviate information security issues for party and government network systems and financial network systems; effectively control the wanton propagation of illegal and deleterious information in mutually connected networks; and raise China's information security technology to a new level[19].

There are several information security issues affecting China. First, China must correct flaws in its information security management system. There is no clear division of labor on information security management among various functional departments, and it is difficult to adapt the present management model to a networked era. Second, there is an inadequate legal system in place to handle information security issues. Critical laws designed to protect the information infrastructure have not been formulated, and the construction of a legal system lags far behind what is required. Third, more research is required in the area of core information security technologies and products. China is limited by its microelectronics industry, which is of a lower quality and standard. Many cryptographic, digital signature, identity authentication, firewall and monitoring system requirements are not fulfilled as a result. Finally, there are many hidden information security dangers associated with China's financial system. The use of lower-grade PCs and the lack of information security methods in supervisory management still make people nervous[20].

Qiushi magazine warned that China must develop an information security system that is strong and adaptable, independent of foreign control, and must combat superstition, rumors, slander, pornography and hackers that corrupt people's minds and threaten national security. Of particular concern is the Internet as a subversive tool. The U.S.'s overreliance on technology causes it to forget that the last line of defense in information security remains people. China must also enhance the use of positive propaganda on the Net, and strive to run a number of web sites that attract people. China must also build up its information security infrastructure including network control centers, appraisal and certification centers, and virus prevention centers. This will help accelerate the promotion of China's information security sector[21].

State Council member Shen Weiguang, reportedly China's first information warfare analyst, wrote that China needed a national information security commission to strengthen centralized leadership over information warfare research. Such a commission should assume total responsibility for China's information security and provide an information security and information warfare studies institute at the highest level. Soldiers no longer have a patent on war. There now are information invasion, information firepower surveillance, information deterrence, and information pollution problems to contend with in information space. According to Shen[22], China must establish information security standards and rules of conduct, and issue information security assessments and inspection standards.

CONCLUSIONS

The purpose of this article was to examine the U.S., Russian, and Chinese understanding of information security. The examination shows that in both definitions and in discussion, Russia and China differ markedly in their idea of information security than does the U.S.. Russia and China are worried about ways to influence the mind and emotions of its citizens. The U.S., on the other hand, treats the mind as a separate issue governed by different terms: human rights, perception management, and so on.

The U.S. does not use the same lexicon in its documents (seldom if ever using "information security" in conjunction with the logic of its citizens), and has been affected by a different situational context.

Governed by such different paradigms, it is easy to fathom how discussion and understanding of the term "information security" in the U.N. could prove to be very difficult. However, with just minor compromises, common ground can be found. It is important to do so now. A decade ago, cloning and stem-cell research were futuristic problems for scientists but today they are problems scientists must learn to control. The speed of development of information technology implies that a futuristic concept such as the electronic manipulation of the mind may be closer than we want to believe. If that happens, the idea of information security will take on a whole new dimension. It is wise to prepare now for the future and develop some universal procedures for scientists to consider.

REFERENCES

1. White Paper, The Clinton Administration's Policy on Critical Infrastructure Protection: Presidential Decision Directive 63, 22 May 1998, downloaded from the Internet.
2. "International Public Information Presidential Decision Directive PDD 68, 30 April 1999, downloaded from the web site of the Federation of American Scientists on 3 July 2001.
3. Joint Publication 3-13, *Joint Doctrine for Information Operations*, 9 October 1998, p. GL-7.
4. Mubin Abdurakhmanov and Dmitry Barishpolets, "From the Dictionary 'Geopolitics and National Security," *Military News Bulletin*, Vol. Vii, No 10, October 1998, p. 13.
5. V.I. Parfenov, "Protecting Information: Terms and Definitions," *Questions of Protecting Information*, No. 3-4, 1997, p. 1 of an insert.
6. Shen Changxiang, "Information Security—an Important Contemporary Issue," *Jiefangjun Bao*, 4 April 2001, p 11 as translated and downloaded from the FBIS web site on 4 April 2001.
7. White Paper, 22 May 1998.
8. Ibid.
9. See http://www.psycdef.se/english
10. ITAR-TASS, "Conference on Information Security, Mass Media held in Russia," 23 October 2000.
11. PLA Daily News, 9 November, 2000.
12. "Security Means Ensuring 'Winning the War,'" *Jiefangjun Bao*, 27 October 2000 p. 1 as translated and downloaded from the FBIS web site on 27 October 2000.
13. Li Xuanqing and Liu Mingxue, "PLA's Security Work Emphasis Shifted to Building of Information Security," *Jiefangjun Bao*, 27 October 2000 as translated and downloaded from the FBIS web site on 27 October 2000.
14. Ibid.
15. No author, title, *Jiefnagjun Bao*, 12 March 2000, p. 4 as translated and downloaded from the FBIS home page on 14 March 2000.

16. PLA Daily News, 9 November 2000.
17. Ibid., Shen.
18. Yang Guangliang, "Paying Attention to Information Security," *Liaowang*, No 23, pp. 37, 38, 4 June 2001 as translated and downloaded from the FBIS web site on 12 June 2001.
19. "PRC High Technollogy Research Development Plan, www.863.org.cn, 6 June 2000 as translated and downloaded from the FBIS web page on 25 October 2000.
20. Ibid.
21. He Dejin, "Raise Network Security Awareness and Build Information Protection Systems," *Quishi*, 1 November.
22. Shen Weiguang, *The Third World War—Total Information War*, 2000 as translated and downloaded from the FBIS web page.

DISCLAIMER

The views expressed in this report are those of the author and do not necessarily represent the official policy or position of the Department of the Army, Department of Defense, or the U.S. government.

The Foreign Military Studies Office (FMSO) assesses regional military and security issues through open-source media and direct engagement with foreign military and security specialists to advise army leadership on issues of policy and planning critical to the U.S. Army and the wider military community.

Please forward comments referencing this study to:

FMSO
ATZL-CTL MR. THOMAS
101 MEADE AVENUE
FT LEAVENWORTH KANSAS 66027-2322

COM: (913) 684-5957
DSN: 552-5957
FAX: (913) 684-4701

E-MAIL: THOMAST@LEAVENWORTH.ARMY.MIL

11. POLLUTION IN THE CASPIAN SEA

THE CASPIAN SEA OBSERVATION AND FORECASTING SYSTEM PROJECT PLAN

ILKAY SALIHOGLU
METU Institute of Marine Sciences, P.O. Box 28, Erdemli, 33731, İçel, Turkey

INTRODUCTION

The fragile ecosystem of the Caspian Sea is affected by natural and anthropogenic impacts. The rising level of the sea and associated secondary contamination resulting from flooding is affecting the environment and the economies of the surrounding countries. The anthropogenic activities of the coastal states in relation to the prospecting, extraction and transportation of oil contribute to contamination with an increasing trend in the near future. The health and well-being of the Caspian Sea, with its important natural resources, has significant impact upon the bordering states (Fig. 1) of Azerbaijan, Iran, Kazakhstan, Russia and Turkmenistan. The Caspian Sea is unique because of its totally enclosed geometry, as it endures external influences without being relieved through some communication with the ocean, and therefore in many ways, its behaviour is similar to a lake. Furthermore, the Caspian Sea is a relatively undisturbed habitat and an enduring natural reserve for its indigenous and other species. With recent changes, its ecosystem could face irreversible damage. The strategy for its protection and preservation brings into focus the need for the adequate prediction of further environmental changes in the system. In this way it appears that there is a great deal to be learned from the Caspian Sea. On the other hand, the Caspian Sea deserves world wide attention as one of humanity's last reserves, and in its present state of a threatened sanctuary, would greatly benefit from experiences obtained elsewhere.

The project reviewed the objectives and the accomplishments of all Caspian Sea programmes, whether completed or ongoing, and compiled the information on existing gaps of knowledge in the region. Utilizing this information, the Project ended with a Caspian Sea Science and Implementation Plan. The Project Plan (NATO/CCMS Pilot Project "Review Of Environmental Projects Of The Caspian Sea For The Planning Of Future Activities" Report No: 239, 1999) basically covers the recommendations for future work aimed at resolving the missing elements in environmental research in the region. The plan identifies the specific anthropogenic and natural causes of environmental problems in the region, and the gaps of knowledge to quantify key variables and processes to be studied. Integrated assessments and scientific investigations of the environmental changes in the Caspian Sea region are identified as the starting point

360

for tools leading to successful predictions and for fruitful scientific collaboration and management.

Fig. 1. Bordering States of the Caspian.

DESCRIPTION OF THE PROJECT PLAN

The Caspian Sea is unique because of its totally enclosed geometry, as it endures external influences without being relieved through some communication with the ocean, and therefore in many ways, its behaviour is similar to a lake. Furthermore, the Caspian Sea is a relatively undisturbed habitat and an enduring natural reserve for its indigenous and other species. With recent changes, its ecosystem could face irreversible damage. The strategy for its protection and preservation brings into focus the need for the adequate prediction of further environmental changes in the system.

As a result of their smaller inertia, semi-enclosed seas, and especially enclosed inland water bodies such as the Caspian Sea, are more sensitive to Global Change, compared to the global ocean. The factors affecting the Caspian are:

- Natural variability and climatic changes,
- Ecological deterioration in regions of great economical value *i.e.* the Volga river delta and the adjoining North Caspian Sea,
- The fate of caviar or sturgeon's eggs,
- Anthropogenic forcing,
- The sea-level changes of greater concern are the exposure of toxic waste burial sites by rising groundwater levels,
- The comb jelly Mnemiopis that has invaded the Caspian and is destroying its fisheries,
- (One of the key issues) Caspian Sea oil: Development of transnational export routes increased the risk of oil pollution.

Because of its specific geographical characteristics, anthropogenic changes created more acute biological, ecological, economic, and social problems than other seas. In this sense, it has a very good natural laboratory for studying human-induced stress on an ecosystem, and may serve as a model of a temporally evolving ecosystem for the world's oceans.

Scientific Rationale of the Plan
The scientific rationale of the Plan is that the existing environmental/oceanographic activities in the Caspian have limited duration and scope. New monitoring programs that use real-time or near real-time physical, chemical, and biological observations and modeling studies are needed to assess and monitor the current state of the basinwide contamination and resulting ecosystem impacts.

Overall Goal of the Project Plan
The overall goal of Project Plan is to provide operationally useful information on changes in the state of marine resources and ecosystems, which will:

- Improve our knowledge of the physical and biogeochemical systems of the Caspian;
- Provide a basis for the assessment of the state and trends in the marine environment regarding the effects of anthropogenic activities;
- Identify causes and solutions of pollution problems; and
- Assist decision-making activities of regulatory and management agencies while stimulating creativity and excellence in research.

Scope of the Project Plan
The scope of the Project Plan is to design and implement a system of observations, data assimilation, and modeling studies to monitor and forecast the state of the Caspian's coastal ecosystem to provide information needed to manage the environment.
The Project Plan specifically includes observations and predictions of the:

- Ecological variables at different trophic levels that underpin exploitable marine resources;
- Sustainability of critical marine habitats;
- Regime shifts and changes in recruitment to fish populations;
- Changes in marine diversity of the coastal zone;
- Impacts from anthropogenic stress on the health of the marine ecosystem;
- Capacity of the ecosystem to transform and store particulate and dissolved organic matter;
- Effects of changes in external forces on the structure and functioning of the ecosystems.

The Project Plan consists of independent subprograms that complement each other. The sub-programs are:

- Pollution monitoring,
- Interdisciplinary modeling,
- Fishery-stock assessment,
- Oil-spill monitoring and control, and
- Database management.

Specific Focus of Project Plan

The Project Plan specifically focuses on a Database-management program where the main objectives of the database-management program are to provide better services to the users, increase the quality of data products with better use of existing data, provide more user-oriented products, and decrease production costs by sharing the work load.

Another main objective of the program is to observe, monitor, and predict ecological variables that relate to living marine resources and control water quality and ecosystem-related health issues. Accomplishing these tasks requires a coherent program with observational, monitoring, and modeling elements.

Need. The need for the Caspian Sea Observation and Forecasting System are the tools leading to successful predictions and for fruitful scientific collaboration and management.

Goals. The goals of the Project Plan are:

- to increase the understanding of the paleoclimate of the Caspian Sea;
- to increase the understanding of its sensitivity to environmental change;
- to increase the understanding of physics and biochemistry for the purpose of identifying sensitivity;
- to determine need for biological warning systems.

Tools. The tools for the achievement of the goals are:

- data analysis,
- observations,
- modeling.

Terms of References are:

- creation of metadata on institutions' and scientists' holdings of various data sets,
- evaluation of water linkages and variability through data analyses ,
- evaluation of sensitivity of the system to natural and man-made changes,
- evaluation of observation systems,
- predictive modeling suited to the Caspian,
- planning of specific process studies,
- evaluation of measures aimed at preventing involuntary import or exportation of alien species.

Parameters to be measured are:

Physics: Temperature, Salinity (CTD-profiles, Nansen bottles, and bathythermograph), Fluorescence, Light attenuation, PAR, Secchi disk depth, Sea level, River volume fluxes, River suspended sediment load.

Chemistry: DO, H_2S, Total sulfides, Thiosulfate, Sulfite, Sulfur-o, TSS, BOD5, COD, DOC, POC, TOC, PO_4, t-P, POP, NO_3, NO_2, NH_3, Urea, PON, t-N, silicate Alkalinity, Carbonate Alkalinity, pH, RedOX potential, Cd, Hg, Pb and Mn, POPs and PH, Humic matter, Radioactive contaminants, Oils.

Biology: Chlorophyll-a, Phaeopigments-a, Primary production, Bacterioplankton, Benthos, Microphytoplankton, Nanophytoplankton, Picophytoplankton, Total phytoplankton, Macrozooplankton, Mesozooplankton, Noctiluca, Total zooplankton, Invertebrates, Fish, Birds, Mammals.

Additional: Meteorological variables, Spectra of transparency, Currents.

364

Links to Other Projects

Project	Area of Interest
World Bank/PHRD Bioresource Network Pilot Projects	Ecotoxicology with emphasis on Caspian Sea
WHO	Not well defined yet
WMO	CASPAS Program
UNEP	Caspian Sea Framework convention
IAEA	Monitoring
NATO/SfP	1- INM-MHI - IMS-METU
2 - WHOI (USA).	
NATO/LG	A Linkage Grant between Turkish, Russian and German scientists on modeling of the Caspian Sea circulationv
EU/TACIS	Capacity building, Infrastructure Project implementation, Copernicus and INCO
National Programs	
Private Programs	

This program can provide data and predictions to meet societal needs.

Potential users and beneficiaries:

- Shipping, oil, and gas industries;
- Port and harbor authorities;
- Commercial fisheries;
- Mariculture operations;
- Reinsurance, tourism, and recreation industries;
- Government agencies;
- Managers;
- NGOs
- The marine-science community.

CASPIAN ENVIRONMENTAL PROGRAMME (CEP): ACHIEVEMENTS AND CHALLENGES

HAMID GHAFFARZADEH
Caspian Observation Programme, Baku, Azerbaijan

Digest: CEP, as an internationally supported inter-governmental regional environmental initiative, has considerably contributed in establishing a regional dialogue to safeguard the heavily stressed environment of the Caspian Sea. It has been successful in engaging the international and the national experts in an undertaking to diagnose the major environmental issues of the Caspian, to identify their root causes and to suggest remedial and preventive measures. CEP is however impacted by daunting challenges, most of which are of a political and economic nature. To succeed, CEP would need to continue beyond the present phase with enhanced international financial and technical assistance. Most problems are of a transboundary nature, which require that assistance is allowed to cross over visible and non-visible political boundaries.

INTRODUCTION

The CEP is a regional umbrella programme born out of and reflecting a desire for regional cooperation to safeguard and rehabilitate the threatened environment of the Caspian Sea. The five littoral countries, namely Azerbaijan, Iran, Kazakhstan, Russia and Turkmenistan, have all endorsed the programme and actively participate in its implementation. The total budget made available to the programme is close to $18 million of which $8.4 million has been funded by the Global Environment Facility (GEF), $5.7 million by the European Union Technical Assistance for CIS programme (TACIS), and $2.9 million by the World Bank. The United Nations Development Programme (UNDP), the United Nation Environment Programme (UNEP) and the private sector, namely the oil and gas industry, have funded the balance. The littoral states have provided the baseline investment estimated to worth $160 million and the in-kind contribution of approximately $1.4 million. The CEP as an initiative was agreed to by major stakeholders in June 1995. The programme documents were developed through a fully consultative process involving the littoral countries and the international partners. The programme started activities in May 1998, and its present phase is expected to continue till end of 2002. Tacis, UNDP, World Bank, UNEP and the United Nations Office for Project Services (UNOPS) are collaborating in the execution and implementation of the programme. Littoral countries are the major owners of the programme.

The CEP aims at the environmentally sustainable development and management of the Caspian environment, including living resources and water quality, so as to obtain the utmost long term benefits for the human populations of the region, while protecting human health, ecological integrity and the region's economic and environmental sustainability for future generations. The ultimate goals of CEP include a) development of a *regional environmental dialogue* through, inter alia, establishment of a Steering Committee (SC), a Programme Coordination Unit (PCU), a number of Caspian Regional Thematic Centers (CRTCs) around the Sea, creation of Caspian Concern Groups and finally formulation and endorsement of a Caspian Framework Convention; b) undertaking **a *transboundary problems/solutions analysis*** through completion of a Transboundary Diagnostic Analysis (TDA) of water related environmental issues and their root causes, formulation and endorsement of a Caspian Strategic Action Plan (SAP) and National Action Plans (NCAPs) and c) *resource mobilization* through identification and early preparation of investment projects which meet CEP criteria of high priority actions.

REGIONAL DIALOGUE

The CEP has succeeded in establishing a regional environmental coordination mechanism which engages international partners and the national decision makers and experts in a constructive regional dialogue on the Caspian environmental issues. The dialogue has been made possible through the establishment of SC, PCU, CTRCs, and the Concern Groups.

The Steering Committee which meets at least once a year is a forum for strategy and policy discussions and decisions. SCMs are often attended by the highest ranking environmental national authorities dealing with the Caspian issues, representatives of the International Partners, CRTCs and private sector. The annual SCMs review the programme's progress, asses its achievements, try to identify barriers to its progress, and direct it. So far the SC has met four times. SC is practically the only institutionalized structure for environmental strategy and policy dialogue around the Caspian Sea.

The management and technical backstopping responsibilities of the CEP are carried out, led by and coordinated from the PCU located in Baku. The PCU, which is jointly staffed by Tacis and UNDP, acts a hub-web of managerial decisions and for environmental exchange of information around the sea. It also functions as an information referral centre. PCU has formed strong links with the governmental agencies as well as with the industry and the NGO community around the sea enabling it to act as a leading partner and facilitator in the ongoing environmental dialogue on environmental issues.

The CEP has assisted in the creation of ten CRTCs around the Caspian. These act as centres of technical excellence, each one specializing in a specific theme. The centres coordinate regional activities related to their specific area of expertise through organizing meetings, workshops, seminars and through research activities. Each country hosts two CTS: Pollution Control and Data and Information Management in Azerbaijan; Emergency Response and Integrated Coastal Area Management and Planning in Iran; Biodiversity Protection and Water Level Fluctuations in Kazakhstan; Legal, Regulatory and Economic Instruments and Fisheries in Russia, and Desertification Control and Sustainable Development and Human Health in Turkmenistan. CRTCs have had varied degree of success but in general have proven to be excellent means of regional environmental cooperation.

The other component of the regional dialogue mechanism is the Public Participation Activities (PPA) element. The CEP has appointed one PPA and has formed at least one Caspian Concern in each of the littoral countries. The groups include representatives of local authorities, NGOs, private sector, academia and media. The PPAs and the groups' aim is to enable the communities voice their Caspian environmental concerns and also to be kept informed of what happens to the environment of the Sea. PPAS are invited to the major events of the CEP including the SC meetings.

The regional dialogue would have been incomplete without the participation of the oil and gas industry. The industry has actively supported CEP activities through information sharing, funding of activities and partnership in research projects. Emergency response activities, pollution monitoring and biodiversity protection have been amongst themes that have attracted considerable private industry support.

In order to institutionalize the dialogue, the CEP has assisted the efforts of the littoral states to develop and collectively agree on a Framework Convention for the Protection of the Caspian Sea. The objective is the protection of the Caspian environment from all sources of pollution and the protection, preservation, restoration and sustainable and rational use of the resources of the Caspian Sea. After seven regional meetings and lively discussion, the text of the convention is near completion except for the 'institutional arrangements'. What holds up the finalization and collective endorsement Convention are however political issues which we shall discuss later in this paper.

TRANSBOUNDARY PROBLEMS/SOLUTION ANALYSIS

To find out what the transboundary issues and problems are and what measures should be undertaken to address these problems and issues, the CEP is undertaking a Transboundary Diagnostic Analysis (TDA) which will be complimented by formulation and endorsement of five National Caspian Action Plans (NCAPs) and a Regional Strategic Action Plan (SAP). The TDA is now nearly completed after three years of hard work, numerous studies and research activities and meetings. The process has brought together a number of highly renowned international experts and national experts to diagnose what the major transboundary issues are, what priority these have, what the root causes are and what impact they make, what Environmental Quality Objective the region should follow and what affordable and technically feasible measures the region should adopt. The NCAP process has recently begun. The aim of the process to develop and obtain national endorsement of strategic plans for the Caspian issues in each of the five countries. The NCAP point of departure form TDA is the 'national' orientation of the latter where both transboundary and national issues are considered and dealt with. The SAP will be developed later in 2002 on the basis of the TDA and NCAPs and will be a blue print for strategic actions in the next 5 to 10 years. The Framework Convention in addition to being a means of dialogue will need to be seen as a mechanism for pursuance of the SAP.

The TDA findings so far have been very important. TDA has identified four major areas of existing issues and two major areas of emerging issues. The existing major issues are:

1. decline in fisheries and economic bio-resources;
2. biodiversity erosion;
3. degradation of and damage to coastal habitats and amenities and
4. decline in overall quality of environment.

The emerging major issues have been identified as:

1. environmental emergencies and
2. intrusion by alien species.

The *decline in fisheries* has been most notable in the decline in sturgeon stock. Although not much reliable stock, reproduction and catch data are available, there is a general consensus that the sturgeon is on the verge of extinction if no immediate and effective measures are taken. The catch which was reported to be around 49 thousand tons at the turn of the 20th century in Volga is now reported to be less than one thousand. On the Iranian side where the catch data are more reliable and more indicative of the stock, the catch has fallen by two thirds in a decade. The decline is so serious that the CITES has threatened the littoral countries except Iran with a total ban on caviar import and the countries have accepted a voluntary moratorium on catch this year. The decline in fish stock is also reported to be considerable in Kilka. A 30% reduction is reported in the southern part while reports indicate noticeable reduction in the north. Other fish stock also show decline. Many reasons are cited and disputed for the decline. Dam construction on Volga which damaged the spawning capacity, river pollution and poaching are listed as major contributing factors, but the root causes are basically the lack of environmental consideration amongst politicians and planners; lack of an agreed regional fisheries plan as well poverty in the coastal communities. It should be borne in mind that that fisheries directly and indirectly provide jobs and income to many of the 12 million residing around the Caspian.

Biodiversity erosion is a threat in the bio rich Caspian. The sea and its immediate surrounding coastal area can boast of being a unique ecological world heritage. The sea is home to some 150 species of fish including some rare sturgeon species and the fresh water seal. The coastal and the river estuaries provide breeding and spawning habitats not only for rare fish, but also for migratory birds such as black crane which travel from Siberia to Africa. The forests in the southern part are the last remaining of their kind across the globe. This rich biodiversity is now severely threatened by mostly man made factors. Sturgeon are being fished to extinction. There are reports of seal death in large numbers. Some suggest that in the last few years, one quarter of the seals have died although this can not be confirmed. Wetlands are being turned into farmland or drained to be converted into residential land. Deforestation is massive. In the southern part, two thirds of the forest have disappeared in less than a generation. It must however be stressed that the biodiversity, although in danger, is still fairly rich. In response to the increasing concern with regards to the biodiversity erosion and the possible impact of pollution on both biodiversity and fish and bio-resources, the CEP has initiated an ecotoxological study of certain species including seals. The early results point to a combination of factors including change in water temperature as well as water contamination.

The *decline in the environmental quality* is serious. While the overall industrial pollution is not as critically high as some environmentalists would have feared it to be, the sea is far from being pristine. Due mostly to the decline in industrial production in the CIS countries, the inflow of industrial contaminants has fallen in the northern part. A large number of factories which used technologically environmental unfriendly methods are out of production for lack of funding. Nevertheless, the potential danger that the factories would go back to the obsolete technology simply to provide job opportunities is always there. The region lacks the investment that is required to renovate and modernize the industry. A rise in production, which is politically desirable, will certainly put additional pressure on the Sea. Dam construction in Volga and Kura have also helped to sediment a considerable quantity of contaminants in reservoirs behind dams, thus helping to prevent outflow of contaminant into the sea. This however has not prevented inflow of certain contaminants, nor has it been of any use for industries that are built downstream of dams. The Caspian Sea has a number of 'hotspots', which are areas of concentrated industrial pollution. Although data and information on contamination and pollution is scant and not easily accessible, it could nevertheless be suggested that the areas around Bay of Baku, the estuaries of Volga, Terek and the inundated oil fields of Kazakhstan are amongst these hotspots. In addition to these, one should include microbiological hotspots near the major urban areas where raw sewerage is flowing into the sea untreated. In the densely populated southern coastal areas and around Baku, this phenomenon is observed. Nutrient enrichment and the consequent eutrophication is also a major menace in the Caspian coastal waters including wetlands. In the south this phenomenon is more pronounced, due to the liberal usage of heavily subsidized chemical fertilizers and pesticides combined with urban sewerage and debris washed into the in-flowing rivers from upstream deforested watersheds. In fact the nutrient enrichment could soon overshadow industrial pollution in many coastal areas.

Damage to and degradation of the coastal areas, habitats and amenities varies in type and degree along the coastline, but it is considerable all over and critical in certain areas. Land degradation, including desertification and deforestation processes, are very much noticeable in Turkmenistan and Iran. Deforestation in Iran is now acknowledged to be a major contributing factor to the frequent flash floods that lead to large scale financial and often human losses on the Iranian side. Man-made factors including high population growth rates and poor land use management are diagnosed to be root causes. Desertification in Turkmenistan is increasing the pressure on water-tables and on the quality of water, thus seriously impacting food chain. Again man-made factors play a major role. Unwise use of wetlands, heavy river pollution and river impoundments, including dams, have seriously impacted natural habitats, ecosystems, fish and birds spawning and breeding grounds. The sea level fluctuation, a Caspian phenomenon which is not fully understood , is also impacting the coastal habitats and amenities. While the rise, close to 1.5 meters in the 1997-1998 period has helped to revitalized some wetlands and has helped biodiversity, its impacts on the infrastructure and amenities has been very serious as large areas of farmland, residential quarters, roads, bridges and port facilities have been inundated. This phenomenon has also contributed to contamination of the sea through inundating waste dumping sites, oil- stone mines and oil drilling and distribution systems.

Introduction of Mnemiopsis leidyi (ML) *a species alien* to the Caspian in now emerging as a major issue. First noted in mid nineties the ML is now observed in sizeable quantities all over the sea. Assumed to having been intruded through ballast water of vessels travelling from Black Sea through the Volga-Don canal, the ML is impacting the food chain through competition with Kilka. ML decimated the Back Sea pelagic stock in the nineties and it is feared that it will have same impact on the Caspian if no effective measure are taken. Already the fisheries authorities across the sea are reporting a considerable fall in Kilka catch which might be attributed to the ML. It must be noted that the Caspian Sea is connected to both the Black Sea and the Baltic Sea through river canals and species alien to each of these seas could fairly easily intrude from the others with unknown impacts.

The other emerging issue is the *environmental emergencies* including major oil spills. The coastal areas of Azerbaijan and Kazakhstan are already dotted with oil drilling, extraction and transport equipment and facilities inherited from the Soviet time. Very often these leak and at times there are spills, although so far no major spill has occurred. Vessel transport of oil and its derivatives across the sea has so far been limited and no major accident has been reported although with the aging fleet, the threat of spills is real. The emerging danger is however the impact that the new oil extraction and transport activities might have on the sea. The potentially large oil and gas resources, particularly near Azerbaijan and Kazakhstan coastal areas, have attracted oil concerns and there are competing plans to cross the sea with many pipelines. The industry is concerned with the environmental impacts of its activities and is making all efforts to minimize risks. Because of the seismic nature of the sea and technological and human errors, the possibility of a major spill however looms.

RESOURCE MOBILIZATION

Resource mobilization has been mostly pursued through the development of Priority Investment Portfolios (PIP) by the World Bank. The PIP intends to assist in the identification and early development of feasible investment projects that address a national priority, have a positive transboundary impact on the Caspian environment and are socially and economically beneficial. The goal is to raise funds for initiatives that the TDA/SAP exercise will have identified as priority interventions. The Bank has fielded a number of project identification missions to the regions and has established a dialogue with the national authorities to identify and prepare PIPs. Simultaneously the Bank is pursuing a complimentary initiative to assist the region in implementing small investment environmental projects that yield short-term results during the lifetime of the CEP. Under this initiative, namely the Small Matched Grant Programme, CEP matches investments by the applicants up to a certain level. The SMG is now in its first year and is considering a number of environmental projects from the region. SMG is mostly oriented towards local authorities, the private sector and NGOs.

The CEP management has also approached the oil and gas industry to seek financial resources and has succeeded in obtaining close to half million dollar from these sources. The Exxonmobile, Shell and BP has been the major contributors. In a separate initiative the CEP explored possibility of establishing a Caspian Environment Fund but the initiative was not pursued for lock of interest from some of the littoral countries.

CHALLENGES

In addition to the issues and problems listed above, the CEP, like most other environmental initiatives in the region, faces a number of serious challenges which are rooted in the social, economic and political conditions of the region.

Lack of and access to reliable data and information has been a major hindrance from the first day of CEP. Certain types of information and data were never systematically collected over a long period in the region. It is hard to come by any information on prices that reflect market values. No long term information has been collected on nutrients. Fisheries data are scant as are pollution data. The collapse of the Soviet Union led to serious a disruption in regional data and information collection and has left a huge information gap. The data and information which has been and/or is being collected is not always reliable. The Soviet era data and information were at time not totally reliable due to the inadequate collection methodology or due to political reasons which required the data to reflect a picture different from reality on the ground. The reliability of present day data and information has improved, but again the old methodologies and lack of modern data and information collection systems and equipment impact the quality. Even in cases where reliable data and information do exist the accessibility remains a challenge. Quite often the required information is withheld for political reasons or it is treated as a commercial item. Whenever information or data is seen to suggest omission or negligence of a certain country or agency to protect the environment, then that piece of information becomes inaccessible or is deemed unreliable. Inaccessibility to data and information has made the regional exchange of the same very difficult if not impossible.

A related challenge is the *underdeveloped social society* in the region. Dominated for decades, if not centuries, by despotic regimes, the region lacks the open, democratic and accountable social structures that are required to ensure free flow of information, transparency, meritocracy and fair competition. The communities have not been allowed, let alone encouraged, to organize themselves to express their environmental concerns. Lack of transparency has created a fertile ground for corruption and even organized crime with the Caviar industry being a well known example. Large, laden and paternalistic bureaucratic structures delay decisions and disfranchise communities, particularly the marginal ones such as small fishermen and environmentalists.

Priority assigned to environmental issues varies across the region but it is basically low all around and at all levels. This poses a serious challenge when politicians and planners allocate the often scare resources to more pressing needs. Environmental concerns are secondary when confronted with the need to create jobs or get the economy moving. The concept of sustainable development is paid lip service to is simply brushed away. The environmentalists do not appear to be held in high esteem by politicians. The lack of dialogue between the development practitioners and the environmentalist is a major issue. The picture become gloomier when the lack of respect for the environment is noted at all levels of the society. To most people the environment is not a high priority concern. The low priority of environment combined with the lack of environmental knowledge in the region has resulted in the lack of or inadequate environmental strategies, policies and measures. It is therefore no surprise to note the lack of regional or national Caspian environmental strategies and policies.

Lack of adequate financial and economic resources to tackle the overwhelming social and economic problems is very serious and perhaps the priority issue in most of the region. Although thought to contain the world's third largest oil and gas deposits and home to most of the world's caviar producing sturgeon, most of the region is basically poorly equipped for the 21st century in terms of immediately available financial resources. The northern part is still going through the painful transition to the market economy. The south is recovering from a revolution and a war and is in transition towards a less governmental dominated economy. Financial resources might become available to the countries in future but for a fairly long time not much will be channeled to the people dealing with environmental issues. To the planners there are more pressing needs to be dealt with immediately. The inefficient and unaccountable bureaucracies will not help the environment cause much either. These structures will absorb a considerable portion of the resources themselves.

The other challenge is the conflicting political agendas in the region. The legal regime of the Sea became a major issue after the dissolution of the Soviet Union. The five countries have since been engaged in a dispute, which at time reaches crisis levels, over who owns what part of the sea and whatever it contains. Border lines are drawn according to national wishes backed by favorable international norms and practices. The engaged parties believe the dispute to be worth over hundreds of millions of dollars, and as such no one is prepared for a compromise. At least not for the time being. The dispute over the legal status is seriously impacting the dialogue regarding the Framework Convention with the majority of the littoral countries wishing to link an agreement on the Convention to the settlement of the legal status. This dispute is further complicated by the traditional rivalry in the Caspian for political dominance and by the arrival of new political forces in the region competing for strategic positions and for hydrocarbon resources. These new forces take sides in the region and further complicate the issues. The question of oil and gas pipelines from the Caspian to the world market is an example of this complicated picture which need not be dealt with here. What needs to be highlighted is however, the fact that these political divisions and differences spill over into the environmental arena. They act as invisible barriers to the flow of much needed technical and financial resources to certain parts of the regions. While the environmental issues are of a transboundary nature, the assistance is regrettably hindered by national and political boundaries.

CONCLUDING REMARKS

The Caspian Sea is a world environmental heritage under serious stresses which are often rooted in social, economic and political conditions. The Caspian problems are not yet beyond solution. The Caspian can be saved.

The CEP has been successful in establishing a constructive regional dialogue to diagnose the root causes of the problems and to come up with remedial and preventive measures. The CEP has also been trying to mobilize resources to assist in the implementation of the suggested measures. In short, its achievements have been considerable.

The challenges to safeguard the Caspian environment have been and will remain daunting. To ensure success it is essential that what the CEP has started is continued. The processes established, including the dialogue process, are not yet sustainable and free of political influence. The national ownership of the environmental dialogue and environmental activities is highly important. The community involvement is equally important.

The social, economic and political conditions of the region however requires the continuation of the international financial and technical assistance in the short to medium term. The assistance should deal with environmental issues and their roots regardless of the visible and invisible political boundaries. The global initiative embodied in CEP needs to be supported.

A REVIEW OF CASPIAN SEA ENVIRONMENT, CLIMATE VARIABILITY AND AIR-SEA INTERACTION

EMIN ÖZSOY

Institute of Marine Sciences, Middle East Technical University, P.O. Box 28, Erdemli, İçel 33731 Turkey

ABSTRACT

The Caspian Sea is located in a region of high environmental stress resulting from anthropogenic effects of modern times. On the other hand, the region also shows signs of large climatic changes throughout its ancient and modern history, best reflected by the sea-level changes. Prominent interannual/interdecadal signals and large scale controls are evident. Only a better understanding of these changes through combined use of observations and modelling can provide the essential knowledge base and foresight for predicting the expected changes in the future.

ENVIRONMENTAL CONCERNS

The Caspian Sea has had an important place in the history of the east as well as the west, being the center of many cultures, migrating human populations, and a host of legends. It is in the general region where earliest human civilizations were created, and were strongly driven by climatic changes (Dolukhanov, 1994). Today the Caspian Sea is of global interest, because of its unique environmental state: the enclosed Caspian Sea is a sensitive system because of its relatively small inertia not being able to absorb the changes imposed, but is also a natural testbed for studying ocean-atmosphere processes that determine the role of climate variability relative to anthrophogenic forcing. The lessons learned from the present environmental threats that result from combined natural and anthropogenic effects in the Caspian Sea could thus serve as a warning for other regions of the world.

Five independent countries (Azerbaijan, Russia, Kazakhstan, Turkmenistan and Iran) share coasts of the Caspian Sea (Fig. 1), but no settlement yet exists about the establishment of marine boundaries between the riparian countries.

Since the later part of the 19th century, oil has been a major activity in the region. Vast oil and gas reserves in the region, after the collapse of the preceding socio-political systems in the 1990's, has drawn the attention of international oil companies as well as riparian and foreign countries, which presently have concentrated their efforts for the development and marketing of Caspian oil.

The Caspian Sea oil reserves, valuable resources as they are, create environmental concerns in the region, which extend far away from their origin, as a result of the need to export through marine routes. The resolution of the immediate environmental threats to the Turkish Straits resulting from the export of Caspian oil through the Black Sea and the Straits is a significant case creating issues that need to be further addressed by international policy and law (Plant, 1996; 2000).

Fig. 1. The Cental Asia region, showing the Caspian and Aral 'Seas' and the political boundaries of riparian states.

The Caspian Sea is a unique reservoir of relic flora and fauna of euryhaline species which are mostly endemic, so that it deserves to be a lake more than a sea (Dumont, 1998). Commercial fish production includes the worlds largest population of sturgeon and black caviar (90% of world stocks), which is of great economical importance for the riparian counties. It is one of the most productive large water bodies of the planet. As a result of the deterioration of the environment, the most treasured product of the Caspian–its caviar–has been virtually finished (Dumont, 1998).

The northern part of the Caspian basin produces one-fifth of the former Soviet countries' agricultural total crop yield, and one-third of their industrial output. In the last 40 years, a string of dams and hydropower plants have been built along the Volga river, to supply electric power to the industries and to irrigate the crops. These uses have lowered water levels enough to severely impair the Volga's capacity to dilute waste and run-off, upsetting the natural balances of salinity, temperature, and oxygen in water downstream.

The environmental crisis in the Caspian Sea resulting from anthropogenic forcing, and accompanied by natural variability and climatic changes, is manifested by dramatic changes in its ecosystem and resources (Glantz and Zonn, 1997). Ecological deterioration has begun in regions of great economical value such as near the Volga river delta and the adjoining North Caspian Sea (Kosarev and Yablonskaya, 1994). The lower reaches of the Volga are heavily developed with numerous unregulated releases of chemical and biological pollutants. Although existing data is sparse and of questionable quality, there is ample evidence to suggest that the Volga is one of the principal sources of transboundary contaminants to the Caspian. Because the Caspian Sea has a drainage area extending from north of Leningrad to south of Tehran, it receives large amounts of wastewater, originating partly from industrial areas of Russia. As a result, fisheries in the region have collapsed.

In addition to these immediate environmental signals, there is a longer term threat, which may lead to a rather rapid, precipitous collapse of the ecosystem: the threat of eutrophication in an enclosed sea. Unfortunately, a similar example has been demonstrated in the Black Sea. For the last few decades there were signs of deterioration, for example, the disappearence of a number of indigenous species from the ecosystem, not sufficiently recognised as a signal. It is in the neighboring Black Sea that the eutrophication process is believed to have led to the sudden collapse of the fisheries in the 1990s.

The sea-level changes in recent times have had important implications for the economic well-being of the coastal populations. Considering flatlands around the periphery, and the petroleum production and shipping facilities in the region, the sea-level changes are of great consequence. Of greater concern are the exposure of toxic waste burial sites by rising groundwater levels.

CLIMATE VARIABILITY

The Caspian Sea is one of the remotest members of the family of interior seas in isolation from the world ocean as one proceeds from west to east following the chain formed by the western and eastern basins of the Mediterranean Sea, Black Sea, Caspian Sea and Aral Sea. The entire area comprising these water bodies is trapped between the continents of Europe, Africa and Asia, and located downstream of Atlantic and European weather events. All three Seas are neighbored by high mountain chains, vast continental flatlands, deserts and fertile lands, in a transition region between the Atlantic and Indian Oceans. While the Black and Caspian Seas have high production supported by nutrients supplied by large rivers, the eastern Mediterranean lacks this supply, and is otherwise known as the 'blue desert' (Fig. 2). Ocean-atmosphere-land interactions in this environment of contrasting marine and continental climates, complex land and sea bottom topography, and energetic mid-latitude atmospheric motions make the region prone to extremes. As a result, the feedbacks to the global system could be disproportionately large in comparison to the size of the region (Özsoy, 1999).

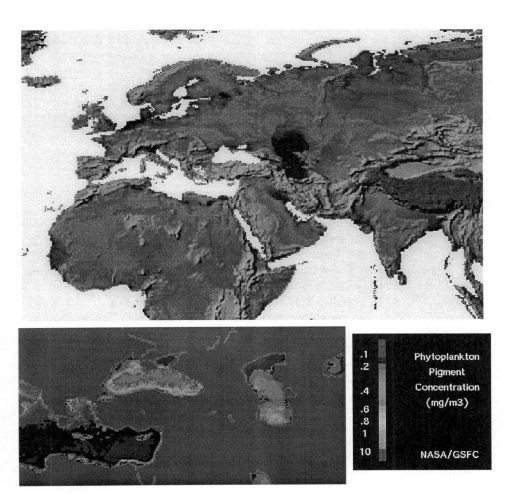

Fig. 2. Top: Topography of Euro-Asian-African continent. The Caspian Sea, currently ~28m below sea level, shown in blue, bottom: CZCS average pigment concentration (mg/m³) in the eastern Mediterranean, Black Sea and Caspian Sea regions. Data after NASA/GSFC.

Because of their smaller inertia, enclosed seas respond faster to climatic forcing compared to the global oceans, and therefore are more sensitive to environmental degradation, which seems to be amplified by cultural and socio-economic contrasts such as demonstrated in the Caspian Sea region.

378

Not far from the Caspian Sea, and indeed very much coupled with it, lies the Aral Sea and the surrounding lands, where the population faces one of the worst environmental disasters that the world has ever seen. The present crisis deeply influences Central Asia and the socio-economic state of its population. On the one hand, the present catastrophe appears to be a consequence of man-made environmental degradation and pollution. On the other hand, longer term changes with equally drastic magnitude could result from natural climatic variability, which appears to be almost as important as the anthropogenic effects. Regional climate modelling with coupling between the atmosphere and the Aral Sea indicates the same result (Small et al., 1999). The changes following 1960s have lead to desiccation of the Aral Sea, and the desertification and pollution of the region at large (Micklin, 1991; Glantz, 1999).

LARGE-SCALE CONTROLS

The region is one of the foremost areas of the world where interannual and long term climatic variability connected with global patterns is relatively large. In the Mediterranean region such variability is well known, and appears to be coupled with the Indian Monsoon system, and through it to the El Nino / Southern Oscillation (ENSO) (Alpert and Reisin, 1986; Ward, 1996).

Weather in Europe, extending well into the Mediterranean and Eurasian regions, is to a large extent determined by conditions in the North Atlantic. The North Atlantic Oscillation (NAO) (Hurrell, 1995) accounts for about one third of the hemispheric interannual variance, especially in terms of surface temperature and evaporation-precipitation anomalies in the European and the Eastern Mediterranean regions (Hurrell, 1995, 1996; Marshall, 1997). It has been linked to sea level changes in the Caspian Sea (Rodionov, 1994), to Danube river runoff directly influencing Black Sea hydrology (Polonsky et al., 1997), as well as to surface winter temperatures, precipitation and river runoff in Turkey (Cullen, 1998).

Large scale control in the region is also well expressed in terms of long range atmospheric transport patterns. Simultaneous transport of aerosol dust from the Sahara desert into the Mediterranean and tropical Atlantic regions (Li et al., 1996; Andreae, 1996) has been demonstrated to depend on interannual patterns of the North Atlantic Oscillation (NAO) (Moulin et al., 1997). An exceptional case during the first half of April 1994 (Özsoy et al., 2001)s has been shown to be triggered by upper air jet interactions and meridional circulations on a hemispherical scale, corresponding to an increase in the positive phase NAO index.

SIMILARITIES IN REGIONAL COOLING PATTERNS

There is a significant degree of synchronism displayed between the Levantine, Black and Caspian Seas, in terms of the air and sea surface temperatures displayed in Figure 3. This is because of the proximity of the three regions, but also a result of the possible large scale controls discussed above, and by Özsoy and Ünlüata (1997) and Özsoy (1999).

Comparisons with the NAO and SO indices and with solar transmission (an indicator of volcanic dust in the atmosphere) shows some cold years linked with negative values of the SO Index (*e.g.* 1982-83, 1986-87, the 1990's), often cited as ENSO years (*e.g.* Meyers and O'Brien, 1995), and some years (1983, 1986-87, 1989-90, 1992-93) to be characterized by high NAO indices.

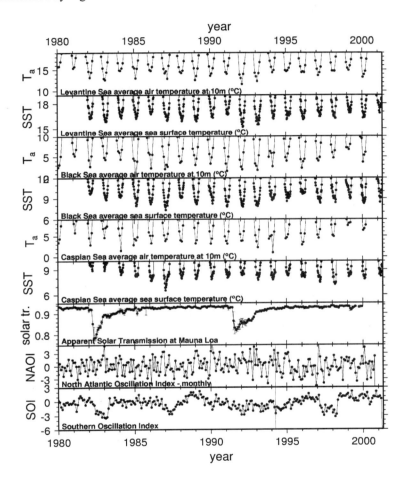

Fig. 3. *Time series of average air (ECMWF/ERA re-analyses, 6 hr forecast temperature at 10m height) and sea surface temperatures (ESA ERS1/ATSR derived monthly averages) for the Levantine, Black and Caspian Seas, Solar transmission at Mauna Loa, and the climatic indices NAOI (seasonal averages) and SOI, for the last two decades.*

380

Relatively cooler winters were detected in the years 1982-83, 1985, 1987, 1989, 1991 and 1992-93 in Figure 3. Some of the cold years correspond to well known cases of intensified convection and deep water formation in the Marmara Sea, the Rhodes Gyre region and Levantine basins of the eastern Mediterranean, and the Black Sea in 1987, 1989, 1992 and 1993 (Özsoy, 1999). There are also some surprises: the year 1987 is one of the coolest years in all three seas, but there is no corresponding decrease of air temperature in the Caspian Sea. Secondly, while the air temperature displays various anomalous years in the Caspian Sea, the sea surface temperature does not respond to it very effectively, except the year 1987, which is not very consistent with the pattern observed in the other two seas. The formation of cold water in the Caspian Sea occurs especially in the northern parts in specific regions, and therefore the sea surface and air temperature averaged over the basin does not appear well correlated.

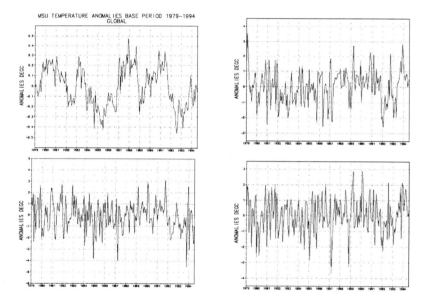

Fig. 4. Average lower tropospheric temperature anomalies for (a) the globe, (b) the Rhodes Gyre region of the Levantine Sea (33°-37°N and 26°-32°E), (b) southwest Black Sea (41°-43°N and 27°-31°E), (c) northern Caspian Sea (45°-47°N and 48°-53°E). The measurements were obtained from the Microwave Sounding Unit (MSU) for the 0-5.6km layer of the troposphere (grid resolution 2.5°) and anomalies calculated over the base period of 1979-1995 are produced by the Global Climate Perspectives System (GCPS) (http://www.ncdc.noaa.gov/onlineprod/prod.html).

Cooling in the lower troposphere (Fig. 4a) occurred globally in 1992-1993 (Spencer and Christy, 1992), following other cooling periods of 1982, 1985-86, 1989 in the last two decades. Significant drops of temperature drops of 1-2°C occurred in the entire region extending from the Rhodes Gyre of the Levantine Sea to the southern Black Sea and the northern Caspian Sea (Figs. 4b-d). An event of global climatic significance (Fiocco *et al.*, 1996), the eruption of Mount Pinatubo volcano in June 1991, resulting in stratospheric warmings (Angell, 1993), decreased solar energy inputs (Dutton and Christy, 1992; Dutton, 1994) and anomalous temperatures (Halpert *et al.*, 1993) in the entire northern hemisphere in 1992-93. Anomalous cold temperatures appeared in the Middle East in very similar spatial patterns during the winters of 1992 and 1993, covering the Black Sea, eastern Mediterranean, and African regions in both years (Özsoy and Ünlüata, 1997). In Turkey, the winter of 1992 was the coldest in the last 60 years (Türke_ *et al.*, 1995), and in Israel, it was the coldest in the last 46 years (Genin *et al.*, 1996).

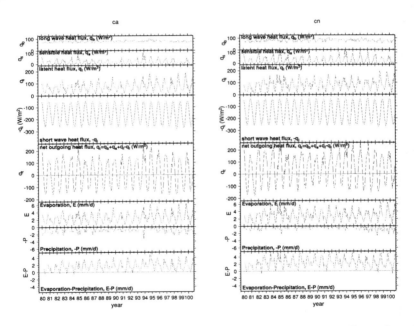

Fig. 5. The components of the sea surface heat and water fluxes based on ECMWF 6 hr forecast fields calculated at 6 hr intervals and averaged at monthly intervals over (a) the entire Caspian Sea and (b) the northern Caspian Sea. The ERA15 reanalysis data set is used for 1979-1993, and the forecast products are used for 1994-2000.

SURFACE FLUXES

To study the effects of climate variability, the surface fluxes computed from uniform quality, decadal atmospheric reanalysis data ERA15 and the forecast data for recent years have been obtained from the ECMWF at 1° nominal resolution, at 6 hr intervals of 6 hr global forecasts, and have been monthly averaged over the sea domain. The air-sea sensible (q_s) and latent heat (q_l), as well as the longwave (q_b) and incoming shortwave (q_i) fluxes as well as evaporation and precipitation are shown in Figure 5 for the studied period.

There are large interannual modulations of the seasonal net fluxes of heat and mass at the surface. The northern Caspian basin has greater latent heat losses, more evaporation and less precipitation than the average for the Caspian Sea.

The comparison of fluxes in the three different seas shows coincidence between the active periods of Black Sea and Levantine Sea, and a lesser degree of sychronism between them and the Caspian Sea (Özsoy, 1999). On the other hand, the Black Sea and Caspian Sea momentum fluxes are larger, more seasonal and more variable than the Levantine Sea, where the larger events come in interannual pulses. The sensible heat flux increases from the Levantine Sea to the Black Sea, reaching a maximum in the Caspian Sea.

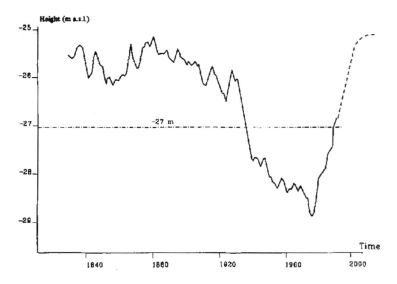

Fig. 6. Recent sea-level changes in the Caspian Sea.

SEA LEVEL CHANGES

The sea-level, besides being a good indicator of climatic fluctuations, is a sensitive measure of hydrometeorological driving factors in enclosed and semi-enclosed seas. In the Caspian Sea, sea-level changes depend on the regional hydrometeorological regime linked to global climate (Radionov, 1994). The sea level changes with climatic and anthropogenic components are of great economic and environmental importance for the surrounding countries. Interestingly, the sea-level change influences the residence time of the deep waters, and therefore has a direct bearing on the health of the Caspian Sea. Abrupt changes in sea-level have occurred twice since the 1830s, as well as earlier in history, as the fate of Khazars occupying its shores in the 10th century stand witness. The sea level has first dropped from a -25 m in 1930s down to -29 m by 1978, and has risen to the present -27 m after 1977. Hydrogen sulfide has been detected in deep waters prior to 1930 when sea-level was high, as a result of insufficient ventilation, limited by the decreased volume of dense water formed on the ice-covered, shallow northern Caspian shelf (average depth ~2m) under the influence of the large, variable inputs of the Volga river (Kosarev and Yablonskaya, 1994).

In the Caspian and Aral Sea region, the most important element controlling the climate variability, directly manifested in sea-level changes, is the hydrological cycle. The region has had a history of large changes in the combined drainage system of the two water bodies, which in geological time included the Black Sea (Dumont, 1998; Boomer et al., 2000). According to past data extracted from geological and archeological records (WMO, 1997), going thousands of years back, there were even larger excursions in the water level, much larger than the present range of sea-level changes in the 20th century.

The present level of understanding of sea-level changes links it rigorously to the water budget of the Caspian Sea. There are new findings that link the rapid changes in sea-level to ENSO / El Nino (Bengtsson, 1998), based on global modelling of the hydrological cycle and data analyses. Further studies (Arpe et al. 1999) that indicate multi-decadal fluctuations and a further positive trend in the sea-level rise based on global warming scenarios.

CIRCULATION AND HYDROGRAPHY

Ongoing studies of circulation and hydrography of the Caspian Sea (Sur et al., 1998, Ibrayev et al., 2000) are based on the accumulation of knowledge by past scientific investigations in the region, although new efforts are being made to prescribe surface and lateral boundary conditions based on realistic data, such as the heat, momentum and water fluxes obtained from the most up to date sources of information. Satellite data (Fig. 7) indicate basic physical mechanisms at work, which are also predicted by numerical models. With these developments there has been a significant improvement in the predicted features of the circulation, ice development, convection as well as in the simulation of sea-level changes. Further studies are planned to be carried out soon through multi-national research.

384

Fig. 7. Sea surface temperrature in the Caspian Sea on (a) 1 August 1997 and (b) 7 April 1994, showing typical summer and winter conditions (Sur et al. 1998). In the first image upwelling is observed along the eastern coast of the Sea adjacent to the deserts. In the second image cold water formed in the shallow northern basin is advected along the westerrn coast and recirculated by the currents as it is intercepted by the Apsheron peninsula.

REFERENCES

1.   Alpert, P. and T. Reisin (1986). An Early Winter Polar Air Mass Penetration to the Eastern Mediterranean, *Mon. Wea. Rev.*,**114**, 1411-1418.
2.   Andreae, M. O. (1996). Raising Dust in the Greenhouse, *Nature*, **380**, 389-390.
3.   Angell, J. K. (1993). Comparison of stratospheric warming following Agung, El Chichon and Pinatubo volcanic eruptions, *Geophys. Res. Lett.*, **20**, 715-718.
4.   Arpe, K. and E. Roeckner (1999). Simulation of the hydrological cycle over Europe: Model validation and impacts of increasing greenhouse gases, *Advances in Water Resources*, **23**, 105-119.
5.   Bengtsson, L. (1998), Climate modelling and prediction, Achievements and challenges, WMO/IOC/ICSU Conference On The Wcrp Climate Variability And Predictability Study (CLIVAR), UNESCO, Paris, 2-4 December 1998.

6.   Boomer, I., N. Aladin, I. Plotnikov and R. Whatley (2000). The Palaeolimnology of the Aral Sea: A review, *Quatern. Sci. Rev.*, **19**, 1259-1278.
7.   Cullen, H. (1998). North Atlantic Influence on Middle Eastern Climate and Water Supply, *Climatic Change* (submitted).
8.   Dolukhanov P. (1994) Environment and Ethnicity in the Ancient Near East. Avebury.
9.   Dumont, H. J. (1998). The Caspian lake: History, Biota, Structure and Function, *Limnol. Oceanogr.*, **43**, 44-52.
10.   Dutton, E. G. and J. R. Christy (1992). Solar Radiative Forcing at Selected Locations and Evidence for Global Lower Tropospheric Cooling Following the Eruptions of El Chichon and Pinatubo, *Geophys. Res. Lett.*, **19**, 2323-2316.
11.   Dutton, E. G. (1994). Atmospheric Solar Transmission at Mauna Loa, In: T. A. Boden, D. P. Kaiser, R. J. Sepanski, and F. W. Stoss (editors), Trends 93: A Compendium of Data on Global Change, ORNL/CDIAC-65, Carbon Dioxide Information Analysis Center, Oak Ridge, Tenn., USA.
12.   Fiocco, G., D. Fua and G. Visconti (1996). *The Mount Pinatubo Eruption, Effects on the Atmosphere and Climate*, NATO ASI Series, Kluwer Academic Publishers, 310 pp.
13.   Genin, A., Lazar, B. and S. Brenner (1995). Vertical Mixing and Coral Death in the Red Sea following the Eruption of Mount Pinatubo, *Nature*, **377**, 507-510.
14.   Glantz M. H. (editor) (1999). Creeping Environmental Problems and Sustainable Development in the Aral Sea Basin}, Cambridge University Press, 304 pp.
15.   Glantz, M. H. and I. S. Zonn (1997). *The Scientific, Environmental and Political Issues in the Circum-Caspian Region*, NATO ASI Series, 2, Environment - vol. 29, Kluwer Academic Publishers, Dordrecht.
16.   Halpert, M. S., Ropelewski, C. F., Karl, T. R., Angell, J. K., Stowe, L. L., Heim, R. R. Jr., Miller, A. J., and D. R. Rodenhuis (1993). 1992 Brings Return to Moderate Global Temperatures, *EOS*, **74**(28), September 21, 436-439.
17.   Hurrell, J. W. (1995). Decadal Trends in the North Atlantic Oscillation: Regional Temperatures and Precipitation. *Science*, **269**, 676-679.
18.   Hurrell, J. W. (1996). Influence of Variations in Extratropical Wintertime Teleconnections on Northern Hemisphere Temperatures, *Geophys. Res. Lett.*, **23**, 665-668.
19.   Ibrayev, R. A., E. Özsoy, A. S. Sarkisyan and H. _. Sur (2000). Seasonal Variability of the Caspian Sea Circulation Driven by Climatic Wind Stress and River Discharge (submitted).
20.   Kosarev, A. N. and E. A. Yablonskaya (1994). *The Caspian Sea*, Backhuys Publishers, Haague, 259 pp.
21.   Li, X., H. Maring, D. Savoie, K. Voss and J. M. Prospero (1996). Dominance of Mineral Dust in Aerosol Light Scattering in the North Atlantic Trade Winds, *Nature*, **380**, 416-419.
22.   Malanotte-Rizzoli, P. and A. R. Robinson (editors) (1994). *Ocean Processes in Climate Dynamics: Global and Mediterranean Examples*, NATO ASI Series C:

386

Mathematical and Physical Sciences, vol. 419, Kluwer Academic Publishers, 437 pp.

23. Marshall, J. and Y. Kushnir (1997). A 'white paper' on Atlantic Climate Variability, an unpublished working document.

24. Meyers, S. D. and J. J. O'Brien (1995). Pacific Ocean Influences Atmospheric Carbon Dioxide, *EOS*, **76**(52), 533-537.

25. Micklin, P. (1991). The Water Management Crisis in Soviet Central Asia. The Carl Beck Papers in Russian and East European Studies, No. 905. University of Pittsburgh Center for Russian and East European Studies, 120 pp.

26. Moulin, C., C. E. Lambert, F. Dulac and U. Dayan (1997). Control of atmospheric export of dust from North Africa by the North Atlantic oscillation. *Nature*, **387**, 691-694.

27. Özsoy, E. and Ü. Ünlüata (1997). Oceanography of the Black Sea: A Review of Some Recent Results, *Earth Sci. Rev.*, **42**(4), 231-272.

28. Özsoy, E., (1999). Sensitivity to Global Change in Temperate Euro-Asian Seas (the Mediterranean, Black Sea and Caspian Sea): A Review, The Eastern Mediterranean as a Laboratory Basin for the Assessment of Contrasting Ecosystems, editors: P. Malanotte-Rizzoli ve V. N. Eremeev, NATO Science Series 2, Environmental Security, **51**, Kluwer Academic Publishers, Dordrecht, s. 281-300.

29. Özsoy, E., N. Kubilay, S. Nickovic, and C. Moulin (2001). A Hemispheric Dust Storm Affecting the Atlantic and Mediterranean (April 1994): Analyses, Modelling, Ground-Based Measurements and Satellite Observations, J. Geophys. Res. (in press).

30. Plant, G. Navigation Regime in the Turkish Straits for Merchant Ships in Peacetime, Safety, Environmental Protection and High Politics, Marine Policy, **20**, 15-27, 1996.

31. Plant, G. The Turkish Straits and Tanker Traffic: An Update, Marine Policy, **24**, 193-214, 2000.

32. Polonsky, A., E. Voskresenskaya and V. Belokopytov, Variability of the Northwestern Black Sea Hydrography and River Discharges as Part of Global Ocean-Atmosphere Fluctuations, in: E. Özsoy and A. Mikaelyan (editors), *Sensitivity to Change: Black Sea, Baltic Sea and North Sea*, NATO ASI Series (Partnership Sub-series 2, Environment, 27), Kluwer Academic Publishers, Dordrecht, 536 pp., 1997.

33. Robinson, A. R., Garrett, C. J., Malanotte-Rizzoli, P., Manabe, S., Philander, S. G., Pinardi, N., Roether, W., Schott, F. A., and J. Shukla (1993). Mediterranean and Global Ocean Climate Dynamics, *EOS*, **74**(44), November 2, 506-507.

34. Robinson, A. R. and M. Golnaraghi (1994). The Physical and Dynamical Oceanography of the Mediterranean Sea, in: Malanotte-Rizzoli P. and A. R. Robinson (editors), *Ocean Processes in Climate Dynamics: Global and Mediterranean Examples*, NATO ASI Series C: Mathematical and Physical Sciences, vol. 419, Kluwer Academic Publishers, pp. 255-306.

35. Rodionov, S. N. (1994). *Global and Regional Climate Interaction: The Caspian Sea Experience*, Water Science and Technology Library, v. 11, Kluwer Academic Publishers, Dordrecht, 256 pp.

36. Small, E. E., Sloan, L. C., Hostetler, S., Giorgi, F. (1999). Simulating the water balance of the Aral Sea with a coupled regional climate-lake model *J. Geophys. Res. – Atmos.*, **104**, 6583-6602.

37. Spencer, R. W. and J. R. Christy (1992). Precision and Radiosonde Validation of Satellite Gridpoint Temperature Anomalies. Part II: A Tropospheric Retrieval and Trends During 1979-1990, *J. Climate*, **5**, 858-866.

38. Sur, H. _., E. Özsoy and R. Ibrayev (1998). Satellite - Derived Flow Characteristcs of the Caspian Sea, in: D. Halpern (editor), Satellites, Oceanography and Society, Elsevier Oceanography Series, 63, Elsevier, 376 pp.

39. Türke_, M., Sümer, U., and G. Kılıç, 1995. Variations and Trends in Annual Mean Air Temperatures in Turkey With Respect To Climatic Variability, *International Journal of Climatology*, **15**, 557-569.

40. Ward, N. (1995). Local and Remote Climate Variability Associated with Mediterranean Sea-Surface Temperature Anomalies, European Research Conference on Mediterranean Forecasting, La Londe les Maures, France, 21-26 October 1995.

41. WMO (World Meteorological Organization (1997). Integrated Programme on Hydrometeorology and Monitoring of Environment in the Caspian Sea Region (CASPAS), WMO-No. 873.

POLLUTION IMPACT TO STURGEON FROM URAL DELTA, KAZAKHSTAN

IGOR V. MITROFANOV

Institution of Zoology, Tethys Scientific Society, Almaty, Kazakhstan

ANIMALS AND SAMPLING

Great sturgeon, Russian sturgeon and Starred sturgeon have annual long-distance migrations from Volga and Ural deltas to the South Caspian Sea and back. All three sturgeon species feeding on fish, shellfish and benthic invertebrates, but Great sturgeon prefer fish (up to 90%), Russian sturgeon – shellfish and polychaete, and Starred sturgeon – other benthic impetrates including small crabs.

Six Great sturgeons (*Huso huso*), two Starred sturgeons (*Acipenser stellatus*) and fourteen Russian sturgeons (*Acipenser guldenstaedti*) were investigated. Organs of sampling were liver, spleen, kidney, and gill. All sturgeons were taken from the commercial fishing in the Ural maritime area (Kazakhstan). Time from taken fishes from the water to dissection was no more two hours. Majority of fishes were still alive, with rare heart activity. After dissecting organs were placed in formalin immediately.

PESTICIDE ANALYSIS IN FISH LIVER

Pesticides and arochlors widely distributed in the Caspian region were determined in the liver of all investigated fishes. In the field 1 g of liver (exact weights) were ground up, placed in different glass vials and 10 ml n-hexane added. Samples were held at ambient temperature for return to the laboratory. Samples were extracted with *n*-hexane. Hexane was removed and replaced by a new portion of 10-ml *n*-hexane and stirred for 10 minutes. This was repeated three times. All hexane fractions (one from fixing and three from extraction) were pooled in one sample. 2-ml sulfuric acid (concentrated) was added to the extract. The mix was intensively stirred for 15 minutes. The sample was then allowed to stratify and the sulfuric acid removed. This was repeated several times until the extract was clear (usually 2-3 times). For removal of traces of sulfuric acid, the extract was washed three times with 15-ml of distilled water. For removal of the traces of water 2 g of Na_2SO_4 (anhydrous) was added to the extract. The extract was decanted and evaporated to 2 ml: If necessary volume was adjusted by a new portion of hexane. Measurement of the concentration and type of pesticide was carried out on the gas-liquid chromatograph CHROM-5, using standard solutions of known substances for comparison.

HISTOPATHOLOGY

Tissue sections from the organs listed above were stained routinely with hematoxylin and eosin (Luna, 1968) and examined for histopathological abnormalities. The severity of abnormalities were scored on a scale of 0-3: Liver: macrophage aggregation – 0, fat degradation and regeneration – 1; Cloudy swelling of cytoplasm –2; necrosis – 3. Kidney: macrophage aggregation, tubula vacuolation, and parasites – 1; glomerulla vacuolation – 2. Spleen: macrophage aggregation – 0; fat degradation and regeneration – 1; vacuolation –2. Gill: aneurysm and parasites – 1; macrophage aggregation and epithelia proliferation within gill – 2. Neoplasms, fibrosis, granulomas, spongiosis hepatis and parasites were noted as being present or absent for each organ.

IMMUNOHISTOCHEMISTRY

Fixed samples were paraffin-embedded, sectioned at 5 μm, deparaffinated and stained using a peroxidase-antiperoxidase detection system (Signet Laboratories, Dedham, MA) with monoclonal antibody 1-12-3 that specifically binds CYP1A proteins (Park et al., 1986). Staining techniques were modified from previously described (Smolowitz et al. 1991) as follows: (1) tissue sections were incubated for 2 hours (150μl at t=0', 60') either with MAb 1-12-3 (2.7 or 1.5 μg/ml 1% BSA/PBS) or with non-specific mouse myeloma protein (1.5μg/ml 1% BSA/PBS) as an antibody control (A comparison of the positive and negative controls incubated with either 2.7 μg/ml or 1.5 μg/ml did not reveal noticeable differences in levels of CYP1A expression) and (2) tissue sections were incubated for 20 minutes in 3-amino-9-ethylcarbazole to develop color. CYP1A expression was quantified according to the occurrence of cells staining (0 = no staining, 1=rare, 2=many cells, 3=all cells) and intensity of stain (0=no staining, 1=very mild, 2=mild, 3=moderate, 4=strong, 5=very strong). The product of occurrence and intensity was termed the CYP1A index and ranged from 0 to 15.

Two types of controls were used: (1) Matching serial sections for all tissues were stained with a nonspecific IgG (purified mouse myeloma protein UPC-10, IgG2A, Organon Teknika, West Chester, PA) to ensure observed staining was specific to CYP1A proteins and (2) sections of winter flounder liver known to express CYP1A strongly, and sections of scup liver known not to express CYP1A (as determined by EROD activity and immunoblotting) were used as positive controls (incubated with MAb 1-12-3) and negative controls (incubated with non-specific mouse myeloma) for the staining technique and as a reference for scoring.

RESULTS

α, β, and γ HCH isomers were founded. α-HCH are the most common among HCH. It was determined in 5 Great Sturgeons and 8 Russian Sturgeons. β-HCH was determined in one Great sturgeon and 4 Russian sturgeons. Only once β-HCH was founded in absence of α-HCH. This is almost true for γ-HCH. Usually concentration of all isomers HCH is not very high and varied from 0.09 µg/g to 0.74 µg/g. Only once concentration of γ-HCH was 1.03 µg/g in the liver of Russian sturgeon. No α-HCH or β-HCH was founded in this fish.

DDT and/or its isomers were founded in all investigated fishes. DDT and DDE are more common then DDD. Usually concentration of DDT is several times higher, then DDD or DDE. Policholbiphenile A50 was determined in all six Great sturgeons, one Starred sturgeon, and 10 Russian sturgeon (13 Russian sturgeon were investigated). It concentration varied from 2.01 µg/g to 12.95 µg/g. It seems, that concentration of all pollutants are higher in Great sturgeon comparative to Russian sturgeon.

Table 1. Concentration of organochlorides in the liver of fishes (µg/g, wet weight) Caspian sea 2000, (Kazakhstan part, near the Ural Delta)

Species	A50	DDT	DDD	DDE	γ-HCH	β-HCH	α-HCH
Great sturgeon	4,53	67,92	16,46	19,67	-	-	0,18
Great sturgeon	12,95	11,58	-	3,94	-	0,40	0,26
Great sturgeon	5,56	4,42	3,12	16,18	-	-	0,25
Great sturgeon	5,18	3,47	-	5,12	-	-	0,32
Great sturgeon	6,83	8,69	-	4,23	-	-	-
Great sturgeon	4,13	7,53	5,16	26,49	0,35	-	0,74
Russian sturgeon	4,14	3,17	-	2,16	1,03	-	-
Russian sturgeon	2,54	4,54	-	1,16	0,12	-	0,10
Russian sturgeon	3,89	10,42	2,45	1,18	-	0,40	0,65
Russian sturgeon	5,55	-	0,65	0,79	-	0,50	0,35
Russian sturgeon	-	-	-	-	-	-	0,31
Russian sturgeon	5,54	-	2,89	1,39	-	-	0,79
Russian sturgeon	2,13	2,01	-	2,01	0,11	-	0,09
Russian sturgeon	2,21	15,20	-	0,52	-	0,19	0,10
Russian sturgeon	-	-	-	-	-	-	-
Russian sturgeon	-	6,08	-	-	-	-	-
Russian sturgeon	2,01	11,15	-	0,63	-	-	0,18
Russian sturgeon	2,07	-	-	0,81	-	-	-
Russian sturgeon	4,34	9,10	-	1,71	-	0,36	-
Russian sturgeon	-	7,13	-	-	-	-	-
Starred sturgeon	2,17	4,86	-	0,85	-	-	-
Starred sturgeon	-	8,10	1,29	1,28	0,28	-	-

Liver:

Fat degradation. Vacuole with glycogen and lipids are volume the whole cell. Nuclei are not evident, cytoplasm are only near cell's wall and become colorless. Cell's walls are regular. It is found out in two Great sturgeons and three Russian sturgeons, and don't found out in Starred sturgeon

Macrophage aggregation. Macrophages usually founded near big blood vessels, hemopoetic cell, and among hepatocytes. It is founded in all fishes.

Regeneration. Regeneration hepatocytes very often displace in narrow stripes with width in two-three cells. Usually it is small area. Nuclei are increased in size and have intensive color. Cytoplasm also has intensive coloration and contains no vacuoles. Usually it is more basophilic comparative to surrounding hepatocytes. It is found out in two Great sturgeons.

Necrosis. Hepatocytes are distracted, all cell's structures destroyed, cytoplasm is almost colorless. Cell's walls are partly destroyed. Two loci of necrosis are found in one Great sturgeon. One of the loci is big enough.

Cloudy swelling. Hepatocytes lose all vacuoles, it become more eosinophilic and granular. Inside cytoplasm more condense parts appear. It is found only once in Great sturgeon.

Spleen

Macrophage aggregation. Macrophages in spleen contain lot of hemosiderin and their appearances confirm misbalance in hemopoetic process. It is found in one Great sturgeon and two Russian sturgeons.

Vacuolation. Nuclei are of normal size and shape and displace to cell walls. It is found in one Great Sturgeon.

Regeneration. Cells have increased nuclei, cytoplasm are paler. It is found in one Great Sturgeon.

Fat degradation. Big lipid vacuoles appear in hemopoetic cells. Nuclei are driven to the walls. One or several vacuoles take more than a half of cell's volume.

Kidney

Macrophage aggregation. It is found in three Great sturgeons and one Russian sturgeon.

Renal tubular vacuolation (RTV). Some cells of canalicules are vacuolated. Nuclei and cytoplasm are driven to the walls. Usually it is only one vacuole and not very big. It is found in one Great sturgeon, one starred sturgeon and five Russian sturgeons.

Glomerulus's destruction. External tunica of glomerulus is destroyed, cells lose connection with each other, intercell's distances increase to half cell's diameter. All cell's structures are normal. Some cells can be vacuolated. Vacuole is not big situated near the wall. It is found in one Russian sturgeon, but without tunica destruction and with vacuoles.

Gills

Aneurysm of secondary lamellas. Dilatation of lamellas and stagnation of blood in it. It is found in three Great sturgeons, one Starred sturgeon, and two Russian sturgeon.

Hyperplasia of interlamellar epithelia. Interlamellar epithelial cells proliferate and volume all the space between lamellas. It is found in two Russian sturgeons.

Proliferation of epithelia. Lamellar and interlamellar epithelial cells proliferate simultaneously. It is found in one Great sturgeon.

DISCUSSION

Comparative to 1999 there are more pathological changes in Great sturgeon liver. Fat degradation, regeneration, and necrosis were observed in the liver of Great sturgeon only in 2000. In kidney of Great sturgeon we found the same changes as in 1999, but macrophage aggregation become more common, and renal tubule vacuolation on the contrary become rarer. In the spleen macrophage aggregation were founded only in 50% of fishes (100% in 1999), and vacuolation and regeneration were founded only in 2000. Aneurysms of secondary lamellas (gill) become more common in 2000, also proliferation of gill epithelia, but there were no parasites founded in the gill and other organs. Indices of pathological changes become higher for Great sturgeon in 2000 for all organs. This means, that environment become worth for Great sturgeon.

On the contrary to Great sturgeon, in Russian sturgeon there are less pathologies comparative to 1999. We found the same pathologies as in 1999, but it become rarer. The only exception are the gills. Percent of fishes with aneurysm of secondary lamellas are similar in 1999 and 2000, but proliferation activity of epithelium cells becomes significantly higher. This means, that environment for Russian sturgeon become a little bite better or the same as in 1999.

It is difficult to compare results of 1999 and 2000 as only two specimens of Starred sturgeon were investigated in 2000. But it seems, that situation is equal to Russian sturgeon, either becomes a little bite better.

Table 2. Histopathological changes in sturgeons (per cent of investigated specimens).

organ	pathology	Great sturgeon		Russian sturgeon		Starred sturgeon	
		1999	2000	1999	2000	1999	2000
liver	M	100	100	100	100	100	100
	Cl	9.09	16.67	-	-	-	-
	Fd	-	33.33	21.43	21.43	6.67	-
	R	-	33.33	-	-	-	-
	Nc	-	16.67	-	-	-	-
kidney	M	28.57	50.00	64.71	7.14	58.33	-
	RTV	28.57	16.67	58.82	35.71	26.67	50.00
	V	-	-	-	-	6.67	-
	Pr	-	-	-	-	20.00	-
	G	-	-	-	7.14	-	-
spleen	M	100	50.00	100	7.14	87.5	-
	V	-	16.67	9.09	-	12.50	-
	R	-	16.67	-	-	-	-
	Fd	-	-	-	21.43	-	-
gill	A	36.36	50.00	12.50	14.28	30.77	50.00
	Pl	-	16.67	-	14.28	30.77	-
	M	-	-	-	-	25.00	-
	Pr	9.09	-	-	-	-	-

M = macrophages

Nc = necrosis

Pr = parasites

RTV – tubula vacuolation

Fd – fat degeneration

R – regeneration

A – aneurysm

G – glomerulla destruction

Cl – cloudy swelling of cytoplasm

Pl – proliferation of gill (lamellar and/or interlamelar) epithelia

394

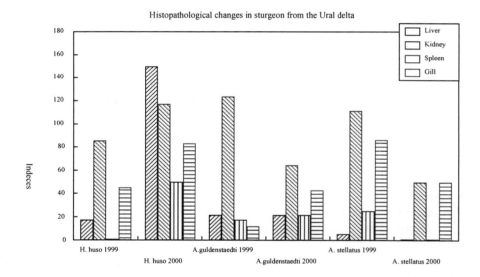

Histopathological changes in sturgeon from the Ural delta

MATHEMATICAL MODELING OF OIL POLLUTION IN THE NORTHERN CASPIAN SEA

ANATOLY VASILIEV
State Oceanographic Institute , Moscow, Russia

In the first part of the presentation, different information observation platforms for meteorological and oceanological data in the Northern and Central Caspian and the large programme for receiving of that data were briefly described. The most suitable for this activity is a CASPCOM (Coordinating Committee on Hydrometeorology and Monitoring of Pollution in the Caspian Sea), which was established in 1994 by the heads of hydrometeorological services of all Caspian States with the active support of the World Meteorological Organization (WMO). Since the hydrometeorological observation network and environmental monitoring network is the main task for the Committee, the conceptual bases of the Integrated Programme on Hydrometeorology and Monitoring of Environment in the Caspian Sea Region (CASPAS) was prepared and adopted by the Committee and WMO (N873) in 1997. The experts from all Caspian countries, as well as from WMO, UNEP, UNESCO and IAEA took part in the preparation of CASPAS. The first phase of CASPAS Programme – the project "Integrated Project for a Monitoring and Information System for the assessment and forecasting of the state of environment and pollution in the Caspian Sea region (IPM&IS)" is planned in 2002 with the financial support of EU and TACIS. One of the most important target of this project is delivering of equipment and devices for complex monitoring to laboratories of hydrometeorological services of all Caspian States by Italy.

As an application of meteorological and oceanological data from different observation platforms for a case study, the modeling oil pollution in the Northern Caspian waters was described. For the description of oil spreading over the area and changes of oil patch with time, the two models developed in the State Oceanographic Institute (SOI) were used.

The first model by Anatoly Vasiliev, named ALAPS-MC, deals with the permanent monitoring data including wind and ice conditions and provides the fields of thermodynamic parameters of sea, such as distribution of temperature, salinity and density, current structure, local and mesoscale turbulence, sea level and ice covering.

The second model by Sergei Ovseenko, named SPILLMOD, is based on the calculations of oceanological fields from the first model and provide the modeling and forecasting of oil spill behavior, that is oil spreading on the sea surface, oil transportation by wind and currents, diffusion, evaporation, emulsification, dispersion into water column, sanding and interaction with devices and dispersants. The model allows to determine zones of a possible presence of oil pollution, probability of damage to the coast in typical and extreme hydrometeorological conditions, and efficiency of actions on localization of oil pollution.

Both models were applied in a case study of an oil spill at the Kazakhstan shelf (Southern-Western Tazigali) in spring 2001. The amount of oil was estimated in a very wide range from 300 to 5000 tons. After ice melting, all oil appeared in the marine waters and drifted forced by wind and currents. The calculation of water movement under specific conditions, such as wind, ice, Volga and Ural riverine etc., were done during a period from 15 March till 7 May during an interval of 6 hours. Spatial interval was lower then 2 n.m. The results of oil distribution for the 5000 tons patch clearly showed the covering of the whole Volga delta after one month. In conclusion, it was postulated that all the Volga delta and even Dagestan coastal waters are the risk zones after large oil spills in the Kazakhstan shelf.

THE CONDITIONS OF THE HAZARDOUS MANAGEMENT SYSTEM IN AZERBAIJAN

RASHID AJALOV
Caspian Environment Programme, Baku, Azerbaijan

THE NATIONAL CONTEXT

Industrial and Associated Environmental Background
The main historical driver for the industrial development of Azerbaijan was its oil and gas reserves, and in addition to major petrochemical industries, a thriving engineering sector developed in support of this. The 'traditional' markets for Azerbaijan's industrial products were developed through its association with the Soviet Union, but were not exclusively within the Soviet Union. In recent years, Azeri industrial enterprises have developed a range of export destinations through Europe, the Middle East and western to central Asia, including CIS countries. Plans for increasing exports and potentially widening future export destinations are a contributory factor to the Government's desire to implement procedures that meet the requirements of international conventions such as the Basel Convention.

A period of economic and political instability following independence from the former Soviet Union resulted in a major decline in the principle industries (including oil extraction) from 1991 to around 1995. Subsequent recovery has largely resulted from the beginnings of renovation of the oil industry and development of an associated gas industry, through a series of production sharing agreements with companies from Europe, North America, the Middle East, Russia and Japan. Although industrial production has not yet mirrored the general economic up-turn, expectations are that this will be seen in the near future.

HAZARDOUS WASTE MANAGEMENT

Until the last decade, the influence of industry on the environment in Azerbaijan has been primarily related to the country's political status. As a consequence of the major oil reserves in the region, the Apsheron Peninsula in particular was developed as a major oil and related chemical production centre for the former Soviet Union, with all production in the hands of the State. For the most part, the Soviet industrial strategy did not make allowances for environmental protection, and the management of waste was controlled primarily, if not exclusively, by logistic and financial influences. Furthermore, Soviet investment declined in the latter half of the twentieth century, and the continuing use of out-dated technologies at inadequately maintained industrial enterprises resulted in inefficient production, high emission rates and production of large amounts of often hazardous wastes.

Following independence, the major Soviet market for Azeri production was lost, and the funding available for industry reflected the economic situation of the new Republic. Again, environmental protection issues were generally not given any priority, as the main concerns were associated with the survival of industry in a declining economic situation. That is not to say that there was no concern over environmental issues; rather that there were insufficient resources available to develop appropriate remedial actions.

In the late 1990s, the country realised considerable economic improvements, including the initiation of renovation of the oil industry through participation with international interests. Improved technologies (for example, resulting in reduced fugitive emissions of volatile organic chemicals (VOCs) during oil production) and the requirements of satisfying an international market for oil, have resulted in better control of production in terms of environmental protection, although there is still considerable room for further improvement.

Although other industrial production has not yet mirrored the up-turn in the oil sector (and the development of natural gas exploitation), expectations are that this will be seen in the near future. The country continues to undergo a transition towards a market-based economy, and is pursuing international interests to support industrial redevelopment. For example, a programme of investment projects is being pursued by UNDP and UNIDO in order to update and revitalise the state-owned Azerkhimia chemical production enterprises in Sumgayit. International backing and plans to develop a major export market are expected to also bring a greater commitment to environmental protection.

Furthermore, environmental consciousness has grown and environmental policy reform is gradually being introduced into national economic development programmes. Therefore, a greater focus on waste is being developed, and the government is pursuing the opportunity to implement a national hazardous waste strategy before major industrial redevelopment takes place.

Environmental consciousness has also grown and environmental policy reform is gradually being introduced into national economic development programmes. However, the implementation of improved environmental protection strategies has been hampered by the country's economic position as well as a degree of inertia in industry and regulatory authorities. Therefore, the main improvements that can be claimed are decreases in levels of continuing environmental pollution only as a result of a down-turn in industrial activity.

Classification of hazardous wastes is based on the former soviet system comprising four classes of hazard; Class I being the most hazardous.

There are some controls on land use planning in relation to hazardous waste management facilities and an environmental impact assessment (EIA) is required.

An assessment has been made of the probable quantities and types of hazardous wastes which are likely to be found in the country, as a base for strategic predictions of future waste–needed to determine the scale of resources required for their management and regulation. This assessment has been based on a combination of data obtained in country and a 'best fit' with Slovakia adjusted to take account of the different GNP of each country. Future waste types consequent on economic development of the country are estimated to be those associated with the chemical and petrochemical industries, agriculture and fisheries; there may also be some associated with the manufacture of textiles, machinery and metallurgy.

During the former Soviet era, little or no attention was given to pollution control, and wastes were stockpiled on the source site awaiting future re-use, recovery or disposal. Specific examples of this comprise the Baku iodine plant, the Baku machinery factory and the chemical factories.

The main industry in Azerbaijan is extraction/mining of raw materials and production of intermediate products (petrochemical industry). Cleaner technologies for the production and manufacturing of these materials have been developed in the country but cannot be applied because of the depressed economic state of the country. The oil companies each adopt their own methods for hazardous waste management, which in most cases involves the development of their own infrastructure (e.g. dedicated landfills).

There is currently little or no hazardous waste management industry in Azerbaijan. Most industrial waste producers (of which the oil producers are the majority) handle their own waste management. There is some recycling of oil sector wastes and some metals recycling. There are a number of landfill sites—for both domestic wastes and hazardous wastes.

RESPONSIBILITY FOR PAST ENVIRONMENTAL DAMAGE

During the privatisation process, the issue of liability for past environmental damage always arises. At the present time, there is no standard model for how this should be allocated and it is subject to negotiation in each individual case.

Essentially there are two models:

- Enterprises can be sold for a relatively high price with the State retaining liability for former environmental damage.
- They can be sold for a lower price with the purchaser accepting liability for all environmental damage on the site(s) acquired.

Waste Framework Directive

The Waste Framework Directive (75/442/EEC, updated by 91/156/EEC), in addition to requiring site licensing and the preparation of waste management plans, introduced a number of important principles, namely:

- The Polluter Pays Principle.
- The Proximity Principle.
- The Principle of Self-sufficiency.

A Community Strategy for Waste Management - 1989

This strategy statement identified, probably for the first time, the now well-known "Waste Management Hierarchy" of:

- Waste prevention.
- Recycling and reuse.
- Optimisation of final disposal.

The EU's Strategy was recently reviewed and revised. The revised strategy places greater emphasis on the adoption of measures to ensure that the Waste Management Hierarchy is applied in practice, in particular the integration of the principle of 'producer responsibility' – making manufacturers of products responsible for ensuring their environmentally sound treatment/disposal at the end of their life – into future EU policy measures and legislation.

Fifth Environmental Action Plan - 1993-1997

This document, entitled "Towards Sustainability", reinforced the hierarchy and defined the priorities within "final disposal" as:

- Combustion as fuel.
- Waste-to-energy.
- Landfill.

It placed great emphasis on waste minimisation and also established some non-statutory targets as follows:

- Stabilise municipal solid waste (MSW) production at the 1985 level of 300 kg per capita/annum.
- Recycle 50% of paper, glass and plastics.
- 90% reduction of dioxins from incineration.
- Banning certain wastes from landfill sites.

Hazardous Waste Directive

The SCE wishes to adhere to as many of its provisions as are practicable. The important provisions are listed below:

- Hazardous waste is defined by reference to a list of wastes/components AND a set of hazardous properties.

- The Competent Authorities must draw up plans for the management of hazardous waste and make these plans public.
- Producers of hazardous waste and establishments and undertakings transporting hazardous waste, must keep certain information and make it available to the Competent Authorities.
- Hazardous waste must be accompanied by a consignment note in the prescribed form.
- On every site where hazardous waste is landfilled, the waste type, quantity and location must be recorded.
- Undertakings which dispose of, recover, collect or transport hazardous waste do not mix different categories of hazardous waste nor mix hazardous waste with non-hazardous waste, unless a permit is obtained.
- Hazardous waste must be properly packaged and labelled in accordance with international and EC standards.
- In cases of emergency or grave danger, hazardous waste must be dealt with so that it does not constitute a threat to the population or the environment.
- Producers of hazardous waste must be subject to appropriate periodic inspections by the Competent Authorities.

Many of these provisions are already incorporated into Azeri legislation.

Likely Future Trends

It can be expected that, over the next few years, greater pressure will be placed on both waste producers and local authorities (through their responsibility for MSW) to increase the amount of recovery and recycling and to reduce the dependence on landfill, particularly for biodegradable materials, through a series of measures, examples of which are given below:

- Further restrictions on the materials permitted to be landfilled.
- More stringent emission requirements for incinerators.
- Extension of the principle of producer responsibility, for example to motor vehicles and consumer durables.
- Compulsory waste minimisation targets.
- The imposition of higher targets for recycling and recovery.

Other External Influences

A number of international conventions have been developed in relation to the management of hazardous wastes and materials, in addition to the Basel Convention. Whilst Azerbaijan is a signatory to only some of these, others are relevant for future consideration. These are considered here, recognising in particular that the major waste issues in Azerbaijan that will be of concern internationally will include:

- Oil and gas industry wastes.

- Banned organochlorine pesticides.
- Radioactive wastes.
- Mercury-contaminated materials.
- Metallurgy sludges and slags

1. Convention On The Prevention Of Marine Pollution By Dumping Wastes And Other Matter (London 1972). Signed by Azerbaijan, but superseded by MARPOL 73/78.
2. International Convention For The Prevention Of Pollution By Ships And Related Protocol (MARPOL 73/78). Not signed by Azerbaijan. Establishes international regulations to control the emission of polluting substances into the marine environment. For example, requires ports to have reception facilities for waste oil and oil-contaminated bilge water and prohibits the disposal of hazardous, sanitary, clinical and plastic waste in all waters. MARPOL covers discharges from vessels of any type operating in the marine environment (including fixed or floating platforms such as oil rigs). Later resolutions incorporated the phasing out of industrial waste dumping and incineration at sea.
3. Prior Informed Consent (PIC) Convention On Hazardous Chemicals (Rotterdam, 1998). Not signed by Azerbaijan. The aim of PIC is to promote a shared responsibility between exporting and importing countries for protecting human health and the environment from the harmful effects of certain hazardous chemicals.

Monitoring

Due to under-staffing, the hazardous waste inspectors are overloaded and it is not possible for them to do more than irregular inspections of the best-known large polluters. Other facilities are inspected rarely.

The inspection is based on a review of contracts and other documents on hazardous waste management and a review of the facility to check handling of hazardous waste. Current waste legislation does not cover all aspects of hazardous waste management and the lack of established standards constrains more effective inspection of hazardous waste generators, transporters and disposers.

Foreign companies working in Azerbaijan do not always comply with the existing regulations. Some of them refuse to allow inspectors to enter their property (although the legislation clearly gives them the right to do so) and often do not report data on hazardous waste to the Statistical Office.

It is difficult to expect improvement in the capacity to undertake monitoring without mobilising the existing laboratory capacity and increasing the number and qualifications of analytical staff. Furthermore, it is essential to establish a strategy, standards, and procedures for hazardous waste management in order to clarify what monitoring is actually required.

Enforcement capacity
Although the existing system of waste legislation includes various tools such as fines, compliance orders or even closure of a facility, there is insufficient enforcement of waste legislation.

There is one general and several specific reasons for this. Generally, people of Azerbaijan (but not only Azerbaijan, this is a typical feature for all post-communist countries) are not very used to the new legal system, where only some specific duties are directly and strictly required and enforced. This may be described as a move from a system where "everything which is not allowed is prohibited" to "everything which is not prohibited is allowed"

The specific reasons are typical for countries in transition: new laws are issued quickly, but lack sufficient back-up by supplementary regulation and allocation of additional resources to ensure implementation. Waste management facilities and waste generating companies, mostly state owned, also lack resources to achieve compliance with the law, often do not know how to achieve compliance and do not even have sufficient funds to pay penalties. Finally, under the pressure of the economic crisis, they have other priorities than environmental protection and improvement of hazardous waste management (such as paying the wages bill each month).

The basic need to improve enforcement capacity is to develop and implement a comprehensive but clear, acceptable and transparent system of hazardous waste regulation, which is fully supported by the state administration.

Future Waste Types Consequent on Economic Development
It may be deduced that, as the Azeri economy develops, the composition of future hazardous wastes will roughly mirror those which are produced in Western Europe and North America—at least in the range of wastes which are produced, if not their proportions. On the basis of a survey of waste producers recently undertaken by the UK Environment Agency, the composition of the wastes which might be hazardous in significant quantities was approximately as follows:

Table 1.

Waste Type	Proportion
Foundry sand and moulds	22%
Water with mixtures of organic and inorganic contaminants	19%
Oil sludges and/or oil water mixtures	19%
Clinical waste	9%
Water with soluble inorganic chemical contamination	5%
Sodium chloride	5%
Coated or chemically treated timber	5%
Calcium sulphate	4%
Toxic metal compounds	4%
Flue dust	4%
Solvents and oils	3%
Animal fats, oils, waxes and/or grease	2%

Of course, there is a wide range of more specialised materials in the UK in much smaller (but significant) quantities in addition to the above. For example, the quantity of hazardous wastes being incinerated exceeds some 80,000 tons p.a.

The balance between different sectors in Azerbaijan, however, will naturally be different from the UK, with a stronger emphasis on waste from the oil and organic chemical sectors. Expansion of the agricultural sector may lead to the production of fertilisers and pesticides. It is also likely that there will be very significant quantities of asbestos (which can be observed in Sumgayit) requiring disposal for some time to come.

The most likely waste types which will be produced in quantity will be:

- Oil sludges and residues.
- Oil/water mixtures.
- Used catalysts.
- Tarry residues – with and without halogens.
- Used solvents – with and without halogens.
- Solid organic chemical residues.
- Other organic chemical wastes – with and without halogens.
- Asbestos.
- PCBs and other transformer oils.
- Pesticide residues.
- Metal finishing wastes (from machinery manufacture).

INTERNATIONAL BEST PRACTICE FOR OIL AND GAS PRODUCTION WASTES

Waste Management Hierarchy
The detailed approaches adopted by national authorities responsible for regulating oil exploration and production organisations vary between countries.

The primary waste-related goal is minimisation of the amount of waste generated, and the secondary goal is maximum reuse and recovery of wastes to reduce the volume—and toxicity—of waste requiring treatment or disposal.

The scope of reuse and recovery of wastes is significant. For example:

- The components of drilling muds can be separated and re-used or burned as a fuel.
- In modern refineries, a large proportion of waste is treated or recycled internally and is recovered or burned as fuel.
- In Europe, priority is being attached to the regeneration of waste oils for use as lubricating oil (in line with the EC Waste Oil Directive).

Wastes not amenable to recovery and reuse and therefore requiring treatment and disposal off-site include spent catalysts, oily sludges and water treatment sludges.

Waste Disposal Technologies
Accepted waste disposal technologies currently in common use include:

- Storage in underground repositories or deep well injection (specifically reinjection into the field).
- Landfilling (typically after stabilisation of the waste to prevent hazardous components leaching into groundwater resources).
- Discharge to surface waters, particularly for less hazardous wastes from off-shore oil production facilities (whilst more hazardous wastes must first be treated – e.g. to reduce the oil-in-water content below 15 mg/l – and if cannot be treated must be hauled to shore and managed in the same way as wastes generated there).

However, the EC Landfill Directive will tighten controls on hazardous wastes, including banning of liquid waste, and explosive, corrosive, flammable and highly flammable waste from all new landfills from July 2001. Therefore in Europe, the option for landfilling oil-related wastes will be removed, and this should be seen as a declining option internationally.

Waste Reduction
The main industry of Azerbaijan is extraction/mining of raw materials and production of intermediate products (petrochemical industry). The country has great potential to reduce its current hazardous waste generation by implementation of technologies utilising by-products, which are currently wasted. The non-oil industrial sectors also have the potential to achieve significant reduction of hazardous waste production by the replacement or up-grading of existing production facilities. This will, of course, require substantial investment for which privatisation may offer the solution.

The country has a substantial research capability in its universities for the development of new methods for hazardous waste reduction, aiming mainly at utilisation of by-products from the chemical industry. Due to economic problems, however, the local market is not able to implement these new technologies and they are often sold abroad. The Azerbaijan State Oil Academy has already implemented some of their research results in Turkey and implementation in the USA is under discussion.

Recycling
Generally speaking, there appears to be very little recycling of wastes of any kind in Azerbaijan, let alone hazardous wastes. The level of recycling is only likely to increase if the cost of disposal increases significantly. Until that time, only selected types of waste will be recycled. I have, however, identified the beginnings of an interest in developing such activities both in the oil sector and other sectors of industry. There are differences, however. In the oil sector, the recycling is done by external companies (other than the waste producer), while in other sectors the development of recycling seems to begin within the companies (as a side activity of the waste generator).

12. PERMANENT MONITORING PANEL REPORTS

COMBINED MEETING OF THE PERMANENT MONITORING PANELS ON WATER AND DEFENCE AGAINST FLOODS AND UNEXPECTED METEOROLOGICAL EVENTS

SOROOSH SOROOSHIAN, ROBERT A. CLARK
Department of Water Resources, University of Arizona, Tucson, Arizona, USA

The combined meeting of the two Permanent Monitory Panels was held in Erice in the San Francesco Center on 19 August 2001.
Participants in the joint meeting included:

Soroosh Sorooshian, University of Arizona, USA
Robert A. Clark, University of Arizona, USA
William Sprigg, University of Arizona, USA
Dumitru Dorogan, Ministry of Waters and Environmental Protection, Romania
Don Scavia, National Centers for Coastal Ocean Science, USA
Donald Boesch, University of Maryland Center for Environmental Science, USA
Peter Douglas, California Coastal Commission, USA
David Brookshire, University of New Mexico, USA
Francois Waelbroek, Jvelich Fusion Center, Belgium
Aaron T. Wolf, Oregon State University, USA

TOPICS DISCUSSED

The meeting opened with a discussion by Soroosh Sorooshian on the role of the World Federation of Scientists (WFS) Permanent Monitoring Panels (PMPs). He pointed out the obvious interaction between the PMP on Water and those on related water quality/quantity issues. The following diagram illustrates the synergistic relationship between water and various PMPs sponsored by the WFS.

The current Erice sessions will include discussions on coastal zone problems sponsored by the PMP on Defence Against Floods and Unexpected Meteorological Events and national and international water conflicts which are related to the various PMPs reviewed by this panel.

A number of topics were discussed by panel members, including:

- Possible future Erice activities.
  - – Tsunamis (observations and warnings).
  - – Climate-related water/weather extremes.
- Interdisciplinary overlap of PMPs.
- Concern about the distribution of scientific information to the public and the effectiveness of public education. The establishment of websites for teachers related to WFS PMPs was discussed. This concern also extends to distribution of scientific information to the press.
- The apparent increase in extreme events (droughts/floods) as a future subject for WFS discussions.
- Arsenic pollution (e.g., the Bangladesh problem) was considered another topic for future discussion (or has it been solved?).
- Progress on various PMP-related problems, including:
  - – Transboundary water.
  - – Pollution (atmosphere/water).

- Panelists felt that preparing video presentations on various PMP topics for distribution to the public is an important World Laboratory (WL) role.
- Examination of the findings of various intergovernmental panels - such as climate (IPCC, Intergovernment Panel on Climate Change) and severe weather – are also pertinent for panel consideration.
- Concerns were expressed by various panel members related to problems such as those related to the Black and Caspian Seas and their impacts on various topics subject to WFS PMP consideration.
- Discussions included other possible topics for presentation at the Erice annual seminars, including:
  - Climate studies (change).
  - Has progress been made in activities related to the numerous subject areas of these PMPs?
  - Have water quality issues, such as arsenic, been resolved?
  - Desertification as related to these PMPs.
  - Roles of carbon dioxide and ozone on climate change.
  - Data for science.
  - Prediction of extreme events, e.g., dust storms, tropical cyclones, cold/extremely hot weather, thunderstorms.
  - Education in meteorology/hydrology.

MOTHER TO CHILD TRANSMISSION OF HIV—ANTIRETROVIRAL THERAPY AND THERAPEUTIC VACCINE: A SCIENTIFIC AND COMMUNITY CHALLENGE

JOINT WORKING GROUP REPORT OF AIDS/INFECTIOUS DISEASES PMP AND MOTHER AND CHILD PMP

G. Biberfeld (Solna, Sweden), P. Biberfeld (Stockholm, Sweden), F. Buonaguro (Naples, Italy), N. Charpak (Bogota, Colombia), G. de Thé (Paris, France), M. Ferreira Rea (Sao Paulo, Brazil), G. Gray (Soweto, South Africa), Ch. Huraux (Paris, France), A. Lindberg (Marcy l'Etoile, France), N. M. Samuel (Guindy-Chennai, India), G. Scarlatti (Milan, Italy), S. Tlou (Gabarone, Botswana), Ph. Van de Perre (Montpellier, France), Zeng Yi (Beijing, China), R. Zetterström (Stockholm, Sweden)

Since our last Erice workshop in August 2000, (Acta Paediatrica 89:1385-6.2000) significant progress has been achieved in both HIV antiretroviral therapy (ART) and in HIV vaccine research and development, with the emerging concept of therapeutic vaccine considered as a possible complement to ART. We reviewed these two areas in the context of mother to child transmission, and appreciated the increasing political awareness concerning the HIV epidemic now considered as a planetary emergency, which led to the engagement of UN and of G8 to provide funds to implement access to ART in low income and most severely affected countries.

In spite of these efforts, UNAIDS estimates that in 2000 around 36 million people were living with HIV infection, 95% of them in the developing world, that at least 5.3 million being infected during the year 2000 alone, and that more than 22 million AIDS patients died since 1985.

Mother to child transmission of HIV (MTCT) represents a particularly dramatic aspect of this HIV epidemic with an estimated 600.000 of newborns infected yearly, 90% of them in Africa. Mother and child health, being a key factor for any sustainable development, avoidance of MTCT HIV must be a priority and become achievable in many countries.

PART ONE

Antiretroviral Therapy

Access to care

- In many parts of the world, HIV is already the leading cause of adult and child mortality. The HIV pandemic compromises the gains made in recent decades in terms of quality of life and life expectancy[1].
- Due to civil society pressure, national initiatives (generic drugs) and high level negotiation with pharmaceutical companies, the cost of ARV for developing countries has been reduced considerably (about one tenth of the initial sale price). A recent decision of the United Nations to prioritise allocation of sufficient resources for HIV, tuberculosis and malaria control should hopefully make a comprehensive package of prevention, care and research including ARV more readily available in the near future.
- However, in resource-limited countries, human, financial and logistic resources needed for programmes of HIV care may compete with other sectors of the health care systems.
- There is a wide discrepancy in access to care including ARV according to different settings in the developing world. The range of people having access to adequate ARV therapy is from a few hundred in many countries to several tens of thousands such as in Botswana, South Africa, Thailand, India and Brazil among others[2-4].
- Important prerequisites for access to care implementation involve the whole health system and structure (accessibility, social acceptability, VCT (Voluntary counselling and testing) structures and competence, education of health professionals...), as well as economic and political commitment.
- A new comprehensive and socially acceptable concept of taking care of households instead of individuals is emerging in some countries such as South Africa[2], Botswana[3] and India[4]. This concept could decrease the economic fragility of affected households and mitigate the impact of HIV on vulnerable children and orphans.
- Innovative and appropriate technologies are developed or are already available, such as plasma and salivary rapid test for HIV diagnosis in adults as well as CD4 counts by alternatives to flow cytometry[5] and modified p24 antigen measurement technologies for monitoring of therapeutic efficacy[6]. These techniques could render monitoring and diagnosis more accessible and affordable in resource-limited settings.
- New drugs that may be active against HIV, including new ARV families (integrase inhibitors, inhibitors of viral assembly, cytokine as adjuvants, etc.) and drugs from the traditional pharmacopoeia (China, India, some African countries) are currently under evaluation[5]. These compounds may improve the efficacy of current ARV regimens.

414

Perinatal and postnatal mother-to-child transmission of HIV
- Antenatal care and VCT are entry points for prevention and care.
- Disclosure of the test results to the husband/partner and significant others varies considerably from one area to another (50-80% in Soweto, South Africa, 17% in Dares Salaam, Tanzania, 15% in Namakal, India, less than 10% in Abidjan, Côte d'Ivoire and Bobo-Dioulasso, Burkina Faso) and is a frequent limiting factor for maternal interventions (ARV, feeding practices,...)[2-4,8].
- Efficacy of short regimen of perinatal prophylactic ARV in reducing MTCT lessens over time if breastfeeding is prolonged and may even be lost[8-9].
- Maternal CD4 count is a strong predictor of the efficacy of perinatal prophylactic ARV. Indeed, in a combined analysis of two clinical trials evaluating short zidovudine regimens in the perinatal period, no efficacy was demonstrable at any time during the follow up in women who had less than 500 CD4 cells per μ at delivery[8,9]. However, in women with more than 500 CD4 cells per μ at delivery, breastfeeding had only a minimal, non-significant, impact on transmission.
- Interruption of maternal ARV administration around the time of lactation may increase short term breast milk viral load and, putatively, infant transmission[8].
- Breastfeeding by HIV-infected women has been reported to be associated with an excess maternal mortality in a clinical trial performed in Nairobi, Kenya[10].

Plea For Action
- The success of HIV prevention and care programmes depends on good access to and the performance of the primary health care system. The strengthening of the necessary infrastructure and human resources to deliver HIV prevention and care is of the utmost priority.
- Access to care/ARV should not be restricted to ARV therapy alone but should be considered as a continuum of medical and psychosocial support. Voluntary counselling testing and care should be regarded as components of a comprehensive package of prevention and care.
- ARV and drugs for prophylaxis and treatment of opportunistic infections should be made available, affordable and sustainable and distributed in an equitable way.
- HIV prevention and care programmes should include also availability of reliable and inexpensive tests to diagnose and monitor the treatment of HIV infection and associated conditions, as well as appropriate training for health care workers in management. Urgent recommendations are needed for criteria for initiation of therapy, scheduling, switching, interrupting and monitoring regimens. In order to ensure success of such programmes, joint decision making involving the whole therapeutic team and the household/family are mandatory. Health care workers should be provided with training on

occupational hazards, appropriate equipment and management of all accidental exposures.

- Pilot country programme of this nature will help to establish policy and allocate adequate resources.
- In order to improve MTCT intervention programme coverage, the disclosure by women of their HIV test result to husband/partner and significant others should be encouraged by learning from successful experiences (such as in South Africa and India), with respect to local socio-cultural mores. Disclosure of HIV status may further encourage husband/partner to get tested and improve overall efficacy of prevention programmes. The social consequences of disclosure should be carefully elucidated in all settings prior to implementation.
- In programmes of prevention of MTCT, CD4 counts, as well as other surrogate markers still to be validated, may become a critical criterion for adapting prophylaxis, maternal treatment and appropriate infant feeding options. More research is required for ensuring transmission risk reduction in mothers who benefited from a perinatal prophylactic ARV but have no acceptable alternative to breast feeding, including ARV regimens covering the lactation period.
- By no means should an HIV-infected pregnant woman eligible for ARV therapy be deprived of adequate ARV therapy for herself, where available.
- Access to appropriate family planning services must be guaranteed.
- A possible association between an excess maternal mortality in HIV-infected mothers and breastfeeding should be urgently scrutinised in existing data sets (retrospectively) and in new research projects.
- Considerable efforts are still needed to optimise safety of all potential feeding practices by appropriate education of both health professionals and mothers and identify adequate standardised indicators to assess infant feeding practices (formula feeding, animal milk, exclusive breastfeeding, early cessation of breastfeeding, etc.).
- More research is urgently needed on differential transmission risks associated with breastfeeding practices, pathophysiology of breast milk transmission and viral/host relationships related to MTCT. Social sciences should also contribute considerably to our understanding of HIV transmission by breastfeeding.

PART TWO

Prophylactic Versus Therapeutic HIV Vaccine

Since 1985, attempts to develop an efficacious HIV vaccine have been as numerous as have the failures. The barrier has been the lack of sufficient scientific knowledge. A major problem in the development of an HIV vaccine is the high variability of HIV. There are two types of HIV: HIV-1 and HIV-2. The HIV-1 is further divided into 3 groups (M, N and O) and each group is subdivided into several subtypes. In addition there are sub-subtypes and inter-subtype as well as inter-group recombinants. The distribution of the various HIV-1 subtypes differs in different parts of the world. Furthermore there are virus variants which differ in their phenotype according to the chemokine receptor usage, the major of which being CCR5 and CXCR4.

Can we make a vaccine? It is still not certain when it will be possible to develop an effective and safe preventive HIV vaccine for use in humans since no efficacy trial has yet been completed. However, knowledge gained over the last decades in studies of the natural infection of humans, in particular long-term survivors, combined with pre-clinical non-human primate vaccine studies have led us to believe that an efficacious vaccine can be developed. Furthermore, approximately 70 phase I/II clinical trials done in humans using different vaccine constructs and modalities of immunization have shown that the tested vaccine candidates were safe and able to induce specific immune responses of varying intensity. An ideal sterilizing prophylactic vaccine should be safe and induce: i) cross-neutralizing antibodies against primary HIV-1 isolates from divergent HIV subtypes; ii) strong and broad CD4+ T-cell responses; iii) poly-epitopic, cross-clade reactive CD8+ CTL responses; iv) mucosal immune defenses; and v) long-term protection.

Prophylactic vaccine approaches. There are several types of possible HIV vaccine candidates including live attenuated virus, whole inactivated virus, recombinant produced subunits, synthetic peptides, live recombinant vaccines and viral DNA. Live attenuated SIV vaccines have been the most efficient in eliciting protective immunity in the SIV/macaque model. However, these vaccines were found to induce disease in macaques and this approach is not applicable in humans for safety reasons. Envelope subunits of HIV-1 elicit antibodies which neutralize laboratory strains but fail to efficiently neutralize field isolates, and usually do not induce cytotoxic T lymphocytes (CTL) responses.

Recent experiences indicates the advantages of using mixed modality immunization, i.e. immunizing with several HIV-1 antigens (env, gag, pol, nef, tat, etc.) either as genetic information in DNA (plasmids) or in live expressing vectors (pox, adeno, salmonella, BCG, etc.), or as virus-like particles, recombinant proteins, peptides, peptide-conjugates. Typically an individual will be primed with one construct (DNA or vector) and boosted with another construct (vector or proteins). Several regimens are currently in phase I/II studies.

Therapeutic vaccine approaches. Before 1996 it was not realistic to have an effective therapeutic vaccine. This is because the HIV-1 infection seriously reduces the number and impairs the function of CD4+ cells, which are central to the immune system. HAART (highly active anti-retroviral therapy), which besides limiting the virus replication and improving the quality of life gives a partial, if not full, restoration of the immune system, allows the development of a therapeutic vaccine controlling HIV infection. An initial phase I study, using a pox vector coding for env and gag protein plus recombinant gp160 protein, delayed viral rebound for four months in 2 out of 4 patients who stopped HAART treatment after termination of vaccine alone. These encouraging results initiated a series of phase II trials with different mix-modality regimens. Data from these trials will be available in the second half of 2002.

Can we make a mother-child vaccine? Current vaccine trials focus on the adult population but do not yet address the children. Even if an efficacious prophylactic vaccine was available there may be an insufficient time to elicit a protective HIV-1 immune response in the newborn against the perinatal HIV infection. Therefore the pregnant woman is the obvious target to vaccinate.

However, pregnant women are usually excluded from vaccination. Indeed it is only the tetanus toxoid that is given to pregnant women in the third trimester in regions where neonatal tetanus is a serious threat. The obvious need of HIV vaccination in pregnant women raises new challenging ethical questions, in particular whether the benefit to the child and mother is greater than potential risks. Therefore we need to evaluate the use of replicating versus non-replicating vectors as well as the use of DNA in experimental animal systems, before starting a clinical trial. Furthermore it will be necessary to design immunization protocols for the pregnant mother, preferably during the third trimester, followed by a prophylactic vaccination for the newborn. Although the primary goal is to prevent HIV-1 transmission to the fetus/child, it is imperative that the mother will continue to be a vaccine recipient.

When can mother-child vaccine be available? As soon as risk factors are evaluated and regulatory approval is obtained phase I trials can be started. However, needed pre-clinical animal studies and the difficult ethical, safety and liability concerns with respect to the pregnant woman are likely to slow development. Therefore it is unrealistic to expect that a comprehensive vaccine program will be available in the coming decade.

REFERENCES

1. Adetunji, J. Trends in under-5 mortality rates and the HIV/AIDS epidemic. Bull World Health Organ 2000; 78: 1200-6.
2. Gray, G. AIDS pediatric epidemics in South Africa. World Federation of Scientists. Planetary Emergencies Conference. Erice, Italy, August 2001.
3. Tlou, S. Mother-to-child transmission of HIV in Botswana. World Federation of Scientists. Planetary Emergencies Conference. Erice, Italy, August 2001.
4. Samuel, N.M. AIDS in India. World Federation of Scientists. Planetary Emergencies Conference. Erice, Italy, August 2001.
5. Lyamuya, E.F., Kagoma, C., Mbena, E.C., Urassa, W.K., Pallangyo, K., Mhalu, F.S. and Biberfeld, G. Evaluation of the FACSCount, TRAx CD4 and Dynabeads methods for CD4 lymphocyte determination. J Immunol Meth 1996;195:103-112.

418

6.   Bush, C.E., Donovan, R.M., Manzor, O., Baxa, D., Moore, E., Cohen, F., Saravolatz, L.D. Comparison of HIV type 1 RNA plasma viremia, p24 antigenemia, and unintegrated DNA as viral load markers in pediatric patients. AIDS Res Hum Retroviruses 1996;12: 11-5

7.   Zeng Yi,. AIDS in China. World Federation of Scientists. Planetary Emergencies Conference. Erice, Italy, August 2001.

8.   Van de Perre, P. Mother-to-child transmission of HIV with special emphasis on breastfeeding transmission. World Federation of Scientists. Planetary Emergencies Conference. Erice, Italy, August 2001.

9.   Wiktor, S.Z., Leroy, V., Ekpini, E.R, et al. 24-month efficacy of short-course maternal zidovudine for the prevention of mother-to-child HIV-1 transmission in a breast feeding population. A pooled analysis of two randomized clinical trials in West Africa. XIII International AIDS Conference. Durban, South Africa. July 2000. [Abstract TuOrB354]

10.   Nduati, R., Richardson BA, John G, et al. Effect of breastfeeding on mortality among HIV-1 infected women: a randomised trial. Lancet 2001; 357:1651-5.

CONTACT INFORMATION

WFS / PMP on AIDS & Infectious Diseases
Secretariat : Pr Guy de Thé, Institut Pasteur, Retrovirus Department, 28 rue du Dr Roux, 75015 Paris   Tel: (33) 1 45 68 89 30 – Fax :(33) 01 45 68 89 31 –e-mail: dethe@pasteur.fr

SUMMARY OF THE WORKSHOP ON GEOPHYSICAL AND GEOLOGICAL PROPERTIES OF NEOS: "KNOW YOUR ENEMY"

J.M. GREENBERG
University of Leiden, Raymond & Beverly Sackler Laboratory for Astrophysics,
2300 RA Leiden, The Netherlands

W.F. HUEBNER
Southwest Research Institute, San Antonio, Texas 78228-0510, USA

ABSTRACT

The first international workshop dedicated to the determination of geophysical properties and physical characterization of NEOs was held in Erice in June 2001. The goals were to discuss the science and technology requirements in these fields for the coming decades. Four quantities were identified for which measurements from NEOs are critically needed: (1) the mass, (2) the mass distribution, (3) material strengths, and (4) the internal structure. Global (whole body) properties, such as material strengths and internal structure, can be determined best from the analyses of penetrating waves: Artificially initiated seismology and multifrequency reflection and transmission radio tomography. Seismology provides the best geophysical (material strengths) data of NEOs composed of consolidated materials while radio tomography provides the best geological data (e.g., the state of fracture) of nonconducting media. Thus, the two methods are complementary: Seismology is best for stony and metallic asteroids, while radio tomography is best for comet nuclei and carbonaceous asteroids. The three main conclusions are: (1) remote sensing for physical characterization should be increased, (2) several dedicated NEO missions should be prepared for geophysical and geological investigations, and (3) it is prudent to develop and prove the technology and learn how to make geophysical measurements on NEOs now.

PREAMBLE

The first international workshop dedicated to the determination of geophysical properties and physical characterization of NEOs was held in Erice June 17 to June 25, 2001. It was supported by the European Space Agency (ESA), the Italian Ministry of University Scientific Research and Technology, the Italian Space Agency (ASI), the USA National Aeronautics and Space Administration (NASA), the Japan Institute of Space and

Astronautical Science (ISAS), the Japan Society for the Promotion of Science (JSPS), and the Sicilian Regional Government. Participants came from China, Germany, Italy, Japan, Netherlands, Poland, and USA. Presentations covered the areas of NEO research in Earth collision avoidance and mitigation, properties of asteroid surfaces and internal structure, geophysical properties, space missions to determine physical properties, laboratory and computer simulation experiments, and comet – asteroid – meteorite – dust links.

We know little about geophysical properties of NEOs. Many parameters are needed to develop credible mitigation or collision avoidance strategies. Current theoretical and laboratory activities are insufficient to provide plausible orbital deflection techniques. *In situ* exploration has been and will be essential. Remote observations will greatly benefit from dedicated ground and space-based visible and IR facilities. We are still far from developing a credible defense system. The NEO impact hazard is a planetary emergency that deserves a major international collaborative effort.

INTRODUCTION

Observations of the planets and their moons show that asteroids and comets have impacted them throughout the history of the planetray system. The evidence from the craters on the Moon and on Mars is overwhelming. The Earth has also suffered such collisions. Even in recent history, such events have taken place. On 30 June 1908 a small object, probably less than 100 m in size, exploded in the atmosphere over Tunguska, Siberia, NNW of the town of Vanavara (about 101° E, 62° N). Although the object was too small to form a crater, it charred the ground and flattened the trees radially outward from the center below the explosion over an area of about 2000 square kilometers. More recently, in July 1994 astronomers witnessed telescopically the collisions of 21 fragments of Comet Shoemaker-Levy 9 with Jupiter. The comet had been torn apart by the gravitational forces of Jupiter as it orbited very closely to the planet. Many of the fragments colliding with Jupiter left "black eyes" that were larger than the Earth.

Collisions of asteroids and comets with Earth have occurred in the past. About 150 impact structures on the Earth's continents have been identified. There may be many more impacts whose structures have eroded over time. Since about 70% the surface of the Earth is covered by oceans, a straightforward extrapolation indicates that at least 500 collisions of asteroids and comets with Earth must have taken place. In Table 1, we give a few representative craters formed by impacts on Earth. In spite of millions of years of atmospheric erosion, these craters are surviving as the evidence of cosmic collisions. The probabilities of further collisions have been documented in several reports and books (Morrison, 1992; Rather et al., 1992; Canavan et al., 1993; Gehrels, 1994; The *Chelyabinsk-70 Workshop*, 1994; The *Planetary Defense Workshop*, 1995). We will not dwell further on these issues.

Table 1. Representative List of Impact Craters on Earth.

Crater	Location	Diameter [km]	Depth [m]	Age [10³ yr]	Comments
Barringer	Arizona, USA	1.3	180	49	
Wolf Creek	Western Australia	0.9	45	300	19°18' S, 127°46' E
Ries and Steinheim	Bavaria, Germany	25. 2.5	240 100	15,000	Simultaneous impacts
Popigai	Siberia, Russia	100.		35,000	71°30' N, 111°00' E
Chicxulub	Yucatan, Mexico	180.		65,000	21°20' N, 89°30' W
Manicouagan	Quebec, Canada	100.		213,000	51°23' N, 68°42' W
Acraman	Australia	160.		570,000	32°01' S, 135°27' E
Sudbury	Ontario, Canada	200.		1,850,000	46°36' N, 81°11' W
Vredefort	South Africa	300		2,020,000	27°00' S, 27°30' E

Finding near-Earth objects (NEOs) has become an important task for planetary astronomy. The International Astronomical Union (IAU) as well as space agencies of several countries have recognized the importance of detecting NEOs, following up on the detections to ascertain their orbital parameters, and cataloging them. As important as these objectives are, they are only a first step toward mitigating Earth collisions of such objects. It is for the first time in human history that defense against such objects is considered possible. To mitigate an Earth collision, we need to know how and where to apply the required forces on an NEO without splitting it. Forces should be directed through the center of mass of the object, otherwise they are wasted in spinning the object (Huebner, 1999). A decision has to be made whether to use gradual, long-term applications of small forces or sudden and large impulsive forces. The material properties of the object and its internal structure (e.g., the state of fracture) play an important role in such decisions. Thus, determination of geophysical properties and geological structures as well as physical characterization of NEOs are very important. This was the first workshop in which these topics relevant to "know your enemy" were of prime consideration.

PRIMARY GOALS

The primary goals of the workshop were to discuss and determine which geophysical properties of asteroids and comet nuclei must be obtained to better understand and implement NEO collision mitigation. This means how to investigate the geophysical properties and geological structures of NEOs. Specifically, methods should be explored to determine these properties and a database of geophysical and related properties should be established. Methods must be expanded for characterization and search. Dedicated space missions and instrumentation must be developed. A roadmap for determining NEO geological and geophysical properties has been outlined.

The status of the NEO program is assessed by the following four categories:

In progress:	Find NEOs, follow them up with observations to determine and catalog their orbits.
Falling behind:	Physical characterization, which includes remote sensing to determine the sizes of objects and their albedo independently. The rate of characterization by remote sensing is not keeping pace with the rapid increase in the rate of new NEO discoveries (Tedesco et al., 2000).
Discussed here:	Determination of geophysical properties such as mass and mass distribution, moments of inertia, material strengths, internal structure, and relationship of global properties to surface properties. Determination of geophysical properties also relates to the secondary goals discussed below.
Final goal:	Develop techniques for Earth-collision mitigation based on geophysical properties. This has to be left for future discussions.

Secondary Goals

Closely related to the determination of geophysical properties and geological structure are the formation and evolution of small bodies in the solar system: Origins of asteroids, comet nuclei, and transition objects, collision and orbital history of these objects, their relationship to meteorites, etc. were all topics of the workshop. Thus, the science of formation and evolution of the planetary system played an important part in the workshop. However, not only science benefited from the approach to address the collision mitigation problems, so did resources exploitation of asteroids and comet nuclei in near-Earth space. Anchoring of a spacecraft on an object that has unknown surface conditions and very weak gravity is a prime concern to several spacecraft missions under development. Resource extractions (e.g. metals and building materials from asteroids and water from comet nuclei) are closely related topics.

A ROADMAP TO DETERMINE GEOPHYSICAL PROPERTIES AND GEOLOGICAL STRUCTURES OF NEOS

A roadmap was developed for geophysical data acquisition and physical characterization. It consists of four parts: The establishment of a database, experimental and theoretical simulations, small bodies missions, and physical characterization. These four parts of the roadmap are discussed below.

<u>Establish an NEO Database</u>

The NEO database consists of four parts: An observational database, a material properties database, database for missions and instrument development, and a database useful for dissemination of projects and results and for public outreach.

Observational Data: The database should contain detailed data for a representative sample of about 30 asteroids and some comets. This number represents approximately the one-sigma level of the estimated 1000 NEOs larger than 1 km in diameter. However, it falls far short of the one-sigma level of the estimated 25,000 NEOs larger than 200 m that could cause considerable regional damage. While the main emphasis is on near-Earth asteroids, main-belt asteroids may be considered as typically representative. Spectra, light curves, etc. can be obtained from ground-based observations. Dimensions, shapes, craters, masses, densities, spin states, etc. must be obtained from space-based observations and measurements. The observational database should gather, connect, and supplement data from various existing databases. A sample of existing databases is shown in Table 2. Links should be established to many more relevant and existing databases.

Table 2. Some Existing Databases.

Name of Database	Address of Database
Planetary Data Systems Small Bodies Node	http://pdssbn.astro.umd.edu
European Asteroid Research Node	http://earn.dlr.de
NASA NEO Program	http://neo.jpl.nasa.gov/
Asteroid and Comet Impact Hazards	http://impact.arc.nasa.gov
Database for Physical Properties of NEOs	http://earn.dlr.de/nea
Rotation periods of Asteroids	http://www.astro.uu.se/~classe/projects/rotast_eng.html
IAU Minor Planet Center	http://cfa-www.harvard.edu/cfa/ps/mpc.html
Occultations of Stars by Asteroids	http://www.lunar-occultations.com/iota/asteroids/astrndx.htm
Rotation Periods of Asteroids	http://www.astro.uu.se/~classe/projects/rotast_eng.html
NEA Tracking (NEAT)	http://neat.jpl.nasa.gov/
Spacewatch	http://pirlwww.lpl.arizona.edu/spacewatch/
LowellObservatory NEA Search (LONEOS)	http://asteroid.lowell.edu/asteroid/loneos/loneos.html
Catalina Sky Survey	http://www.lpl.arizona.edu/css/
Japanese Spaceguard Association (JSGA)	http://www.spaceguard.or.jp/
Lincoln NEA Research	http://www.ll.mit.edu/LINEAR/
Asteroid Radar Research	http://echo.jpl.nasa.gov/
European NEA Search	http://www.astro.uu.se/planet/earn/euneaso.htm
Uppsala-ESO Survey of NEOs	http://www.astro.uu.se/planet/uesac_eng.html
NEAR-Shoemaker Mission	http://near.jhuapl.edu/

424

In Table 3 we summarize observational data of importance to NEO collision mitigation. Most of these data can be acquired by remote sensing, either ground-based or from spacecraft flybys. For example, some indication of the shape, such as the axes ratios, can be obtained from ground-based observations. Radar can determine the shape in much more detail including surface roughness, but the object must come close to Earth. Spacecraft flyby missions are needed for objects that are more distant. Mass determination requires at a minimum a close and slow spacecraft flyby. The mass is determined from the departure of the trajectory of the spacecraft caused by the gravitational attraction of the object. The density is not measured directly. It is derived from the size, shape, and mass of the object. Completed, ongoing, and planned missions to asteroids and comets will fill many of the gaps in our knowledge. Such missions include Galileo, Cassini, NEAR-Shoemaker, DS-1, Stardust, CONTOUR, Deep Impact, Muses-C, and Rosetta.

Table 3. Observational Properties.

Properties	Asteroids	Comet Comae	Comet Nuclei
Size	•		•
Shape	•		•
Radar shape models	•		•
Axes ratios	•		•
Mass	•		•
Density	•		•
Spin state	•		•
Spin rate	•		•
Spin axis orientation	•		•
Light curve	•	•	•
Albedo	•		•
Polarimetry parameters	•	•	•
Taxonomic classification	•		
Absolute magnitude (H, G)	•		
Spectral type (8 color)	•		
Binzel's spectroscopic survey	•		
Orbital parameters	•		•
Geological properties	•		•
Geophysical properties	•		•
Internal structure	•		•
Family membership	•		
Binaries	•		
Dust bands	•		
Crater counts	•		
Boulder counts	•		
Fracture size & spacing distribution	•		
Gas production rate		•	
Chemical composition	•	•	•
Dust-to-gas mass ratio and $Af\rho$		•	
10 μm emission feature		•	

Dust trails		•	
Meteor streams		•	
Size distribution		•	
Density		•	
Composition		•	

Material Properties Database: Two extreme approaches can be considered for nudging an object out of its orbit. One extreme is to apply a relatively small force for a long time. This method is preferred and can be implemented if a warning of a potential collision is determined decades in advance. It has the advantage that the incoming object is less likely to be fragmented by the collision avoidance measures. Fragmentation could cause a series of smaller collisions with Earth over very wide or possibly global regions.

The other approach is to use a large force for a very short time. The impulse transmitted to the object is the same in both cases. However, the response of the object may differ significantly. When a sudden (explosive) force is applied to an object, a shock wave may be transmitted through the object. This shock wave may cause spall on the far side of the object, therefore reducing the effectiveness of the applied impulse.

Static Data

By static data, we mean data important for applying a gentle push during collision mitigation. In terms of impulse transmitted to an NEO, this means a small force acting over a long time.

Table 4. Static Properties.

Property	Conditions
Mass and mass distribution	
Ice – dust – void mixtures	$T < 200$ K
Density	
Porosity	
Pore radii	
Electric permittivity	
Loss tangent	
Thermal conductivity	As function of temperature
Heat capacity	As function of temperature
Enthalpy	As function of temperature
Sound speed	

Dynamic data

By dynamic data, we mean data relevant to impulsive transmission of an impulse during collision mitigation. This implies a large force acting over a short time, an explosion.

Table 5. Dynamic Properties.

Property	Conditions
Momentum coupling coefficients	For various types of interactions
Strain rates	
Young's modulus	
Poisson ratio	
Yield, flow, or fracture stress in compression	
Yield, flow, or fracture stress in tension	
Hugoniots	
Grüneisen parameter	
Energy and momentum dissipation rates	As function of density, porosity, etc.

Missions and Instrument Development: Several fully instrumented rendezvous missions will have to be carried out. Such mission also may define additional types of measurements that will be needed and explore different techniques for their effectiveness, limitations, spatial resolution, and dynamic ranges to determine and characterize whole-body properties of a variety of NEOs. Lander technology will have to be developed as part of the fully instrumented missions but also as goals for missions of opportunity. After these basic determinations have been made, micro-spacecraft missions may follow to explore detailed and specific properties. In a final phase of NEO exploration flyby measurement techniques using penetrators and instrumented penetrators should be investigated.

Instrument development can be divided into two categories: *In situ* surface and remote sensing instruments and instruments for determining bulk properties. Remote-sensing instruments have been widely used and are well developed. In this category fall spectroscopic and thermal IR measurements that can be used to determine emission properties. The *in situ* surface instrumentation, such as multi-axial accelerometers, sample coring, and penetrators to determine composition and geology will need further development. Penetrators and landers that can work in swarms on and just below asteroid and comet nucleus surfaces may need to be developed.

In the second category are instruments for determining bulk properties. This group encompasses the most important aspects for primary and secondary goals of geophysical and geological exploration of NEO properties. While drilling and digging on an NEO can provide detail about the composition and structure of an NEO, it does so only locally, at one spot and to a very limited depth of typically one to a few meters. Analysis of waves that penetrate an NEO to a depth of several hundred meters or are transmitted through the entire object give whole-body or global information. Among such waves are electromagnetic (radio) waves and sound (seismic) waves. While general background radiation can be useful, artificially induced waves and pulses are more easily analyzed. Radio tomography and seismology are complementary methods to achieve the goals of geophysical determinations of asteroids and comet nuclei. Electromagnetic radiation is limited to nonconducting materials. These are typically fluffy and porous objects such as comet nuclei that are composed of ice and dust and carbonaceous asteroids. Sound waves are rapidly dissipated in these objects. On the other hand, sound waves propagate well in dense objects such as stony and metallic asteroids.

Artificially induced wave pulses can be timed. This makes determination of sound speeds, be they three-dimensional compression or shear waves or two-dimensional surface (Rayleigh) waves, possible. Sound speed is related to material strength parameters. In addition wave shapes can be used in the analysis of material properties (see, e.g., Huebner et al., 2001).

Radio tomography does not provide as much information about material properties as it does about the structure of an object. Radio waves are refracted, reflected (scattered), and absorbed depending on the electric permittivity of the materials. Reflections occur at interfaces between materials. They also reveal fractures and other discontinuities in materials. Thus, radio tomography is very useful in revealing the internal structure of an object. This is very important for distinguishing, for example, monolithic objects from rubble piles. With the use of tomographic techniques, the measured radio echoes can be converted to three-dimensional images of the interior of an NEO. This procedure is similar to ultrasonic imaging in medicine. We distinguish between transmission and reflection tomography. For transmission tomography a transmitter as well as a separate receiver are needed. For example, the transmitter can be on a rendezvous spacecraft orbiting an NEO while a receiver has been landed on its surface. Reflection tomography on the other hand, can have the transmitter and the receiver on the orbiting spacecraft. It has the added advantage that the timing from reflected signal reveals the location of a discontinuity on the object. Multifrequency radio tomography permits penetration of the radio signals to different depth. The most powerful radio tomography is a combination of multifrequency transmission and reflection tomography. The complex permittivity of materials determines the speed of the radio signal propagation and its dissipation in the material.

We encourage complete data analysis of past and present missions to exploit all information that may be useful for NEO collision mitigation. We also encourage development of new missions to small solar system bodies. In particular, to accomplish good data analysis we need missions using multi-axial accelerometers, sample coring, and penetrators to determine thermal properties (heat of fusion, heat of vaporization, heat conduction, heat capacity, etc.). Whole-body properties can best be obtained from analyses of patterns and transmission speeds of waves that penetrate the entire body. Transmission radio tomography, i.e., the determination of the complex electric permittivity, will work best on comet nuclei and carbonaceous asteroids. Reflection radio tomography can reveal reflecting surfaces (e.g., fractures) within a stony body. Seismology is best suited for consolidated (as opposed to highly fragmented or porous) objects. For example in nickel-iron objects, radio waves will be absorbed while seismic (sound) waves are easily transmitted.

Probes need to be developed to measure electric conductivity locally. Spectroscopic and thermal emission properties such as intensity as a function of wavelength, particle sizes, particle density, particle temperature, etc. need to be measured. Gamma, alpha, neutron, and x-ray probes should be used to measure the composition of objects.

Dissemination and Public Outreach: Topics of interest include not only properties of asteroids, comets, and comet nuclei. It is also important to inform the public about potential methods of NEO collision mitigation. The origin and evolution of the planetary system and resources exploitation in near-Earth space are topics of great interest. Possible projects for public outreach on properties of NEOs include public forums, high school projects, semipopular articles and forums, museum exhibits, and multimedia websites.

Experimental and Theoretical Simulations

Theoretical Models: Development of thermal models is important for the analysis of asteroid data. Coupled thermal and gas diffusion models are needed for the interpretation of comet nucleus data. Temperature profiles determine sound speed. Gas diffusion reveals information on porosity.

What are the parent bodies of near-Earth asteroids and how steady is their population? Models to investigate the source regions of NEOs will aid in understanding the evolution of the planetary system. The collision history of asteroids will clarify their injection rates into NEO orbits. The dynamical mechanisms deserve further investigations.

Computer Simulations: Very important are simulations of artificially activated seismic events. What is the best placement of an activator on an asteroid? Since asteroids have no atmosphere and negligible gravity, how does one achieve consistently good coupling of the activator to the body with the least mass and at the least expense? What signals can be expected for various objects? What is the best placement of seismometers relative to activators?

Similar questions arise for radio tomography. How complex is the analysis of refracted, multiply reflected, scattered, and attenuated radiation? Simulations of transmission and reflection tomographic experiments will be most useful to answer these questions.

Numerical integration of orbits of observed NEOs is important to understand their past history and predict future encounters. Closely related to this are models for asteroid–asteroid collisions. In the case of comets, a further complication arises from the non-gravitational forces (the recoil from the outgassing) on the motion of their nuclei.

Laboratory simulations: Laboratory simulations should be carried out to investigate artificially activated seismic experiments. Laboratory simulations of radio tomography can be carried out using scaled objects and wavelengths. These types of experiments, when performed in conjunction with theoretical simulations will be useful tools for mission planning and for analysis of mission data.

Landing and anchoring experiments are a challenge for bodies with little gravity to hold down an instrument package. Much can be learned from the successful landing of the NEAR spacecraft on Eros. Landing and anchoring is part of the Rosetta mission to Comet Wirtanen.

Penetrator technology will play an important role in future missions to NEOs. This technology is in its infancy of development. It needs to be expanded to permit the use of swarms of minipenetrators.

Missions to Small Solar System Bodies

Table 6. Some critically needed data from NEOs.

Quantity to be measured	Acquisition technique
Mass	Spacecraft trajectory
Mass distribution	Spacecraft trajectory
Spin axis orientation	Remote sensing
Spin rate	Remote sensing
Moments of inertia	From mass distribution and spin states
Material strengths	Artificially activated seismology
Internal structure (fracture state)	Seismology and radio tomography
Porosity	From mass and volume determination, drilling, digging
Composition	x-ray, gamma-ray, neutron spectroscopy, drilling, digging

While global (whole body) data acquisition is preferred, some locally obtained data is also desirable. Local data will usually be limited to a depth of one to a few meters. Drilling and digging is the most likely process of obtaining local data. Methods for this approach are being developed for the Rosetta mission to comet Wirtanen.

Whole body (global) data acquisition depends primarily on wave analyses. Wave speeds and wave shapes can be analyzed. Artificially initiated seismology, using impacts and explosives, is most useful for measuring the sound speeds for compressional, shear, and surface waves. Multifrequency transmission, reflection, and combined transmission and reflection radio tomography are being considered. Radio tomography is being developed for the Rosetta mission to Comet Wirtanen.

Comprehensive data analyses of past missions should be carried out in an effort to identify properties most important to successful mitigation. We also endorse new missions to small solar system bodies missions and their comprehensive data analyses. Models and instrumentation has to be developed for geophysical experiments.

Physical Characterization

Ground-based physical characterization does not keep pace with the discovery rate of NEOs (Tedesco et al., 2000). The reason for this is the rapidly increasing rate of discovery of NEOs, and a constant rate in physically characterizing these objects. An increase of remote sensing characterization is most desirable. A most crucial datum needed for assessing the hazard from NEOs is the mass of each object. This information is missing for most NEOs. While mass cannot be determined by physical characterization, determining the size of an object is a first indication of its mass. What is measured is the brightness of the object, which is a combination of its cross section and its albedo. To separate these two quantities, simultaneous visual and IR observations are needed.

Highest priority after discovery and orbit determination should therefore be given to ground-based characterization. This requires:

- Access a system that can acquire simultaneously thermal IR and visible measurements of the largest possible fraction of the NEO population on short notice.
- Observations of a sample of NEOs over large wavelength and phase angle ranges and with sufficient resolution to resolve spin variations and to provide input for detailed thermophysical modeling.

The detailed requirements include:

Primary physical characterization of the general population in order to:
- Obtain a statistically significant database of fundamental NEO properties, namely effective diameter and albedo.
- Measure a number of objects of each taxonomic type.
- Develop a more accurate assessment of the hazard posed by objects in different size and mass ranges.
- Assess selection effects in survey discovery statistics.

Detailed observation of a sample NEOs over a range of:
- NEO sizes, rotation states, taxonomic types.
- Phase angle (> 90 deg. highly desired)
- Wavelengths (taken simultaneously) in the optical regime and the thermal- to far-IR and submillimeter to radio.
- Time to determine light curve amplitude and shape.

Observational basis for thermophysical models of representative NEOs.

431

Observing strategy:
Technical advantages of space-based NEO observing program:
- Detection efficiency better than ground-based facilities
- High duty cycle
- Observations near the Sun
- Mid- and far-IR observations.

Routine rapid access to ground-based or space-based observing facilities for NEO characterization is essential.

SUMMARY

Many types of data were identified for which measurements from NEOs are needed, but the four most important quantities to be measured are: The mass, the mass distribution, the material strengths, and the internal structure. Velocities relative to Earth can be determined quite accurately, but the mass is needed to determine the energy of impact as well as the energy and momentum needed to deflect the object in its orbit. The mass distribution is needed to determine the center of mass of the object. Any applied force should be directed through the center of mass of the NEO to ensure that the energy and momentum transfer are most effectively applied into translational motion and not into spinning the object. Material strengths and material structure determine the response of the object to the applied forces. Global material strengths and structure are best determined from artificially activated seismology experiments and from multifrequency radio tomography.

CONCLUSIONS

Mass, not size is the important quantity determining the destructive energy of an impactor and the energy needed to deflect it. However, size together with taxonomic classification are important indicators of mass. *Increase remote sensing for physical characterization.*

Digging and drilling give local information only. Global geophysical properties (material strength and internal structure) can be determined from wave analyses. Seismology for consolidated matter complements radio tomography for loose, nonconducting matter. *Several dedicated NEO missions are needed for geophysical measurements.*

It is prudent to *develop and prove technology and learn how to make geophysical measurements now.* Instrument and mission development takes time.

REFERENCES

1. Canavan, G.H., Solem, J.C., Rather, J.D.G. (eds.) *Proceedings of the Near-Earth-Object Interception Workshop.* Los Alamos National Laboratory report LA-12476-C, 1993.

2.   *Chelyabinsk-70*. Space Protection of the Earth against Near-Earth Objects. Organized by VNIITF, Russian Federal Nuclear Center, 1994.

3.   Gehrels, T. (ed.) *Hazards Due to Comets and Asteroids*. The University of Arizona Press, Tucson, London, 1994

4.   Huebner, W.F., `Physical and chemical properties of comet nuclei' in *International Seminar on Nuclear War and Planetary Emergencies*, 23rd Session, Goebel, K. (ed.) p. 169-179, 1999.

5.   Huebner, W.F., Cellino, A., Cheng, A.F., Greenberg, J.M., `NEOs: Pysical properties' in *International Seminar on Nuclear War and Planetary Emergencies*, in press, 2001.

6.   Morrison, D., The Spaceguard Survey: Report of the NASA International Near-Earth-Object Detection Workshop. JPL/Cal/Tech report, Pasadena 1992.

7.   *Planetary Defense Workshop*. Lawrence Livermore National Laboratory report CONF-9505266, 1995.

8.   Rather, J.D.G., Rahe, J.H., Canavan, G. *Summary Report of the Near-Earth-Object Interception Workshop*. NASA, Washington, DC, 1992.

9.   Tedesco, E.F., Muinonen, K., Price, S.D. *Space-based infrared near-Earth asteroid survey simulation*. Planet Space Sci., in press, 2000.

PMP REPORT: MISSILE PROLIFERATION AND DEFENSE

PROF. ANDREI PIONTKOVSKY
Strategic Studies Centre, Moscow, Russia

The last thing I want to do on our last day in this beautiful environment is to be a bearer of bad tidings. But unfortunately the bad news is that our group did not manage to produce a joint statement.

As we all know, the joint statement is reached by a consensus of all the participants of a group and during our last seven or eight meetings we always managed to find common ground. This time we tried to do so until the last moment but unfortunately there were very strong disagreements on one of the points which I refer to later. Certainly the main responsibility for this failure rests with the chairman. But in my view it also reflects a profound political problem beyond my control or the control of any other chairman.

Firstly I would like to remind you of one of the points in our famous Erice Statement of 1998. This Statement was signed by all the Russian, American, European participants including all the members of our current group, and also by many top representatives of the American, Russian and British Establishment.

The main recommendation in 1992 was that we reach an agreement to move forward on the entire complex affair of defensive strategic arms control to unambiguously reach highly effective limited national missile defenses while preserving the retaliatory capability of the United States and Russia.   All the Americans enthusiastically signed this document because it was a confirmation of the American position. The Russian participants also signed this document, which was a kind of conceptual and political breakthrough because it completely contradicts the position of the Russian government on this point. Not only did we sign the Statement, but we did our best in our country to change the perception and mentality of Russian politicians and decision makers to induce acceptance of this strategic formula as being beneficial not only for the United States but also for Russia and for the world. Why do I mention these documents so extensively? Because now, when I reach the point in our draft of a final statement which produced controversy, you see that it is just a repetition of the same formula.

The point, which was resolutely opposed by one of the members of our group, was:

> "The panel recommends that the governments of Russia and the United States replace the ABM Treaty by an agreement that permits the deployment of a limited number of ABM launchers and does not jeopardize the mutual deterrent

capability. The Agreement must however provide protection from unauthorized rogue missile launches, a threat which must be taken into account in the near future. The Agreement shall be subject to regular review in order to adjust the agreed upon number of launches as required by changes to the international security environment."

It is the same formula, perhaps somewhat more technically developed, but our American friend, Greg Canavan, resolutely opposed this formula and we could not reach a consensus for the final statement of the group.

I could stop here in describing the poor results of our group, but may I elaborate on this problem outside the context of our working group because it seems to me that, controversy aside, our group fully reflected the problems which will be the substance, the crux of the strategic debates, of the top world leaders in the coming months.

The situation is that, for years, the United States convinced Russia and other countries that it was necessary to modify the ABM Treaty to allow them to protect themselves from the rogue states and other accidental launches. Well, we experts realized that that was a legitimate security demand and we subscribed to it and advocated it in our country and almost succeeded in obtaining agreement.

Then the United States drastically changed their position and all the statements made by top American politicians, especially during the past two or three months, give us to understand that America is no longer interested in an agreement on arms security at all. That they themselves would not like to subscribe to any arms control agreement. They want to be free of any international law-based restraint and able to formulate this new strategic framework as they like.

It is a new approach and what happened inside our group is just a reflection of this new approach. As little as a year ago this formula did not provoke any controversy. I think it is really an unhelpful development, as I said during my talk two days ago, even from an American point of view, because it is very difficult to convince the outside world of such an approach to world stability.

This is the latest July monthly offering of the Institute of Security Status of the Western European Union. The subject is "Nuclear Weapons and Nuclear Debate". It is a collection of pieces by the most famous European strategists including Russians. The Russian piece is the most conservative and the most mild in discussing this new American approach. May I just quote one phrase from the Preface: "It will not be easy to sell American partner countries, even its closest allies, on the idea that strategic regulation should paradoxically become the rule in the international relation future. Yet, without this support, is it not possible that America will become nothing more than a "rich lonesome cowboy?" I feel the same sorrow as the author of this paper and, if I am allowed to make some prediction, I think this position of unilateral stance allows no agreement at all. Agreement is almost unacceptable in the current American political criteria. This position cannot be sustained by American administration for long because it provokes very serious criticism in the United States itself including in the United States Senate.

As for future of our group, I always thought that we were coming here to Erice not to publish additional papers and to increase our quotation index, but to prepare together recommendations for our governments to act upon. Especially in this nuclear field, because our Annual Meeting is called Nuclear War - Global Emergency and we have just concluded its 26th Annual Meeting. So if our task is to prepare a conceptual draft of this new agreement, and if one of the very important players in the world says that he is not interested in any agreement at all, what is the point of our work?

But I do not want to be so pessimistic, I would like to finish on an optimistic note because I think that this position will not be sustained for long. It is a very unconvincing, bizarre mixture of arrogance and ignorance, of hypocrisy and naiveté and it will be rejected within the United States itself and then we will again be able to return to more constructive work.

At the workshop of 19 August and in further meetings of panel members during the subsequent days the following observations and conclusions had been forwarded:

- The Erice statement on ballistic missile defense of 1992 advocated the development of a global missile defense system by like-minded nations against unauthorized missiles launches from all origins and destinations.
- The 1998 statement endorsed the 1992 statement and emphasized the need for protection against theater threats and regional threats. Subsequent events have underscored the threats from rogue, accidental, and unauthorized launches. Thus, it is appropriate for like-minded countries to jointly develop, deploy, and control defenses against these threats.
- While the defenses that have been deployed by Russia and that are under development by the United Sates will provide useful protection for them, global defenses that cover allies and friends would provide greater protection and stability for all as well as the foundation for the common defense of all like-minded nations.
- Defenses could be developed rapidly by the nations and incorporated into a global defense system at a fraction of expenditures for offensive weaponry. They would effectively eliminate the threat of nuclear war by design or accident and devalue ballistic missiles thus reducing proliferation incentives.
- Limited deployment of defenses as currently planned by the United States would not affect the strategic balance between Russia and the United States and would not stimulate new strategic arms races. However, in order to preserve strategic stability and prevent a return to antagonistic strategic relationships, like-minded nations shoud agree that further defense deployments should be accompanied by appropriate reductions in offensive systems.
- The panel considered for recommendation that the governments of Russia and the United States terminate the ABM Treaty and enter into an agreement that permits the deployment of a limited number of ABM launchers and does not jeopardize their mutual deterrent capabilities while providing protection from

rogue, accidental, and unauthorized missile launches that must be reckoned with in the forseeable future. The agreement shall be subject to regular review in order to adjust the agreed upon number of launchers as required by changes in the international security environment

- The panel considered for recommendation the establishment of an exploratory group for discussion of architectural options for the evolution of a global missile defense system by like-minded countries and for developing models for estimating substitution rates between defensive and offensive systems based on a common understanding of security requirements and processes that are at the heart of strategic stability.

WORLD FEDERATION OF SCIENTISTS PERMANENT MONITORING PANEL ON INFORMATION SECURITY

DR. HENNING WEGENER
German Ambassador to Spain (former), Madrid, Spain

The group met on 19 August to continue the preliminary discussion it held last year, with a view to reaching a basic understanding on the thrust of its future work, to adopting a work programme and to deciding on a suitable division of labour among its members.

The discussion confirmed the Panel's view that the global connectivity of, and the increasing societal dependence on, modern information technologies, together with its susceptibility to disruption, increased the fragility of civil society; in addition to violating privacy and causing damage to business information interests, critical societal infrastructure assets were potentially imperiled, and attacks on national defense assets would constitute a major threat to peace and security.

The adoption of strategies to prevent and defend against unlawful attacks on these information infrastructures (depending on the context often referred to as "information warfare", "cyber crime" or "cyber terrorism") and to safeguard information security had thus become a major challenge to societies and states as well as the international community. However, the awareness of these dangers was as yet insufficient, not least in the international business community. It had to be increased world-wide, and available techniques to preserve information security had to be coordinated, completed and made effective as a matter of urgency. For this purpose, a first and major priority of the Panel should be the elaboration of a comprehensive interdisciplinary report (or White Book), to be given wide international distribution. Being issued under the auspices of the World Federation of Scientists, it would enjoy added impact.

For the structure and contents of the Report the Panel adopted the following decisions.

An introductory chapter should highlight the threats emanating from the potential of the aforementioned attacks and provide a succinct status report of action already undertaken on the national, regional and international level. This chapter should avoid duplication with work already performed elsewhere, and should limit itself to cross-references where feasible. However, a glossary of terms relevant to the topic might usefully be included.

It was agreed that the Report should primarily focus on strategies to deter and defend against action directed against the integrity, confidentiality and availability of computer systems, networks and computer data, and that its recommendations should be geared to this purpose.

The introductory sections would be followed by eight chapters of an operational character, each resulting in specific recommendations:

1. Needs and measures in the field of education; strategies to heighten public problem awareness and to propagate already existing methods and techniques.
2. National and international monitoring systems (CERT, FIRST, networks of the UN, ICAO, IAEA, etc.) and possibilities for their increased use, interaction and coordination for the purpose of monitoring IT disruption.
3. Investigative/forensic mechanisms, existing and new (including INTERPOL, EUROPOL, etc.) and their optimum interplay in order to assess attribution of authorship and safeguard evidence.
4. Guidelines for national criminal codes and their application throughout the international community.
5. Techniques for the possible intervention by law-enforcement authorities in the contents of international networks, and the limits to such measures of control (including the problem of encryption) under national legislation.
6. Cooperation on preventive and protective techniques and in introducing common standards for, and management of, information security. Possible support for the introduction of such techniques in developing countries
7. Management of consequences of acts of disruption (emergency assistance, ensuring business continuity, lessons learned).
8. Filling gaps in existing international law, remedying deficiencies in international cooperative practices.

The Report should, if possible, be written, discussed and finalized prior to the International Erice Seminar in August 2002. It could then be made available to delegations to the UNGA in September of that year.

The Panel undertook to assign drafts for the individual subchapters of the Report to its present members, but left open the possibility of soon recruiting a limited number of additional members under the auspices of geographical balance and special expertise.

Looking beyond the work outlined above, the Panel confirmed its intention to continue to monitor closely events in this rapidly evolving field, devoting special attention to the emergence of new threats to information security and innovative ways to defend against them.

WORLD FEDERATION OF SCIENTISTS PERMANENT MONITORING PANEL ON POLLUTION—2001 REPORT

DR. RICHARD C. RAGAINI
Department of Environmental Protection, University of California, Lawrence Livermore National Laboratory, Livermore, CA, USA

The continuing environmental pollution of earth and the degradation of its natural resources constitutes one of the most significant planetary emergencies today. This emergency is so overwhelming and encompassing, it requires the greatest possible international East-West and North-South co-operation to implement effective ongoing remedies.

It is useful to itemize the environmental issues addressed by this PMP, since several PMPs are dealing with various overlapping environmental issues. The Pollution PMP is addressing the following environmental emergencies:

- degradation of surface water and ground water quality
- degradation of marine and freshwater ecosystems
- degradation of urban air quality in large (mega) cities
- impact of air pollution on ecosystems

Other environmental emergencies, including global pollution, water quantity issues, ozone depletion and the greenhouse effect, are being addressed by other PMPs. The Pollution PMP coordinates its activities with other relevant PMPs as appropriate. Furthermore, the PMP will provide an informal channel for experts to exchange views and make recommendations regarding environmental pollution.

PRIORITIES IN DEALING WITH THE ENVIRONMENTAL EMERGENCIES

The PMP on Pollution monitors the following priority issues:

- clean-up of existing surface and sub-surface soil and ground-water supplies from industrial and municipal waste-water pollution, agricultural run-off, deforestation, and military operations
- reduction of existing air pollution and resultant health and ecosystem impacts from long-range transport of pollutants and trans-boundary pollution
- prevention and/or minimization of future air and water pollution

440

- training scientists & engineers from developing countries to identify, monitor and clean-up pollution

MEMBERS OF POLLUTION PERMANENT MONITORING PANEL

The following scientists listed below were appointed by the chairman as permanent members, because of their interdisciplinary expertise in environmental matters:

Chairman Dr. Richard C. Ragaini, Lawrence Livermore National Laboratory, USA
Dr. Lorne G. Everett, University of California at Santa Barbara, USA
Prof. Ilkay Salihoglu, Middle East Technical University, Ankara, Turkey
Prof. Sergio Martellucci, University of Rome, Italy
Dr. Gennady I. Palshin, ICSC-World Laboratory, Ukraine
Prof. Paolo Ricci, University of San Francisco, USA
Prof. Vittorio Ragaini, University of Milan, Italy
Academician Albert Tavkhelidze, National Academy of Sciences of Georgia

ASSOCIATE PANEL MEMBERS OF THE POLLUTION PMP

The following scientists listed below were appointed by the chairman as associate panel members:

Prof. William Sprigg, University of Arizona, USA
Prof. Robert Clark, University of Arizona, USA
Prof. Joerg Drewes, Arizona State University, USA
Dr. Herman Bouwer, US Department of Agriculture, USA
Dr. Vladimir Mirianashvilli, Institute of Geophysics, Georgia
Mr. David Rice, Lawrence Livermore National Lab, USA
Prof. Soroosh Sorooshian, University of Arizona, USA
Ms. Kay Thompson, Department of Energy, USA

HISTORICAL AREAS OF EMPHASIS OF THE POLLUTION PMP

The following areas listed below have been addressed by the Pollution PMP since its beginning in 1997 in order to more effectively monitor generic and hot-spot impacts of pollution in developing countries:

- 1998: Impacts of Sewage Treatment Wastewater Used for Irrigation
  - emphasis on impacts of pharmaceuticals and disinfectant byproducts
- 1999: Memorandum of Agreement (MOA) between WFS and the US Department of Energy
  - cooperative ventures in the Black Sea are the first implementation of the MOA
- 1999: Contamination of Groundwater by Hydrocarbons
  - emphasis on solvents and petroleum

- 1999: Black Sea Pollution
- 2000: Contamination of Groundwater by Hydrocarbons
  - emphasis on MTBE
- 2000: Black Sea Pollution by Hydrocarbons From Oil Spills
- 2001: Caspian Sea Pollution
- 2001: Transboundary Water Conflicts
- 2001: Water and Air Impacts of Automotive Emissions in Megacities
- 2002: Radioactivity Contamination of Soils and Groundwater

The Pollution PMP has addressed these issue areas by sponsoring talks and workshops; writing reports, recommendations, project proposals; and by continued monitoring.

SUMMARY OF ACTIVITIES OF THE POLLUTION PMP IN 2001

January: DOE/NOAA/WFS Tbilisi Workshop on Oil Spill Plume Modeling
- Organized by K. Thompson and R. Ragaini
- Four Black Sea scientists from Ukraine, Georgia (2), and Romania (Dimitru Dorogan) were supported by WL scholarships to participate in the Workshop

August 5: Panel on Water and Air Impacts of Automobile Emissions
- Organized by S. Martelucci
- Held at 31st Course of International School of Quantum Electronics "Global Automotive Laser Applications"
- Summary can be obtained by contacting Professor Martelucci

August 22: International Seminar Talks on Transboundary Water Conflicts
- Jointly organized by R. Ragaini and S. Sorooshian, Chairman of the Water PMP
- Papers are contained elsewhere in this volume

August 24/25: Workshop on Caspian Sea Pollution
- Jointly organized by R. Ragaini and I. Salihoglu
- Workshop summary and papers are contained elsewhere in this volume

August 15: WL Pilot Proposal Approved: "Environmental Pollution Caused by Utilization and Processing of Radioactivity-Bearing Minerals in Vietnam" for 2002-2005
- Co-directors: P. Ricci and D. Lien of the Vietnam Atomic Energy Commission

FUTURE ACTIVITIES PLANNED BY THE POLLUTION PMP

August, 2002: Organize a session on Radioactivity Contamination of Soils and Groundwater at the International Seminar

- Include talks on "Long-Term Stewardship": the US program to stabilize and monitor the Department of Energy sites in the US, which are contaminated with radioactivity, rather than clean them up.
- Also schedule talks on other sites around the world, such as Semipalatinsk, with radioactive contamination.

August, 2002: Organize talks on the current state of the Aral Sea, including the situation of the island containing a former Soviet military chemical/biological agent laboratory.

EXTENDING ACTIVITIES OF THE WORLD FEDERATION OF SCIENTISTS IN UKRAINE

PROF. VALERY P. KUKHAR, DR. GENNADY I. PALSHIN
ICSC-World Laboratory, Ukraine Branch

THE UKRAINE'S SCIENTIFIC POTENTIAL TODAY

- 120 institutions of the National Academy of Sciences,
- \> 100 institutions of the Academy of Agrarian Sciences and Medical Sciences,
- ~ 150 universities and other high education institutions, and 600 research organizations (in Industry, Medicine etc.)

The overall staff employed (both scientific and technical) ~ 180,000 people.

PLANETARY EMERGENCIES, WHICH ARE THE FOCUS OF THE WFS IN THE UKRAINE

- Chernobyl
- Energy
- Water Pollution Issues with a special emphasis on the Black Sea pollution
- Sustainable Agriculture and Food Processing
- New Military Threats in the multi-polar World (Danger of Proliferation of the Weapons of Mass Destruction)

Chernobyl
The Workgroup set up in Ukraine, particularly, includes: Professor Victor Baryakhtar as a Head, and Professors Valery Kukhar, Vyacheslav Shestopalov, Emlen Sobotovich. All are members of the WFS and take part in Erice Seminars.
Main trends of research activities are:

1. Problems of the Chernobyl Nuclear Power Plant
2. Biomedical Issues of Chernobyl Accident aftermath
3. Conversion of the "Shelter" into an environmentally safe object

Decommission of the Chernobyl Nuclear Power Plant –
Ukraine has no experience in decommission of civil graphite RBMK NPP, thus, the scientifically based program has to be worked on outside of the Ukraine.

Related Issues:
- waste and fuel handling;
- equipment for decontamination and disposal (recycling);
- personnel employment.

Biomedical Issues of Chernobyl Accident aftermath –

Related Issues:
- "Samossely" – people willfully settled in the Exclusion and Resettlement Zone as large as 30 km around the Nuclear Plant;
- Health Monitoring among rescue personnel and population exposed;
- Databank development.

A new book under the editorship of Professor Baryakhtar is titled, "30-km Zone. Biological and medical after-effects."

Conversion of the "Shelter" object into an environmentally safe entity –
Behavior of fuel-containing mass inside the 4th unit is the primary concern for scientists. The monitoring shows the compact glasslike masses to transform now in brittle porous compounds and dust. There is water inside and outside the Shelter.

Energy
While importing power sources—oil and gas in large quantities from Russia and Turkmenistan—Ukraine has available a large stock of oil and uranium. Ukrainian scientists work on:

- fuel cycle development for nuclear power plants;
- advanced technologies and equipment for oil extraction from thin and steeply inclined seams;
- various aspects of the unprofitable mines closures (more than 100 or 40% of them have to be closed so far) – workers employment, underflooding of lands etc.;
- development of methane separation from oil;
- Ukrainian Power Industry development concept

Black Sea Pollution–
Investigation of the Black Sea pollution problems as part of the Water-related emergencies monitoring is a great concern of Ukrainian scientists working under the WFS auspices.

The Ukrainian Research Institute of Sea Ecology is led by Professor Mikhailov to implement the Program of the Black Sea Monitoring

Foods-associated emergencies–
Agrarian, Biological and Information Scientists work together on the Agrarian Information Support system development.

Research being made in conjunction of the three fields by Ukrainian scientists has been summarized in recommendations for the Ukrainian Government.

As result, it has been designated at the governmental level to work out a concept and program of the National Agrarian Producer Consulting Network in Ukraine.

The International Symposium "Role of Agrarian Information in the European Integration processes" was held last May with partial support from the Ukrainian Branch of the World Laboratory.

Three research projects in the field of Agrarian Biotechnologies are being developed under the leadership of the Ukrainian WFS members – Professor Vitaly Kordium and Dr. Gennady Palshin.

This research particularly covers the biological method of nitrogen fixation, technology of plant resistance enhancement by specially designed bacteria and development of a gene vaccine against the swine fever disease.

This research also intends studying the risk of their impact on biodiversity.

Kangaroo Mother Care method–
Agreement on the WL Kangaroo Mother Care Center foundation in Kiev (Ukraine) has been signed by Professor Zichichi, Public Health Ministry and Academy of Medical Science officials.

Methodical help provided by Dr. Nathalie Charpak, Director of KMC WL Center in Colombia during her visit to the Ukrainian Research Institute for Pediatrics, Neonatology and Gynecology in Kiev on last March with kind support of the World Laboratory Headquarters.

The initiative has been approved of the National Foundation "Ukraine for Children" working under the auspices of the Ukrainian President's spouse Ludmila Kuchma.

PERMANENT MONITORING PANEL REPORT: LIMITS OF DEVELOPMENT AND SUSTAINABILITY

HILTMAR SCHUBERT
Fraunhofer-Institut for Chemical Technology, Germany

ATTENDANCE AND MEETINGS

The PMP has met on August 19 and 21 with about 15 attendees.

PAPERS PRESENTED

During the morning 4 papers were presented:

- "SUSTAINABILITY OF DEVELOPMENT", by Hiltmar Schubert
- "SUSTAINABILITY: PROPOSAL FOR A PLAN OF WORK", by Geraldo G. Serra
- "LIMITS TO GROWTH; LIMITS TO PLANNING", by K.C. Sivaramakrishnan
- "PLANNING A SUSTAINABLE URBAN DEVELOPMENT", by Juan Manuel Borthagaray

The first two contributions dealt with the items and objectives of the PMP.

"SUSTAINABILITY OF DEVELOPMENT" - Hiltmar Schubert

The objectives contain the exhibits and explanation of special situations and technical and/or social counter- and/or preventive measures to avoid emergencies. In this context global and different regional limits have to be dealt with and the situation of undeveloped and developed countries and countries in transition have to be taken in consideration.

To avoid these limits, a "Sustainable Development" has to be reached. For "Sustainable Development" the definition of the well known "Brundland-Commission" would be very useful:

"Meeting the needs of the present population without endangering the wellbeing of future generations." (1987)

In order to consider the special situation in undeveloped countries and countries in transition, the following sentence was added:

"The aim of sustainable development is the just and equitable improvement of the quality of life for all people."

Sustainability can be attained only by a network of economic, ecological and social components. These components influence each other and have to be balanced (Fig.1):

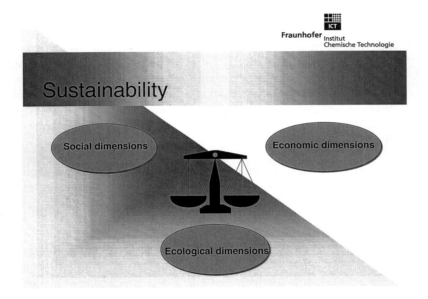

For instance, if an economical way for a "Sustainable Development" cannot be found, ecological and social solutions have no future in a long term. The issues of development concern:

- Population
  - Demographic growth
  - Quality of life
  - Education
- Resources (inputs)
  - Raw Materials
  - Water
  - Food
  - Energy
- Waste (output)
  - Into water
  - Into soil
  - Into atmosphere

For the protection of the human being and for the environment, the "Enquete Commission" of the German Federal Diet stated in 1997 the following rules:

- The rate of usage of renewable resources should not exceed the rate of regeneration.
- Non-renewable resources should be used only in the extent of an equivalent compensation, which is to be found in other renewable resources or by higher productivity (renewable or non-renewable resources).
- The introduction of materials into the environment should take into consideration loading limits of the environment. Also, the very sensitive function of the regulation in the environment has to be taken into account.
- The timing of anthropogenic input or-action into the environment should be balanced with that of the relevant natural processes.
- Dangerous or non-acceptable risks for human health by anthropogenic actions should be avoided.

If practical experiences are taken into consideration, a dynamic interpretation should be used. Referring to resources, the following points underline the dynamic behavior of a "Sustainable Development:"

- Resources have no static quantities.
- Resources are based on common intentions (nuclear energy, oil, agricultural useful areas...etc).
- Preferences of future generations may not match the preferences of today.
- Resources are also determined by technical means in order to utilize these materials (availability).
- The importance of resources will move constantly because of the progress of R and D.
- Human ingenuity has enlarged the basis of resources as a rule.

"SUSTAINABILITY: PROPOSAL FOR A PLAN OF WORK"—Geraldo G. Serra

Prof. Serra presented a general review of the performance of the Limits of Development PMP and a proposal for guidelines for next year's work.

Table 1 – Categories for sustainable development studies.

Use of renewable resources • Water: availability of fresh water, production of drinking water, sewage treatment, recycling and flooding; • Energy: hydroelectric power, wind, solar, geothermal and bio mass; • Soil • Food
Development of substitutes and technologies for non-renewable resources • Energy: fossil fuels • Soil
Development of technologies to reduce emission, recycle or make disposal of pollutants harmless • Water • Soil • Food
Demographic Growth

Prof. Serra proposed also a continuous monitoring of information and interpretation on data concerning the objectives of the PMP, particularly through the organization of a data bank on those subjects and the maintenance of a page on the WFS website to diffuse the proposals of the PMP.

450

"LIMITS TO GROWTH; LIMITS TO PLANNING"—K.C. Sivaramakrishnan

The paper describes, as an example, the emerging of urban corridors starting from Bombay towards the southeast direction over more than 1000 km. The urbanization of these transport corridors represent a typical example of "Limits of Development" because these areas are running out of water and land and suffer from air-, water- and ground-pollution. The poor people living there are getting poorer and rich people getting richer (Fig. 2).

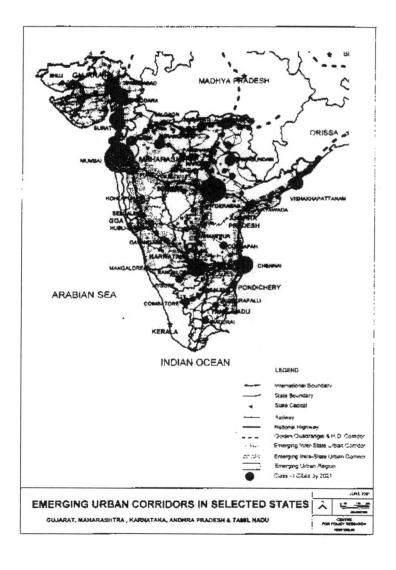

LEGEND

International Boundary
State Boundary
State Capital
Railway
National Highway
Golden Quadrangle & H.D. Corridor
Emerging Inter-State Urban Corridor
Emerging Intra-State Urban Corridor
Emerging Urban Region
Class - I Cities by 2021

EMERGING URBAN CORRIDORS IN SELECTED STATES

GUJARAT, MAHARASHTRA , KARNATAKA, ANDHRA PRADESH & TAMIL NADU

"PLANNING A SUSTAINABLE URBAN DEVELOPMENT"—Juan Manuel Borthagaray

The paper proposed guidelines for city design including the following aspects:

- Economic and Social Issues
  - An Urban Bill of Rights to all Citizens
  - Environmental and Social Values in the Economic Equation
- Scientific and Technological Issues
  - Clean Energy and Substitution of Non-Renewable Resources
  - Water Conservation and Minimization of Solid Waste
- Physical Design Issues
  - Symbolic Centers and Public Spaces to favor Encounter and provide Enjoyment of the City
  - Land Use and Production Techniques to Minimize Transport
  - Densification to favor Pedestrians- and Public Transport
  - Discouragement of individual Motorcars
  - Transport Lanes harmonious with Public Spaces
  - Proximity between Urban Areas, Parks and geographic Scenery

DISCUSSION

During the second part of the meeting other members presented their views on the papers and proposals. Prof. Huo Yu-ping presented his views on the PMP main subject, from a Chinese viewpoint, focusing on the question of mineral carbon consumption and the production of CO_2, as being as important issue to the Chinese development.

Important contributions were presented by Christopher D. Ellis, Margaret S. Petersen, Gerhard Kuies, Teja Valencic, Bertil Galand, Mbarek Diop, Juras Pozela, Leonardas Kairiukstis and Zenonas Rudzikas.

CONCLUSIONS

Emergency Related to Development Limits
To speak with one language, the following definition of "Limits of Development" was agreed:

Limits of Development are the observance of limitations of ecological, economical and social developments in order to avoid emergencies to human beings in a global or regional sense.

The PMP has adopted to keep the title "Limits of Development" for the PMP but adding the expression *"Sustainability"* to it.

The main aspects should be in the same category, as shown in Figure 3 :

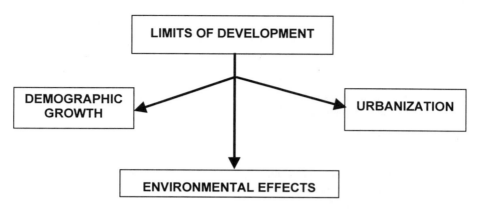

Monitoring the scientific results of research
The PMP has proposed to monitor scientific results on the subjects:

- Use of renewable resources
- Development of substitutes and technologies for *non-renewable resources*
- Development of technologies to reduce *emission,* recycle or make disposal of *pollutants* harmless
- Demographic growth
- Urbanization

Organization of a data bank
The PMP intends to feed in a permanent manner a data bank on relevant findings in the field mentioned above and will produce an annual report evaluating the progress made.

2002 Session Theme
Especially for the year 2002, **Urban Waste** should be the topic. The objective is the monitoring of the state of the art and science in this subject, what means gathering information and interpretation taking into account various countries, development stages and available technologies.

2003 Session Theme
For the year 2003 the State of the Art and Science of the topic **Urban Mobility** is proposed taking the following items in consideration:

- Moving people, material, waste, information and energy
- Organization of urban mobility

WORLD FEDERATION OF SCIENTISTS PERMANENT MONITORING PANEL, CLIMATE, OZONE & GREENHOUSE EFFECT

PROF. WILLIAM SPRIGG
Institute for the Study of Planet Earth, University of Arizona, Tucson, Arizona

The WFS PMP #7 met on the occasion of the 26th Session of the International Seminars on Planetary Emergencies to review and plan its activities. Panel representatives joined in discussions with the water resources and extreme meteorological events panels to plan joint activities.

ACCESS TO DATA & TOOLS FOR RESEARCH

The Climate Panel continues to promote equal access to scientific data, information, and the tools of research may be given to scientists of all nations. A workshop is proposed in conjunction with the next (27th) Session of the International Seminars on Planetary Emergencies to identify and plan ways to encourage data exchange and fill important gaps in international research programs.

The Panel recommends that a task group be established to develop and implement an overall strategy for PMP action, a narrower view than had been recommended earlier.

GREENHOUSE WARMING

Recognizing the need to understand the current wisdom regarding the forces and mechanisms behind global warming, the Panel recommends a special session be held at the 27th Erice Seminar on this subject. This session would explore the uncertainties behind global warming as well as clearly outline the well-established and understood components of the problem.

The Panel recommends that the WFS send an observer to annual meetings of the Intergovernmental Climate Convention to monitor international concerns and proposed actions and to provide input as an NGO. The Panel further recommends that WFS participate at the World Summit on Sustainable Development in Johannesburg in October 2002.

13. TRANSMISSIBLE SPONGIFORM ENCEPHALOPATHY WORKSHOP

STRUCTURAL VARIATIONS OF ABNORMAL PRION PROTEIN IN GERSTMANN-STRÄUSSLER-SCHEINKER DISEASE

PEDRO PICCARDO, BERNARDINO GHETTI
Department of Pathology and Laboratory Medicine, Indiana University
School of Medicine, Indianapolis, IN USA

FABRIZIO TAGLIAVINI, ORSO BUGIANI
Istituto Neurologico "Carlo Besta", Milano, Italy.

INTRODUCTION

A central event in the pathogenesis of prion diseases is a conformational change of the normal prion protein (PrPc) into a pathogenic isoform (PrP^{sc})[1-3]. PrPc has a predominantly α-helical structure; in contrast, PrP^{sc} has an increased ß-sheet structure[4-8]

The methionine/valine (M/V) polymorphism at the *Prion Protein* gene (*PRNP*) codon 129 can influence the clinical phenotype produced by a mutation[2]. In Creutzfeldt-Jakob disease (CJD), a sporadic or inherited disease characterized by dementia and spongiform degeneration in the cerebrum and cerebellum, two main PrP patterns, PrP^{sc} type-1 and type-2, have been described[9-10]. PrP^{sc} type-1 following deglycosylation is characterized by a PK-resistant C-terminal fragment of 21 kDa, while PrP^{sc} type-2 is characterized by a C-terminal protease-resistant peptide of 19 kDa[9]. Gerstmann-Sträussler-Scheinker disease (GSS) is a genetically determined prion disease clinically characterized by ataxia and cognitive impairment[3]. GSS is caused by several mutations in *PRNP*[3,12,13]. The pathologic hallmark of GSS is the accumulation of PrP, with and without amyloid tinctorial properties, in the brain[3]. Our studies have shown that the pattern of PrP^{sc} isoforms in the multiple GSS variants analyzed is different from that seen in CJD[11-12]. Patients with CJD present full-length and N-truncated PrP fragments while patients with GSS present full-length as well as N- and C-terminal truncated PrP peptides[10-12,45,15]. In GSS associated with the *PRNP* F198S mutation (GSS F198S), we showed that the smallest fibrillogenic peptide has an N-terminus at residue G81[16]. Similar studies in patients with GSS associated with the *PRNP* A117V (GSS A117V) mutation show major N-terminal cleavage site at residue G81 and G88-G90[17-18]. In addition, we have reported on the presence of PrP^{sc} isoforms of 27-29, 18-19 and 8 kDa in patients with GSS F198S[11].

In GSS A117V, previous studies have been contradictory in terms of whether or not PK-resistant PrP is present[12,19,20]. The aims of the present study are to determine the biochemical characteristics of PrP in several patients and an asymptomatic carrier from two unrelated USA families with GSS A117V[21,22]. In addition, based on the observation that the pattern of digestion of PrP^{sc} depends on the tertiary structure of the protein, we investigated whether PrP conformational isomers are present in phenotypically different GSS variants. To explore this possibility, we determined the N-terminal cleavage sites of the PrP fragments that accumulate in GSS A117V and GSS F198S.

RESULTS

The experiments were carried out using brain tissue of individuals, from previously described families, with *PRNP* mutations A117V and F198S[21-26] (see Table 1). Western blot analysis for detection of PrP was done in brain extracts, subcellular and purified PrP fractions, as previously described[11-12,15,27-31].

Table 1.

| | Patient | K | AD | Duration | Residue 129 | G | PrP | | | Spong | NFT |
							Neocortex	Striatum	Cerebellum		
GSS A117V	1	SG	61	3 yrs	MV	M	+/++	++	-	-	-
	2	SG	39	3 yrs	MV	F	+/++	++	+/-	-	-
	3	SG	32	7 yrs	VV	M	+/++	+/++	+/-	-	-
	4	SG	33	5 yrs	VV	M	+/++	++	+++	-	-
	5	AS	45	4 yrs	VV	F	++/+++	++	+/-	-/++	-
	6	AS	50	N/A a	MV	M	-	-	+/-	-	-
GSS F198S	7	IK	64	6 yrs	MV	F	+++	+++	+++	-	+++
	8	IK	61	12 yrs	VV	F	+++	+++	+++	-	+++

a: asymptomatic carrier. K: kindred, AD: age at death, G: gender, Spong: spongiform degeneration, NFT: neurofibrillary tangles. Semi-quantitative analysis of lesions (-) absence, (+) mild, (++) moderate, (+++) severe. Subjects 1-6 correspond to 2 families with GSS A117V and patients 7 and 8 to the Indiana kindred of GSS F198S.

PrP in brain extracts of subjects with GSS A117V

Immunoblots of brain extracts from subjects with GSS A117V demonstrated PrP in non PK-treated homogenates, however considerable heterogeneity of PrP^{sc} was observed. PrP^{sc} varied in overall amount and in the relative quantity of the different isoforms, in different brain areas of individual patients and among patients. In many samples obtained from the frontal cortex and cerebellum no PrP^{sc} was detected. Nevertheless, PrP^{sc} was seen in selected samples of the frontal cortex of all patients following digestion with PK10 µg/ml. In cases 1-5, a PrP^{sc} band of 14 kDa was observed, with patients 1-3 and 5 usually showing an additional band of 7 kDa. In some blots, a 7 kDa band was also seen in patient 4. Extracts obtained from patient 3 showed 7 and 14 kDa bands in samples digested with PK at concentrations ranging from 5 to 100µg/ml.

PrP in subcellular fractions of subjects with GSS A117V.
PrP partitioned with membranes in samples obtained from the frontal cortex of all subjects (i.e. 1-6) with GSS A117V. We observed that the membrane fraction obtained from the patients (i.e. 1-5) contained detergent-insoluble and PK-resistant PrP (i.e. PrP^{sc}) fragments of 7 and 14 kDa. PrP^{sc} was not detected in membranes or in the soluble fractions of the sample obtained from the frontal cortex of the asymptomatic carrier (case 6).

Partially purified PrP in subjects with GSS A117V.
Due to the variability of detecting abnormal PrP in brain homogenates and to the fact that previously published protocols are effective in concentrating relatively pure PrP^{sc}, we analyzed partially purified fractions obtained from patients 1-5 and from the asymptomatic carrier (case 6). Following PK digestion, two major bands of 14 and 7 kDa were detected in the patients (i.e., 1-5). A similar pattern was seen when immunoblots were probed with a panel of antibodies raised against the mid-region of PrP (i.e., PrP residues 89-112). No PrP^{sc} was observed in the partially purified sample obtained from the asymptomatic carrier.

Analysis of glycosylation state of PrP^{sc} fragments.
In CJD, PrP^{sc} is composed of a C-terminal glycosylated PK-resistant core. The 7 kDa PrP^{sc} fragment seen in GSS A117V is detected with Mab 3 F4. Considering the apparent size and immunologic profile of this peptide, we speculated that this peptide would not contain the C-terminal N-linked glycosylation sites at residues 181 and 197. To explore this possibility further, we performed enzymatic deglycosylation on samples obtained from all patients (i.e., 1-5), and showed that an identical mobility is seen in PrP^{sc} peptides before and after PNGase F treatment. This finding supports the concept that glycosylation sites are not present in these fragments.

PrP-fragments in non-PK-treated samples
We noticed that in selected samples the low molecular weight PrP fragments (i.e., 7 and 14 kDa) could be seen before PK-digestion. To analyze if the truncated fragments were insoluble in nondenaturing detergent we partially purified PrP in non-PK-digested samples (through ultracentrifugation in the presence of Sarkosyl). In all patients (i.e., 1-5), we observed bands of 27-35, 14 and 7 kDa. The stoichiometry of the low molecular weight fragments in PK- and non-PK-digested samples is different. PK-digestion increases the amount of the 7 and 14 kDa bands. The similar size of the 7 kDa fragments in non-PK and PK-treated samples suggest that a similar or identical cleavage site may be present *in vivo* and *in vitro*.

Determination of the N-terminal cleavage site of PrP fragments in patients with GSS A117V and GSS F198S.

To characterize the primary structure of the 7 kDa PrP^{sc} fragment, partially purified and PK-treated PrP obtained from the frontal cortex of patients with GSS A117V (i.e., 1-5) was analyzed using Edman chemistry. Sequence analysis of the 7 kDa peptide from patient 3 (for 20 cycles) yielded GQGGGTHSQWNKPSKPKTNM as the major N-terminal sequence corresponding to residues 90-109 of PrP. Analysis of the 7 kDa band from the other four patients showed that the major sequence also started with residue 90 of PrP. As stated above, a 7 kDa band was also seen in immunoblots obtained from non PK-digested samples. Sequence analysis of this 7 kDa band from patient 4 yielded GXGQGGGTHSQXNKP as the N-terminal sequence corresponding to residues 88-102 of PrP.

Patients with GSS F198S and GSS A117V accumulate PrP^{sc} fragments that can be cleaved by PK to generate peptides of different mobility (i.e., the small fragment is 7 kDa in GSS A117V and 8 kDa in GSS F198S) in gel electrophoresis. This observation could be due to the presence of conformational isomers that expose different PrP residues to hydrolysis by PK in the presence of detergents. To explore this possibility, we analyzed the 8 kDa PrP^{sc} fragment obtained from the frontal cortex of two patients with GSS F198S. Sequence analysis for 29 cycles showed residues 74-102 of PrP in both cases. This indicates PK cleavage in the second PrP octarepeat region for GSS F198S, different from the PK cleavage site for GSS A117V.

To analyze the possibility that different PrP conformers might be present in different brain areas, we determined the N-terminal sequence of PrP^{sc} purified from the caudate nucleus (an area with absent or small amount of amyloid), and the cerebellum (a region with abundant amyloid) in a patient with GSS F198S. Similar results were observed in both samples indicating G58, G66 or G74 as the N terminus. Although we were not able to determine sufficient residues on the fragments to identify the exact N terminus, based on the size and immunologic profile, the simplest interpretation of the data is that the major N terminus is at residue 74, as found in the frontal cortex.

DISCUSSION

This study demonstrates that PrP^{sc} is present in patients with GSS A117V but was not detected in the samples analyzed from the asymptomatic carrier. The 14 and 7 kDa PrP species were observed in brain extracts and microsome preparations. Whether PrP^{sc} partitions with membranes due to protein-protein, protein-lipid interactions or the formation of sedimentable aggregates in the presence of detergents remains undetermined. PK-resistant fragments of 14 and 7 kDa were also seen in purified PrP^{sc} fractions. In all patients (i.e., 1-5), the sequencing of the 7 kDa peptide in non-PK and PK-digested samples, showed a major N-terminal cleavage site between G87-G88 and W89-G90, respectively.

It is noteworthy that PK-digestion changed the stoichiometry of the various PrP peptides observed in the undigested preparations with increase in the 14 and 7 kDa fragments and disappearance of the high molecular weight isoforms of 27-35 kDa. These findings indicate that these patients (i.e., 1-5) accumulate larger PrP isoforms that can be cleaved to smaller, insoluble, PK-resistant fragments. The detection of low molecular weight peptides in samples that have not been treated with PK, suggest that truncated PrP is generated *in vivo* by hydrolysis of preexisting intermediates. These peptides may be the result of the metabolism of PrP in a pathway associated with amyloid formation, since similar size fragments have been found in isolated amyloid cores of patients with GSS[16-18].

These data differ from that showing the absence of PrP^{sc} in patients with GSS A117V[20]. It has been proposed that a transmembrane PrP isoform plays a central role in the pathogenesis of GSS A117V[20]. We did not investigate the presence of different topological form of PrP. Thus, we cannot determine if the PrP^{sc} present in the patients analyzed by us (i.e., 1-5) corresponds to the transmembrane, extracellular, secreted or membrane bound via a phosphatidylinositol anchor PrP species. It is important to note that the amount of PrP^{sc} in the patients (i.e., 1-5) reported here is significantly smaller than that found in other GSS variants[11-12]. In agreement with the results presented here, recent studies in a patient from an Alsatian with GSS A117V[18] showed that the major component of amyloid fibrils was a 7 kDa peptide with ragged N terminus starting mainly at G88 and G90[18].

The 14 and 7 kDa bands, seen in GSS A117V (i.e., patients 1-5), are detected by antibodies directed to the mid-region of PrP indicating that these fragments include the intact epitope recognized by Mab 3F4 consisting of residues 109-112. It has been shown that in a normal human brain, PrPc is endogenously cleaved at residues H111 or M112[15]. Thus, our results suggest that patients with GSS A117V have an alternative metabolic pathway, leading to the accumulation of N- and C-truncated PrP fragments with abnormal physicochemical properties. The data also suggests that the proteolytic pathway present in these patients with GSS is different from that previously described in CJD[10,15]. In the latter, N-truncated, C-terminally-intact PrP^{sc} fragments accumulate in the brain[10,15].

Studies on amyloid fractions have previously shown that patients with GSS F198S accumulate amyloid peptides of 11 kDa spanning residues 58-150[32]. Further analysis showed that the smallest amyloid subunit in GSS F198S and GSS Q217R corresponds to a 7 kDa fragment comprising residues W81-Y150 and W81-E146, respectively[16]. In addition, preliminary data on a patient of an American family with GSS A117V showed a similar fibrillogenic fragment with a ragged N-terminus corresponding to W81, G82 and Q83 and the C-terminus at E146[17]. Thus, patients with these GSS variants may accumulate amyloid subunits of similar size and primary structure, in spite of the different genotypes and phenotypic presentations.

We hypothesized that the purification of PrP^{sc} may allow the isolation of (i) PrP peptides that have not acquired fibrillogenic properties, (ii) peptides that are not metabolized in an amyloidogenic pathway, or (iii) peptides that are precluded from the extraction procedure used to isolate amyloid cores. Therefore, to expand our studies we purified PrP^{sc} from patients of the Indiana kindred with GSS F198S and determined the N-terminal cleavage site of the 8 kDa fragment isolated from areas with and without amyloid accumulation. In all the samples analyzed the major N-terminal cleavage site corresponded to residue G74 of the octapeptide-repeat region, suggesting that as yet unidentified local factors may contribute to PrP amyloidogenesis. In view of the fact that the smallest amyloidogenic fragment has a mobility of 7 kDa and N-terminus at W81 and PrP^{sc} peptides of 8 kDa have an N-terminus at G74, we speculate that sequential proteolytic cleavage of a precursor PrP^{sc} fragment generates fibrillogenic peptides in GSS F198S.

In conclusion, the data obtained in the GSS variants analyzed in this study demonstrate that N- and C-truncated PrP isoforms of different size and N-termini, accumulate in GSS A117V and GSS F198S. Based on the principle that the pattern of digestion (i.e., protease digestion in the presence of detergents) depends on the tertiary structure of proteins[33], the data suggest that PrP conformational isomers are present in these patients.

The results show that octarepeats 3 and 4 are an integral part of the 8 kDa peptide present in GSS F198S, but not in the 7 kDa fragments detected in patients with GSS A117V. The importance of the accumulation of peptides with different N-termini in these GSS variants is unclear at this time. However, other investigators have suggested that Cu2+ binds to a structure defined by two of the octarepeats in PrP containing the sequence PHGGGWGQ[34]. They proposed that this binding could induce conformational changes in PrP[34]. In addition, short peptides corresponding to the octapeptide repeat motif of PrP have been reported to bind Cu2+[35,26]. Moreover, it has been shown that PrP fragments can be transformed from a predominantly α-helical monomeric form to an oligomeric ß-sheet rich secondary structure[37-38]. In conclusion, we speculate that PrP^{sc} fragments with a distinct structure could have different neurotoxic properties or a tendency to form aggregates, providing a possible mechanism underlying the difference in phenotypic presentation among GSS variants.

ACKNOWLEDGMENTS

Supported by PHS P30 AG 10133, R01 NS 14426. We thank Bradley S Glazier for editorial assistance.

REFERENCES

1. Prusiner, S.B., Scott, M.R., DeArmond, S.J., Cohen, F.E: Prion protein biology. Cell 1998, 93:337-348
2. Gambetti, P., Petersen, R.B., Parchi, P., Chen, S.G., Capellari, S., Goldfarb, L., Gabizon, R., Montagna, P., Lugaresi, E., Piccardo, P., Ghetti, B. Inherited prion diseases. Prion Biology and Diseases. Edited by Prusiner, S.B. Cold Spring Harbor Laboratory Press, 1999, pp. 509-583
3. Ghetti, B., Piccardo, P., Frangione, B., Bugiani, O., Giaccone, G., Prelli, F., Dlouhy, S.R., Tagliavini F: Prion protein amyloidosis. Brain Pathology 1996, 6:127-145

4. Bolton, D.C., McKinley, M.P., Prusiner, S.B.: Identification of a protein that purifies with the scrapie prion. Science 1982, 218:1309-1311
5. Meyer, R.K., McKinley, M.P., Bowman, K.A., Braunfeld, M.B., Barry, R.A., Prusiner, S.B.: Separation and properties of cellular and scrapie prion proteins. Proc Natl Acad Sci USA 1986, 83:2310-2314
6. Caughey, B.W., Dong, A., Bhat, K.S,. Ernst, D., Hayes, S.F., Caughey, W.S.: Secondary structure analysis of the scrapie-associated protein PrP 27-30 in water by infrared spectroscopy. Biochemistry 1991, 30:7672-7680
7. Safar, J., Roller, P.P., Gajdusek, D.C., Gibbs, C.J., Jr.: Conformational transitions, dissociation, and unfolding of scrapie amyloid (prion) protein. J Biol Chem 1993, 268:20276-20284
8. Pan, K.M., Baldwin, M., Nguyen, J., Gasset, M., Serban, A., Groth, D., Mehlhorn, I., Huang, Z., Fletterick, R.J., Cohen, F.E., Prusiner, S.B.: Conversion of a-helices into ß-sheets features in the formation of the scrapie prion proteins. Proc Natl Acad Sci USA 1993, 90:10962-10966
9. Parchi, P., Castellani, R., Capellari, S., Ghetti, B., Young, K., Chen, S.G., Farlow, M., Dickson, D.W., Sima, A.A., Trojanowski, J.Q., Petersen, R.B., Gambetti, P: Molecular basis of phenotypic variability in sporadic Creutzfeldt-Jakob disease. Ann Neurol 1996, 39:767-778
10. Parchi, P., Wenquan, Z., Wang, W., Brown, P., Capellari, S., Ghetti, B., Kopp, N., Schulz-Schaeffer, W.J., Kretzschmar, H., Head, M.W., Ironside, J., Gambetti, P., Chen, S.G.: Genetic influence on the structural variations of the abnormal prion protein. Proc Natl Acad Sci USA 2000, 97:10168-10172
11. Piccardo, P., Seiler, C., Dlouhy, S.R., Young, K., Farlow, M.R., Prelli, F., Frangione, B., Bugiani, O., Tagliavini, F., Ghetti, B.: Proteinase K resistant prion protein in Gerstmann-Sträussler-Scheinker disease (Indiana kindred). J Neuropathol Exp Neurol 1996, 55:1157-1163
12. Piccardo, P., Dlouhy, S.R., Lievens, P.M.J., Young, K., Vinters, H.V., Zimmerman, T.R., Mackenzie, I.R.A.M., Brown, P., Gibbs, C.J., Jr., Gajdusek, D.C., Pocchiari, M., Bugiani, O., Ironside, J., Tagliavini, F., Ghetti, B.: Phenotypic variability of Gerstmann-Sträussler-Scheinker Disease is associated with prion protein heterogeneity. J Neuropathol Exp Neurol 1998, 57: 979-988
13. Panegyres, P.K., Toufexis, K., Kakulas, B.A., Brown, P., Ghetti, B., Piccardo, P., Dlouhy, S.R., Cervenakova, L., A new PRNP mutation (G131V) associated with Gerstmann-Sträussler-Scheinker disease. Archives of Neurology (in press)
14. Parchi, P., Chen, S.G., Brown, P., Zou, W., Capellari, S., Budka, H., Hainfellner, J., Reyes, P.F., Golden, G.T., Haw, J.J., Gajdusek, D.C., Gambetti, P.: Different patterns of truncated prion protein fragments correlate with distinct phenotypes in P102L Gerstmann-Sträussler-Scheinker disease. Proc Natl Acad Sci USA 1998, 95:8322-8327
15. Chen, S.G., Teplow, D.B., Parchi, P., Teller, J.K., Gambetti, P., Autilio-Gambetti, L.: Truncated forms of the human prion protein in normal brain and in prion diseases. J Biol Chem. 1995, 270:19173-19180

464

16. Tagliavini, F., Prelli, F., Porro, M., Rossi, G., Giaccone, G., Farlow, M.R., Dlouhy, S.R., Ghetti, B., Bugiani, O., Frangione, B.: Amyloid fibrils in Gerstmann-Sträussler-Scheinker disease (Indiana and Swedish kindreds) express only PrP peptides encoded by the mutant allele. Cell 1994, 79:695-703

17. Tagliavini, F., Prelli, F., Porro, M., Rossi, G., Giaccone, G., Bird, T.D., Dlouhy, S.R., Young, K., Piccardo, P., Ghetti, B., Bugiani, O., Frangione, B.: Only mutant PrP participates in amyloid formation in Gerstmann-Sträussler-Scheinker disease with Ala>Val substitution at codon 117. J Neuropathol Exp Neurol 1995, 54:416

18. Tagliavini, F., Lievens, P.M.J., Tranchant, C., Warter, J.M., Mohr, M., Giaccone, G., Perini, F., Rossi, G., Salmona, M., Piccardo, P., Ghetti, B., Beavis, R.C., Bugiani, O., Frangione, B., Prelli, F.: A 7 kDa prion protein fragment required for infectivity- is the mayor amyloid protein in Gerstmann-Sträussler-Scheinker disease A117V. J Biol Chem 2001, 276:6009-6015

19. Tateishi, J., Kitamoto, T., Doh-ura, K., Sakaki, Y., Steinmetz, G., Tranchant, C., Warter, J.M., Heldt, N.: Immunochemical, molecular genetic, and transmission studies on a case of Gerstmann-Sträussler-Scheinker syndrome. Neurology 1990, 40:1578-1581

20. Hedge, R.S., Mastrianni, J.A., Scott, M.R., DeFea, K.A., Tremblay, P., Torchia, M., DeArmond, S., Prusiner, S.B., Lingappa, V.R.: A transmembrane form of prion protein in neurodegenerative disease. Science 1998, 279: 827-834

21. Nochlin, D., Sumi, S.M., Bird, T.D., Snow, A.D., Leventhal, C.M., Beyreuther, K., Masters, C.L.: Familial dementia with PrP-positive amyloid plaques: a variant of Gerstmann-Sträussler-Scheinker syndrome. Neurology 1989, 39:910-918

22. Ghetti, B., Young, K., Piccardo, P., Dlouhy, S.R., Pahwa, R., Lyons, K.E., Koller, W.C., Ma, M.J., De Carli, C., Rosenberg, R.N.: Gerstmann-Sträussler-Scheinker disease (GSS) with PRNP A117V mutation: neuropathological and molecular studies of a new family. Neuropathol App Neurobiol 1999, 25:57

23. Ghetti, B., Tagliavini, F., Masters, C.L., Beyreuther, K., Giaccone, G., Verga, L., Farlow, M.R., Conneally, P.M., Dlouhy, S.R., Azzarelli, B., Bugiani, O.: Gerstmann-Sträussler-Scheinker disease. II. Neurofibrillary tangles and plaques with PrP-amyloid coexist in an affected family. Neurology 1989, 39:1453-1461

24. Farlow, M.R., Yee, R.D., Dlouhy. S.R., Conneally, P.M., Azzarelli, B., Ghetti, B.: Gerstmann-Straussler-Scheinker disease. I. Extending the clinical spectrum. Neurology 1989. 39:1446-1452

25. Dlouhy, S.R., Hsiao, K., Farlow, M.R., Foroud, T., Conneally, P.M., Johnson, P., Prusiner, S.B., Hodes, M.E., Ghetti, B.: Linkage of the Indiana kindred of Gerstmann-Sträussler-Scheinker disease to the prion protein gene. Nat Genet 1992, 1:64-67

26. Hsiao, K., Dlouhy, S., Ghetti, B., Farlow, M., Cass, C., Da Costa, M., Conneally, P.M., Hodes, M.E., Prusiner, S.B., Mutant prion proteins in Gerstmann-Sträussler-Scheinker disease with neurofibrillary tangles. Nat Genet 1992, 1:68-71

27. Kascsak, R.J., Rubenstein, R., Merz, P.A., Tonna-DeMasi, M., Fersko, R., Carp, R., Wisniewski, H.M., Diringer, H.: Mouse polyclonal and monoclonal antibody to scrapie-associated fibril proteins. J Virol 1987, 61:3688-3693

28. Sastry, S., Linderoth, N.: Molecular mechanisms of peptide loading by the tumor rejection antigen/heat shock chaperone gp96 (GRP94). J Biol Chem 1999, 274:12023-12035

29. Hope, J., Morton, L.J.D., Farquhar, C.F., Multhaup, G., Beyreuther, K., Kimberlin, R.: The major polypeptide of scrapie-associated fibrils (SAF) has the same size, charge distribution and N-terminal protein sequence as predicted for the normal brain protein (PrP). EMBO J 1986, 5:2591-2597

30. Piccardo, P., Langeveld, J.P.M., Hill, A.F., Dlouhy, S.R., Young, K., Giaccone, G., Rossi, G., Bugiani, M. Bugiani, O., Meloen, R.H., Collinge, J., Tagliavini, F., Ghetti, B.: An antibody raised against a conserved sequence of the prion protein recognizes pathologic isoforms in human and animal prion diseases including vCJD and BSE. Am J Pathol. 1998, 152: 1415-1420

31. Liepnieks, J.J., Ghetti, B., Farlow, M., Roses, .A.D., Benson, M.D.: Characterization of amyloid fibril ß-peptide in familial Alzheimer's disease with APP 717 mutations. Biochem Biophys Res Comm 1993, 197:386-392

32. Tagliavini, F., Prelli, F., Ghiso, J., Bugiani, O., Serban, D., Prusiner, S.B., Farlow, M.R., Ghetti, B., Frangione, B.: Amyloid protein of Gerstmann-Straussler-Scheinker disease (Indiana kindred) is an 11 kd fragment of prion protein with an N-terminal glycine at codon 58. EMBO J 1991, 10:513-519

33. Nakayama, N., Arai, N., Kaziro, Y., Arai, K.: Structural and functional studies of the dnaB protein using limited proteolysis. Characterization of domains for DNA-dependent ATP hydrolysis and for protein association in the primosome. J Biol Chem 1984, 259:88-96

34. Stöckel, J., Safar, J., Wallace, A.C., Cohen, F.E., Prusiner, S.B.: Prion Protein Selectively Binds Copper(II) Ions. Biochemistry 1998, 37:7185 -7193

35. Hornshaw, M.P., McDermott, J.R., Candy, J.M., Lakey, J.H.: Copper binding to the N-terminal tandem repeat region of mammalian and avian prion protein: structural studies using synthetic peptides. Biochem Biophys Res Commun 1995, 214:993-999

36. Miura, T., Hori-I, A., Takeuchi, H.: Metal-dependent alpha-helix formation promoted by the glycine-rich octapeptide region of prion protein. FEBS Lett 1996, 396:248-252

37. Zhang, H. Kaneko, K., Nguyen, J.T., Livshits, T.L., Baldwin, M.A., Cohen, F.E., James, T.L., Prusiner, S.B.: Conformational transitions in peptides containing two putative alpha-helices of the prion protein. J Mol Biol 1995, 250:514-526

38. Zhang, H., Stöckel, J., Mehlhorn, I., Groth, D., Baldwin, M.A., Prusiner, S.B., James, T.L., Cohen, F.E.: Physical studies of conformational plasticity in a recombinant prion protein. Biochemistry 1997, 6:3543-3553.

14. AIDS AND INFECTIOUS DISEASES WORKSHOP

COMMUNICATION STRATEGY FOR THE INTRODUCTION OF ANTI RETROVIRAL THERAPY: A CONCEPT PAPER

DR. SHEILAH TLOU
University of Botswana, Gaborone, Botswana

BACKGROUND

The Government of Botswana has committed to make anti retroviral drugs available to all HIV positive people in Botswana. It has thus taken the lead over many other nations in the region and this has been hailed internationally. In most resource poor countries, the government has not been able to provide free access to Anti Retroviral Therapy (ART) through the public health system. This has created its own complications in terms of issues of equity and adherence. Brazil has been the pioneer in breaking the barrier and making ART freely available through the public health system. Many elements combined to make this possible, not the least of which was the decision to defy international patents and produce the drugs indigenously at a fraction of the cost at which they are supplied by the multinational drug companies. Another critical element in their strategy was a concerted and planned communication effort in mass media and social mobilization activities.

Before ART became available, the treatment available for people who tested HIV positive was mainly treatment of opportunistic infections such as TB through IPT, and palliative care. Therefore ART represents a great hope for:

- People who are HIV positive, their families and caregivers.
- People who will test HIV positive in the future, their families and caregivers.
- Employers (both government and private) who will not lose a sizeable portion of their workforce in the next few years, and thereby be able to recover their investments in training and experience.
- The government and employers who will eventually spend less in hospitalization costs.

There is as a result, a great interest in ART among all groups in Botswana, a country which, unfortunately has the highest rate of infection (38.5%) in the world today. However, ART is not simple. Even if we assume that there is an uninterrupted supply of the drugs which are supplied free; there are complex issues of managing the program which need to be understood and addressed for the program to be a success.

Communication strategies would have a critical role to play in addressing these challenges.

CHALLENGES

In Botswana, the challenges are of a different nature from other countries. It has on the one hand, many positive things in its favor which would create a conducive environment for the launch of ART. Some of these are:

- A well developed public health system which has been able to deliver quality primary health care.
- A high literacy rate (68.9%).
- A committed leadership.
- A small population (1.6 million).
- A strong economy.
- A high level of awareness about HIV/AIDS.

However, many of these factors also make this a challenging exercise, at least in the immediate term. Some of these are:

- Inadequate skilled human resources. ART is resource intensive.
- Low density of population making access for all difficult.
- High mobility of population making monitoring difficult.
- Severe strain on the health system as a result of the epidemic.
- Although literacy levels are high, education is still relatively new. Illiteracy also still exists.
- Although a high GDP per capita places Botswana in the category of middle income countries, inequitable distribution of income and unemployment continue to be problems.
- Mass media options are limited. Television, which can powerfully communicate emotive issues quickly, is very new, and its access as well as programming is still limited. Television was a key intervention in the case of Brazil.
- The planning and implementation mechanism for IEC in the HIV/AIDS program as a whole, is in the process of evolving.
- The need for communication efforts to move beyond the knowledge area to behavior change has only been realized recently, and still remains to be implemented.
- An understanding of the reasons for behaviors of particular target groups (who continue to practice risky behavior), which form the basis of behavior change strategy, is in the process of being developed through ongoing research.

PROPOSAL

Based on a study of these factors, the consultants contracted to develop the launch plan concluded that meeting the current levels of demand for ART requires a large increase in capacity at the national, district, and sub-district levels, and that this process will take time. Therefore, they have proposed a strategy of demand management in the initial phase. This is proposed to be done by targeting specific patient groups. Of the estimated 300,000 HIV positive people in Botswana, approximately 200,000 would qualify for ART using the criterion of symptomatic with CD4 count <200, which is used in some countries. It has been proposed for the first year that of these, ART will be made available to the following four groups of HIV positive people who are symptomatic and have CD4 count <200:

- Adult inpatients,
- TB patients,
- Pregnant women and their partners, and
- Children.

Four locations have been proposed for the launch:

1. Gaborone
2. Francistown
3. Serowe
4. Maun

It has been additionally proposed that the initial launch be only in two towns—Gaborone and Francistown—followed by the remaining two as soon as the first two are set up. On this basis, the total number of people to whom ART will be offered in the first year is estimated to be 16,890 or approximately 44% of national demand in the selected patient categories. (At the national level, these four categories account for 36% of total treatable demand.) The recommendation of groups is based on the following considerations:

- Clinical appropriateness and relevance,
- Political and ethical acceptability,
- Allows the demand to be managed without putting any other population group at risk, and
- Lends itself to high levels of compliance.

ROLE OF IEC IN ART

The success of the ART program will depend in large part to there being effective communication in place to address and prepare all relevant target groups. It is a large complex technical issue that needs to be explained correctly and meaningfully to all target groups, so that they are correctly informed and their expectations are realistic. The experience of other countries such as Brazil shows that effective communication was a critical element in the success of the program.

As has been said, there are great public expectations from ART without a concomitant understanding of its limitations. In addition, there are misunderstandings about who among HIV/AIDS patients would qualify for ART. There is also the potential for ART to detract from the crucial prevention effort. A carefully planned communication effort for ART that effectively educates the public and manages demand without creating a negative reaction or stifling hope, is therefore, urgently required.

APPROACH TO IEC STRATEGY

As IEC is a component on the total launch strategy, it has to be in synchrony with what is being proposed in the overall strategy. The strategy proposed for the launch is demand management as it is felt that the delivery system will not be able to meet full demand to begin with. The IEC strategy will therefore, also have to follow this approach. Demand creation, which is often the primary goal of IEC efforts, is consequently clearly not a goal of IEC in this instance. On the contrary, the initial objective would be to control demand to keep it at manageable levels. A note of caution, therefore, needs to be sounded at the outset about pronouncements and activities that would create demand which would not be met and which could then generate a counter productive backlash of disappointment and possibly resentment.

Based on this, it is recommended that the guiding principle for the IEC strategy should be caution. We need to engender hope. At the same time, we also need to inject realism into this hope. The IEC strategy will have to perform this delicate balancing act.

ISSUES

In the light of this, some of the issues that an IEC strategy for the launch of ART needs to address follow.

Managing public expectations

Following the President's pronouncement, expectations and hopes that the solution to the epidemic is now here, are high. It should be remembered that Botswana is a small country in terms of numbers, and that this can have both positive as well as negative implications. On the positive side, it does not take long to communicate a message to the population. On the negative side, incorrect impressions can also spread fast. It needs to be explained to the public that ART requires a great expansion in infrastructure and that this will take time.

Ensuring that correct information regarding ART is available to providers, clients and opinion influencers

As has been said above, misinformation can spread fast in Botswana. There is already a lot of misinformation about ART that is circulating including:

- ART is a cure for AIDS.
- All HIV positive people will be put on ART.
- When an HIV positive person is put on ART, he/she ceases to be infective. This is particularly dangerous as the patient will continue to infect others as well as expose him/herself to the danger of re-infection, about which not much is currently known.

This misinformation needs to be corrected. It also needs to be emphasized that everyone who is HIV positive does not medically qualify for ART, and that many people can successfully wait for years to start therapy.

Possible shift in focus from prevention

This misinformation has resulted in some countries like the USA, in a shift in focus away from prevention activities and the feeling among some clients that they now no longer need to change their behavior, as the cure is available. It has been reported in some studies that certain groups in these countries have reverted to behaviors which they were beginning to change before ART became freely available. In Botswana, where behavior change has not so far taken place in sufficient degree to impact the epidemic, this is a particularly dangerous possibility. The ART program has therefore become available to all those who qualify for it on medical grounds, only in a phased manner, during which time the infection could spread further. It needs to be emphasized that prevention is still the only way to avoid HIV/AIDS.

Adherence

ART is extremely complicated for the 'lay' person who may not also be very well educated. Although one of the criteria for selection of the proposed groups is likelihood of adherence, this has nevertheless to be ensured through appropriate communication to providers, patients, their families and caregivers. It needs to be emphasized that ART is lifelong. In addition, the drugs could have adverse side effects. The implications of a lack of adherence in terms of development of drug resistances and the danger of re-infection, as well as many other factors such as diet, timings of taking drugs, etc. need to be communicated in a manner in which they are remembered and followed.

Impact on programs

Specific programs will, in the immediate term, be impacted by the launch of ART in different ways, which will need to be taken into consideration while developing communication strategy. Some of these follow.

VCT: The success of ART will ultimately depend on the HIV serostatus of most of the population being known, so that all clients who medically qualify can be put on it. This will eventually also become an important motivation for voluntary testing and, in fact, should encourage it. However, as has been said, to begin with, the promise of putting all qualifying HIV-positive people on ART cannot be met. Testing efforts will therefore have to continue to rely on other motivations for some more time.

PMTCT: The impact of the launch of ART on this program is uncertain. It will have a positive impact on the one hand as qualifying pregnant women and their partners will be put on ART, and it will, therefore, encourage pregnant women to test and enroll in the program. However, it will create a communication challenge as the reasons for all pregnant women not being put on the treatment will have to be carefully communicated to the clients, so that this does not create a negative impact. There is also the issue of whether extending ART to all pregnant women would decrease the transmission rate and whether, therefore, CD4 <200 is a valid criterion for this group.

Orphan care: As all HIV-positive children with CD4 counts <200 will be immediately put on ART, its launch will impact positively on this program. It should act as a motivation for caregivers to register children in the program, and help in overcoming stigma. However, here again the reason for not putting all HIV-positive children on ART will have to be handled carefully.

IPT: As TB patients are one of the recommended groups for introduction, SRT will definitely impact on this program immediately. The exact nature of this impact is not, however, clear at this point.

Stigma

As the availability of ART will ultimately encourage more testing and openness about HIV, it will help to reduce stigma. However, for reasons stated above, this will take some time to happen, and an overarching IEC campaign to reduce stigma and encourage openness and dialogue must also be initiated immediately.

Counseling

Counseling skills are already inadequate to deal with the current level of demand. The launch of ART has to be accompanied by intensive and extensive counseling for all people who are being put on the treatment, their caregivers and families as well as those HIV-positive people to whom it cannot be offered immediately. This will, therefore, place an even greater strain on those already-stretched services. Thus far counseling skills were mainly vested in nurses, who mostly do not have the time to counsel. The learning from the PMTCT program was that these skills need to be augmented by training suitable non-medical personnel. This process has now been initiated, and perhaps needs to be upscaled to meet the counseling needs of ART.

IEC Structure

The planning and implementation structure for IEC for the HIV/AIDS program as a whole is in the process of developing. The roles of NACA and the AIDS/STD Unit and Family Health Division/Health Education Unit of the Ministry of Health (which were before the formation of NACA responsible for IEC for HIV/AIDS), are in the process of being clarified. There are a large number of stakeholders, many of them undertaking IEC campaigns in their respective areas, with at the present moment, not too much coordination. Better coordination is also required between the national level and the district level in IEC planning and management. There are procedural delays in DMSACS receiving funds for IEC activities. All this poses a challenge for the implementation of any IEC strategy for HIV/AIDS, including for ART. To expedite the process and to ensure synergy, it is recommended that a package of materials for ART be produced nationally and distributed to the districts to be used in district level activities.

IEC OBJECTIVES

Short term (one year)

- To correctly inform and prepare the provider system for the proposed rollout of ART.
- To sensitize and correctly inform all stakeholders, policy makers and opinion influencers such as political and traditional leaders, the media and private sector, so that they, in turn, communicate the right information and keep public expectations realistic.
- To correctly inform the public at large about ART, including who would qualify and when.
- To ensure adherence of those put on ART.
- To synergize ART with prevention activities and strengthen them.

Medium term (1-3 years)/ Long term (3-5 years)

There is currently no clear-cut plan for the medium and long term. At this point it seems likely that in the next phase, ART will be made available to all remaining symptomatic patients and eventually to all qualifying HIV-positive people. As these plans are not definitive yet, it is difficult to specify IEC objectives for these phases at this stage. However, it is recommended that an attempt be made to develop at least broad plans beyond the first year as they will help greatly in making the limited reach of the proposed plan for the first year more palatable to the stakeholders as well as to the clients.

If this is not done, the proposal could sound negative and create disappointment and possibly resentment both among some stakeholders and clients who will not receive treatment for at least a year, countering the positive impact of the government's initiative. If stakeholders and clients could be told that constraints in human resource capacity may not make it possible to immediately put all qualifying HIV-positive on ART, but that it would nevertheless happen in phased manner over the next three years (which is spelled out), it will soften the blow to a great extent. This will enable us to hold out hope to those qualifying HIV-positive people to whom ART cannot be immediately offered.

IEC TASKS

- To sensitize and prepare target groups prior to the launch,
- To ensure continuous access of target groups to correct information,
- To maintain morale both of providers and of clients—those on ART and those who are not, and
- To develop a mechanism for feedback which can then become the basis or further communication.

This will essentially be done through:

- Advocacy,
- Addressing the provider system at all levels including medical and non-medical staff and traditional healers, and
- Addressing the client groups

A THREE-TIERED APPROACH

Communication for ART will be required at three levels:

- The national level,
- The district and community levels, and
- The points of care.

TARGET GROUPS/MESSAGES

The target groups for the IEC strategy will have to be determined by the client groups defined for the launch and the IEC tasks as articulated above. The client group is broadly all HIV-positive people. These can be divided into:

- Those to whom treatment will be offered in the first place, and
- Those to whom treatment will not be offered in the first place.

Communication will have to be addressed to both of them. The specific groups to whom ART will be offered initially have been proposed but are not yet finalized. However, it is being taken as an assumption for the moment. Obviously the target groups for communication will also change if the target groups for the program change. At present it is assumed that they are as follows:

- **Adult inpatients**: These are patients who have AIDS defining illnesses and are admitted in hospital as a result. They would mostly have CD4 <200. The target group for the launch of ART among this group and the communication task for each is as follows:

Providers
- Inform them what to expect from ART so that they can appropriately counsel patients.
- Prepare them for additional responsibilities in a situation where they are already stretched.
- Training in counseling for ART for those who will be put on treatment and those who will not.

Clients
- Give them hope.
- Give them correct and detailed information on ART: what it is, how to take it, what to expect from it, how long it is to be taken.

Families/caregivers of patients
- Give them correct information on what to expect from ART.

- **TB Patients**: These are registered patients with active TB who are HIV-positive, are symptomatic and have CD4 count <200.

Providers/clients
- The combination of DOTS and ART would be even more complicated and needs careful explanation both to providers and to clients. Clients and their families must be counseled on what to expect from ART.

- **Pregnant women and their partners**: These are pregnant women and their partners who are HIV-positive, are symptomatic and have CD4 <200.

Providers
- The providers for this group will have a difficult communication task as they will have to explain to the pregnant women why all pregnant women who test positive will not be put on ART. This might be difficult for a 'lay' person without much education to grasp. It would cause some disappointment and may negatively impact on the motivation to test if not handled properly. Providers would therefore have to be trained to handle this situation sensitively and communicate the correct information. They may require aids such as flip charts and videos for this.

478

Clients
- For those clients in this group who will be put on ART, communication has to focus on getting them to enroll in the PMTCT program so that it can strengthen the prevention of transmission through this route. Therefore, it will have to be closely networked with this program.

- **Children:** These are HIV-positive in patients >6 months, who are symptomatic and have CD4 <200.

Providers
- The providers for this group will find the decision to give ART only to qualifying children and the communication of this to parents and caregivers particularly difficult, as children are always the most vulnerable group. It will require careful handling

Clients
- These children are of varying ages. Explaining the therapy to the younger would be difficult, but it does need to be done to get their cooperation. Special pediatric counseling skills need to be developed for this. Adherence could be a problem with some of them.

Advocacy
The advocacy task is an overarching one and needs to address the following groups:

- Policy makers and political leaders, particularly the President, the Minister of Health, and the minister of Local Government,
- MPs,
- Government departments/sectors,
- Private sector: some companies such as Debswana have already started their own ART programs,
- Chiefs,
- Traditional healers: they could actually further spread misinformation about ART to further their own clientele, as they might feel that their 'business' is under threat. Their support to the program must be actively solicited through meetings and dialogue,
- NGOs, CBOs, and FBOs through their networks,
- DMSACS, and
- Other stakeholders.

The launch plan needs to be shared with these groups and their support solicited, so that they, in turn, can communicate this to others. This group would be most easily able to understand the constraints facing the government and the reasons for a limited launch as a result. They should, therefore, be effectively used to further explain this to other groups and the general public so that they are patient and realistic in their expectations.

GENERAL MESSAGE COMPONENTS FOR THE PUBLIC

- ART cannot cure HIV/AIDS and is a challenging lifetime commitment.
- Continue to practice safe sex by using condoms and avoiding high risk sexual behavior.
- Not every HIV-positive person needs ART and it may be in the individual's best interest to wait.
- Support those who are on therapy. It will reach everybody who needs it as soon as possible.

SUGGESTED ACTIVITIES

Advocacy

- Press conference addressed by the President and Minister of Health, which presents a realistic picture of the launch plan. Well designed and informative press kits including a question and answer booklet to be given to members of the media who come for the press conference, detailing launch plans and presenting them in a positive but realistic light.
- A series of seminars at national and district levels to present the launch plan and reasons for limitations to the various stakeholders and target groups who have to actively support the program if it is to be successful. A professionally designed presentation to be developed for this purpose. These groups also need to be consulted on the support they could offer and what support they in turn would require. A coordination mechanism is needed to monitor this process and to take it forward.
- More intensive activity at launch sites. Technical and financial support to concerned DMSACS for this purpose.
- Television and press interviews with President/Minister of Health and other key figures in which they will talk about the plans for the launch of ART.
- ART plans to be discussed at all major national and local events and presented realistically in a manner to correct disinformation.
- Discussion programs with key stakeholders on television and radio in which these plans would be discussed.
- Phone-in programs on radio and television to answer queries and establish a dialogue with communities.
- A newsletter keeping all stakeholders in touch with the latest developments on HIV/AIDS and ART.
- A website, so that all people with internet access can readily access the latest information on HIV/AIDS and ART.

Providers
- Seminars/workshops with different groups of providers to discuss launch plans and expectations from them.
- Training in ART, including counseling for ART specifically, and training of trainers.
- Build capacity of non-medical cadre of counselors such as social workers, community volunteers and FBOs as fast as possible.
- Appropriate training materials to be developed—training modules, videos, question and answer booklet.
- Develop system for periodic counseling for counselors.
- Motivational program for providers.
- Direct mail to establish a continuing line of communication with providers about latest developments in ART.

Clients
- Providing correct information and explaining launch plans through television and radio programs, editorial content in newspapers, posters, leaflets, question and answer booklets and videos at points of care. It should be remembered that points of care will only be at limited locations to begin with.
- Counseling using flip charts, videos and other communication aids for those who will be put on the treatment, and those who will not, as well as for those who are planning to test, explaining the possibilities.
- Developing an attractive, well designed timetable for patients on ART with drug regimen, diet, meal times, etc., including place to record adherence.
- Encouraging and supporting development of support groups for those who are on therapy and those who are not. Maintaining contact with these groups and supplying them correct information.
- Correcting misinformation through meetings with village health committees, community groups, kgotla meetings, etc., and explaining launch plans.

TIMING/PROCESS

These activities need to start as soon as possible as ART is already being discussed by a number of people and needs to be placed in the proper perspective so that unrealistic expectations are countered. It is also suggested that professional services be contracted for those activities which can be contracted out, such as the public relations and direct response activities as well as production of materials. This will remove this burden from government/stakeholders who can then focus on other aspects of the program, leaving such aspects to professionals who will ensure quality, consistency and quick action. This process needs to be coordinated by a central body which will also be responsible for coordinating the entire communication program for ART.

The communication program must be closely tied in to and continuously receive feedback from the rest of the program, so that it can adapt to changing situations. The ART program itself must be closely tied in to the prevention program and continuously receive feedback from it so that it can keep it informed and be informed itself. A mechanism needs to be developed for this purpose. As this will be the first experience of ART through the public health system in Botswana, it will need to be closely monitored and supervised.

15. POLLUTION WORKSHOP

ENVIRONMENTAL POLLUTION IN THE CASPIAN SEA: WORKSHOP SUMMARY, CONCLUSIONS AND RECOMMENDATIONS

RICHARD C. RAGAINI
Lawrence Livermore National Laboratory, Livermore, California, USA

ILKAY SALIHOGLU
METU Institute of Marine Sciences, Erdemli, Icel, Turkey

ABSTRACT

The workshop was held on 24-25 August 2001, and addressed four issues: (1) review of the current knowledge of the Caspian Sea environment; (2) identification of unresolved environmental issues in the Caspian Sea; (3) current and planned environmental activities in the Caspian Sea; (4) possible WFS/World Lab future actions.

WORKSHOP SUMMARY

Friday, 24 Aug.
R.C. Ragaini welcomed the participants to the Workshop and introduced the Agenda. He expressed his views on and impressions on the presentations and discussions for the Caspian Sea Session in the International Seminar. R. Ragaini then briefly explained the aim and scope of the workshop.

The Caspian Observation and Forecasting System was presented by I. Salihoglu. The subject basically covers the recommendations for future work aimed at resolving the missing elements in environmental research in the region. It pinpoints the specific anthropogenic and natural causes of environmental problems in the region, and the gaps of knowledge to quantify key variables and processes to be studied. Integrated assessments and scientific investigations of the environmental changes in the Caspian Sea region are identified as the starting point for tools leading to successful predictions and for fruitful scientific collaboration and management.

H. Ghaffarzadeh, who participated the workshop on behalf of the Caspian Environmental Programme (CEP) located in Baku, described the Caspian Environmental Programme and the regional structure. He gave extensive information on the resources, objectives and aims, achievements on several issues relevant to the Programme, conflicting political agendas and conclusions. The main conclusions of the presentation are:

- Essential to learn more about the Caspian;
- Caspian faces serious issues but there is still hope;
- Financial and technical assistance required for some time come;
- Political and economic issues are seriously impacting the sea;
- We should not allow the environmental voice to be silenced.

L. Shabanova based her talk on the Oil and Gas Pollution of the Caspian which covers terrestrial, limnological and oceanographical aspects. The subject refers to the relevant legislative issues in Kazakhstan and projection of a four years programme for the reduction of pollution in the country. The assessment of the oil and gas pollution indicated the following impact:

- Disturbance of natural landscapes and loss of agricultural lands;
- Groundwater water level rise followed by formation of small saline lakes;
- Mortality of waterfowl and small animals;
- Air pollution as a result of evaporation of hydrocarbons at oil fields;
- Economic losses in regard to immobilization of oil at oil storage.

E. Ozsoy presented the "Sea Level Fluctuations of the Caspian Sea". Briefly he gave a perspective of the historical events. He stressed the importance of the sudden changes in sea level, which occurred twice since 1830 when recording of sea level began. Large changes have occurred during the past decades: starting in 1930, sea level fell from a level of about -25 m down to -29 m by 1978. The latter part of this drop is believed to be associated by construction of hydropower plants along the Volga and other rivers of the Caspian. From 1977 to present, the sea-level trend has reversed, such that at present the ocean level stands near -27 m. The rapid rise and fall in sea level has had catastrophic effects on the landforms and settlements surrounding the Caspian Sea. Ozsoy also stressed the importance of the imbalances in water budget, originally depending on climate, but also modified strongly by anthropogenic regulation schemes, which lead to patterns of inter-annual, inter-decadal or longer-term variations in sea level throughout the history of the Caspian Sea..

Igor Mitrofanov from Kazakhstan described the Ecotoxicology of the Oil Pollution in the Caspian. He devoted his talk on the influence of specific pollutants as oil and pesticides to some Caspian Sea fish species. The species studies were the ones which can be identified as ecologically important indicators and also have economical value. The conclusion of the presentation is that the water quality of the whole Caspian Sea is still good enough to sustain the fish species. However there are some "Hot Spots" where concentrations of different pollutants are high and critical for fish.

N.T. Ibrahimov made two successive presentation complimentary to each other. The first was the description of the "Urgent Environmental Investment Project". The project aims action in four areas (regions) identified with distinctive environmental problems in the Azeri National Environmental Action Plan. Information for the measures introduced in Azerbaijan in order to protect the Caspian against pollution were described.

His second topic was "Economic Instruments for Environmental Management in the Region". In this context the growth of economy-environment relations is criticized, and a general description of the economic situation in the Azerbaijan Republic is made. The importance of economic regulation of environmental protection was mentioned and information on the resources for proper waste management given. He also referred to the recent Waste Framework Directive adopted by the EU and to the Community Strategy for Waste Management–1989 and to the Fifth Environmental Action Plan–1993-1997. The Future Trends on waste management were explained briefly, and this linked to the external influences. The country monitoring policy is criticized and enforcement capacity mentioned. Estimates were made of the consequences on the country economy of future waste types. Waste reduction and the importance of recycling were stressed.

K. Thompson, "Communication Across the Black Sea and Caspian Sea via Internet Technology," briefly explained the aims and the set up of the web site for the Black Sea. Further information on the contents of the web site was provided.

Saturday, 25 Aug.
A. Vasiliev's topic was "Mathematical modeling of oil pollution in the Northern Caspian Sea". The CASPAS program was presented extensively, and relevant information of the Russian Federation activities in the Caspian was demonstrated. State Oceanographic Institutes activities via the Astrakhan branch were also demonstrated.

The model developed at the State Oceanographic Institute of Russia is a strong tool, and it is based on the information gathered from several platforms in the northern Caspian. An application to a case of an oil spill emerging from a drilling accident was demonstrated. The conclusions of the report are:

- An efficient monitoring system covering all the Caspian is needed;
- An early alarm system is essential for the whole Caspian;
- An operational system to prevent pollution is needed.

Activities of the State Oceanographic Institute and Astrakhan departments of Roshydromet were also demonstrated.

A. Korshenko introduced the topic of "Monitoring of marine pollution in the Northern Caspian Sea". Examples and case studies were demonstrated..

L.M. Michaud from the International Meteorological Organization of the UN presented, "Integrated Project for a Monitoring and Information System for the Caspian Sea Region," a short subsidiary presentation. He gave brief information on the historical background, new developments in technology, institutional arrangements of CASPCOM and CASPAS. Mr. Michaud gave further information about his personal experience and impressions on the achievements in the Caspian region.

Prof. Lado Mirianashvili presented a project proposal outline to be implemented for the Caspian (Appendix I).

488

CONCLUSIONS AND RECOMMENDATIONS

CONCLUSIONS
- The Caspian Sea is a World heritage and needs to be preserved for future generations;
- The Caspian Sea faces environmental stresses which includes:
  - industrial pollution;
  - nutrient enrichment;
  - fisheries decline;
  - biodiversity erosion;
  - damage to and degradation of coastal habitats;
  - invasion by Mnemiopsis leidyl and other alien species;
  - environmental accidents including oil spills.
- The Caspian problems are rooted in complex natural, technological, social and economic issues, which need to be fully understood; political and economic issues are strongly coupled with the environment and seriously impact the sea;
- The environmental issues should increasingly be brought to the foreground and the environmental voice should not be silenced by other temporary concerns of the society;
- Priority action against pollution caused by hydrocarbon exploration and production needs to be implemented simultaneously by all riparian countries;
- Biomarkers are essential for the marine and fresh water pollution monitoring and Caspian Sprat (Clupeonella delicatula caspia) and Goby species (N. Gorlap - N. Fluviatilis - N. Melanostomus) are identified for this purpose;
- Hydro-meteorological, climate and geo-dynamical monitoring is essential for any environmental program; it is relevant to the Caspian Sea where the hydro-meteorological and environmental control services of the five littoral countries are collaborating to produce the basic marine meteorological, geodynamical and oceanographic data;
- Continued financial and technical assistance is required for some time to come, as the economic situation requires external inputs for sustained activities;
- The development of genuine scientific collaboration among local institutions and other scientific organizations is the only feasible way to achieve continued investigation of the problems and mitigated measures.

RECOMMENDATIONS
- A permanent Caspian panel with membership including environmental politicians; economists and strategists is recommended;
- A Caspian Symposium to be dedicated to the short and long term problems of the Caspian is recommended;

- The development of a network for efficient data exchange and communication between Caspian Sea riparian countries is essential. Validated data should be freely available;
- Build a central file of data base models, accessible to all, for the Caspian and Black Sea environment on the internet;
- Research funding attention is required to deal with the Mnemiopsis leidyl issue;
- Technical assistance should be made available to all stake holders; no "invisible" border should be a barrier to assistance to deal with "transboundary" environmental issues;
- Tools of operational oceanography must be developed and mitigated to ensure protection of the environment and to sustain it;
- Environmental issues in the Caspian Sea area are of interest to oil and industrial companies, and they should be involved in future workshops for their technical inputs;
- Sea level changes are a key element of the Caspian environmental protection, therefore a system providing abiotic sea level parameters is a priority issue;
- A common system to monitor pathology of fishes, birds, mammals etc. needs to be established;
- Signal monitor locations need to established in the river systems to detect accidental releases;
- The tritium concentrations existing at depth in the Sea need to be identified;
- Establish an international hydrocarbon test site to demonstrate effective hydrocarbon clean up technologies.

RECOMMENDATIONS FOR THE WORLD FEDERATION OF SCIENTISTS
- Decreasing bio-diversity in general is an emergency, and should be handled by the Pollution PMP, and a workshop should be devoted to this issue;
- The WFS/WL should generate and promote close relations with the hydro-meteorological and environmental control services and relevant projects in the Caspian;
- The WFS Pollution PMP should monitor the work of the Caspian littoral states and regional and international organizations to develop systems to minimize the environmental impacts to the Caspian Sea of pollution from oil exploration, production and transportation, and support participation by marine sciences community to these efforts;
- Under the MOA with the U.S. Department of Energy, it is recommended that the Pollution PMP support development of joint Caspian research project proposals to international science funding organizations;
- The WFS should use the existing scientific and intellectual potential of the Caspian littoral countries in the realization of CEP and other projects, to direct available financial and technical funds to support local institutions and scientific groups;

- Decisions must be based on scientific understanding of the system, rather than being based on partial remedies by purely economic and social concerns, therefore it is recommended that the WFS improve the dialogue between scientists and political decision makers in the Caspian region;
- The WFS should create a regional center for the Caspian Sea Region.

Signed: All Participants

Richard Ragaini, USA
Emin Özsoy, TURKEY
Lorne Everett, USA
Vittorio Ragaini, ITALY
Ludmila Shabonova, KAZAKHSTAN
Igor V. Mitrofanov, KAZAKHSTAN
L.M. Michaud, SWITZERLAND
Namik T. Ibrahimov, AZERBAIJAN
Zenonas Rudzikas, LITHUANIA
Rashid Aielov, AZERBAIJAN
lkay Salihoglu, TURKEY
Margaret S. Peterson, USA

Hamid Ghaffarzadeh, CEP/AZERBAIJAN
Anatoly S. Vasiliev, RUSSIA
Mbareck Diop, DAKAR-SENEGAL
Alexander Korshenko, RUSSIA
M. Kay Thompson, USA
Dorogan Dumitru, ROMANIA
Valery Kukhar, UKRAINE
Leonardas Kairiunatis, LITHUANIA
Vladimir (Lado) Mirianashvili, GEORGIA
Yuras Pozela, LITHUANIA
Donald F. Boesch, USA
Robert A. Clark, USA

APPENDIX I: OUTLINE OF THE PROJECT PROPOSAL BY GEORGIA

(a) Objective
1. Complex study of the Caspian Sea comprising:
    a) pollution with oil,
    b) pollution with chemicals,
    c) sea level changes,
    d) atmospheric and hydrological processes,
    e) geodynamics of the adjacent area,
    f) biodiversity

2. Filling in gaps in the results of accomplished Caspian Sea projects, introducing "d"
    and "e"; usidely using remote sensing; involving experience of the international
    scientific community with large involvement of local scientists, since they better
    know the local problems and have access to dates.

(b) Tools
1. Use of existing observation stations and installations (purchased lately in a course of
    international projects), with necessary enhancement.
2. Complex, multiparameter, multinational approach.

(c) Steps
1. Compilation of universal database based on historical data, data obtained in past years
    (available through bulletins regularly published till 1992), news data obtained within
    the frames undergoing projects funded by western agencies (the data cannot be
    refaired by the project executors, since they were obtained due to international
    assistance and are free for world wide use); current data obtained in a course of our
    program.
2. Development of new models with more parameters.
3. Training of scientists involved.
4. Periodical meetings for identification of successes and failures for further perfecting
    of the project.
5. Drawing up recommendations even before accomplishing the project, after accusing
    enough materials and confidence, with the purpose of slowing down degradation
    process.
6. Involvement of policy makers from early stage of the project accomplishment.
7. Drawing final recommendations

(d) Participation
Georgia can be involved in data base compilation (due to skilled GIS specialists),
geodynamical studies (due to experience in these studies), hydrometeorological studies
(complete data bases and skilled specialists).

OIL AND GAS POLLUTION OF THE CASPIAN ECOSYSTEM IN KAZAKHSTAN

LUDMILA SHABANOVA
National Environmental Center for Sustainable Development, Koksheta, Kazakhstan

The Caspian region is an interlacement of economic, political and environmental interests due to considerable fish, oil and gas resources. Extending for 700 miles from north to south and for 170 miles from west to east, the Caspian Sea borders with Kazakhstan and Turkmenistan in the east, Azerbaijan and Iran in the south and Russia in the north and west.

Currently, all the Caspian littoral states produce oil and gas (Fig. 1). There is an expansion of production of these valuable minerals. Production of oil and gas resources tends to increase in this region and approximate estimations of which are 20-180 billion barrels for oil and 160-500 trillion ft^3 for natural gas. However, exploration of oil and gas reserves poses significant risks. These risks are being increased due to the following facts: i) the Caspian Sea is an inland reservoir, surrounded by land, ii) other pressing environmental problems in the region both natural and anthropogenic.

Impact, as a result of oil and gas production in the past and future, should be properly taken into consideration in the context of other environmental risks in the Kazakhstan part (Fig. 2). They are as follows: industrial pollution, discharge of wastewaters, overuse of fertilizers and pesticides in agriculture, municipal wastes and associated products damaging environment as a result of rapid development of the oil and gas complex. According to geological surveys, there are 148 oil fields amounting in total to 2,2 billion tons of oil reserves in the Kazakhstan part of the Caspian Sea. Over the years more than 50 oil fields have been developed, to most of which obsolete technologies and equipment were applied and they contributed greatly to emitting toxicants of low entrapment. Only in 1998, production in the Caspian region of: oil (including gas condensate) - 18,19 mln tons, gas - 5,36 billion m3. Products of primary oil processing produced 4,1 mln. tons.

All these factors produce negative impact and damage environment of the region causing degradation of ecosystems and migration of marine inhabitants, water pollution, flora and fauna depletion, mutations, hydrocarbon pollution, threats to human health.

Air pollution poses a major threat in the region.

Main air pollution sources are 269 enterprises with over 5600 stationary pollutants emission sources. In 1998, the overall emission volume for all these enterprises produced

135,057 (100%) th. tons, from which solid substances -2,281 (2%), gas - 132,775 th. tons. (98%), including: sulfur dioxide - 43,815 (33%), carbon oxide - 56,064 (47%), nitric oxide - 8,265 (13%), hydrocarbons - 17,559 (13%), others - 7,072 (6%) th. (Fig. 3).

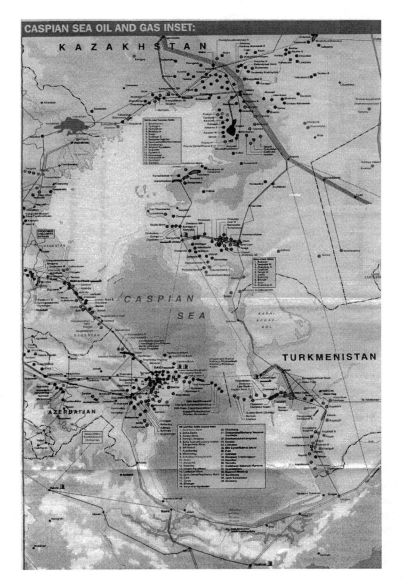

Fig. 1. Caspian Sea and Oil Inset.

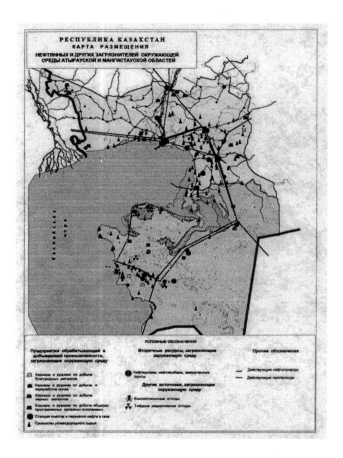

Fig. 2. Republic of Kazakhstan. Map of locations of oil and other environmental pollutors in the Atyrau and Mangystau oblasts.

Concerning the overall emission volume, oil and gas production as well as processing industries contributed greatly to air pollution (Fig. 4). Total emissions of these enterprises produced 95,5% of the overall volume in the Atyrau oblast, among which Tengizshevroil Ltd. - 61%, Tengizmunaigas JSC - 15,5%, Embamunaigas JSC - 8%, ANPZ JSC - 9%, ATEZ -1%.

Environmental load in the region is complicated by a continuous warm period throughout the year increasing inflow of many chemical substances from natural sources into the atmosphere as follows: sulfur dioxide, nitric oxide, hydrocarbons, etc.

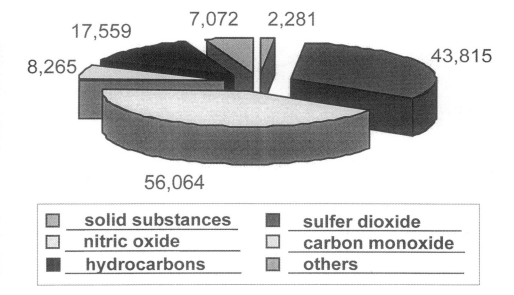

17,559   7,072   2,281

8,265

43,815

56,064

	solid substances		sulfer dioxide
	nitric oxide		carbon monoxide
	hydrocarbons		others

Fig. 3. Volume of emission of pollutants into atmosphere by ingredients (totally in 1998: 135, 057 th. tons.).

496

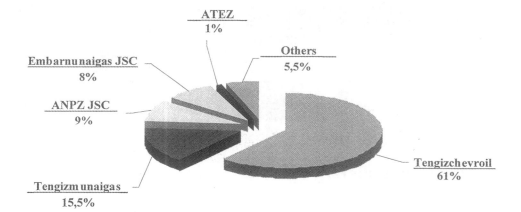

Fig. 4. Contribution of enterprises to air pollution.

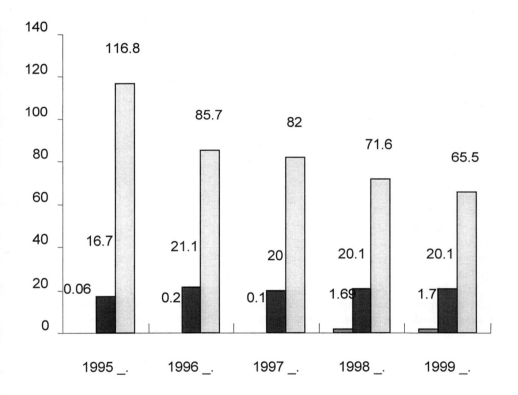

Fig. 5. Quantity of flared gas by enterprises of the Atyrau oblast in 1995-1999.

One of the major environmental problems at oil fields is associated gas recycling when extracting oil.

Currently, over 800 mln. m^3 of associated gas is flared annually in the republic. In 1995, when developing the Mangyshlak oil field, a total of 46293 mln. m^3 of associated oil gas in produced over the republic, from which 37938 mln. m^3 or 81,9% were flared, emitting millions tons of pollutants into the atmosphere.

Despite such a long period of developing the Prorva oil fields in the Mangystau oblast, so far the problem of associated gas recycling has not been solved. Totally, since the development of oil fields, over 5 billion m^3 of associated gas have been flared and 500 th. tons of pollutants emitted into the atmosphere. In 1996, only emissions into the atmosphere were 23 th. tons.

In the Atyrau oblast, 1 billion m^3 of associated gas were produced at the oil fields of West Kamyshitovoye, Rovnoye, Zaburunye, Botakhan, Oryskazgan, East Makat, North Kotyras for the whole period of development, from which 776 mln. m^3 were flared and 138 th. tons of pollutants were emitted into the atmosphere.

There is a progressive decrease of emissions followed by gas flaring, most of which were harmful substances, from flaring at the Prorva oil field (Fig. 5). Further decrease of emissions to 20-25% is possible by constructing a mini-plant to recycle associated oil gas. At present, Tengizmaunaigas JSC, jointly with foreign companies, is planning to construct a mini-plant to recycle associated oil gas and generate electric power with capacities of 25 mln. m^3 of associated oil gas.

Associated oil gas recycling with the purpose of saving energy, presently flared (around 740 mln. m^3 annually) may be referred to energy-saving and ecology as a measure to cease fuel flaring at other installations and to mitigate harmful emissions as well as ÑÎ2.

The Republic of Kazakhstan was one of the first to sign and ratify the UN Framework Convention on Climate Change. At the Kyoto International Forum on Global Climate Change, Kazakhstan supported initiatives related to the establishment of world market carbon crediting. After the successful Kyoto Conference, the Kazakhstan Carbon Initiative Project was awarded the status of governmental and was included in the long-term Strategy for Social-Economic Development of Kazakhstan. The Ministry of Natural Resources and Environmental Protection carries out overall coordination of all activities related to the implementation of the Framework Convention in Kazakhstan.

One of the peculiar features of the Caspian Sea is specific sensitivity of its ecosystem to external impacts. They are as follows: coastal areas being practically at sea-level and shallow parts of Northern Caspian, most of which belong to Kazakhstan (where small sea level fluctuations and even strong winds cause significant changes of water area), ice cover in winter in the Northern Caspian, sea inlandness.

Sea level fluctuations contribute to oil pollution of Caspian ecosystems as a result of underflooding and flooding operating and capped oil and gas fields (Fig. 6).

According to available data, there are 127 flooded wells located only in the Kazakhstan sector, which, in connection with catastrophic sea encroachment, were not closed or conserved properly. Due to corrosion and destruction of plugs, some of them are currently a source of pollution.

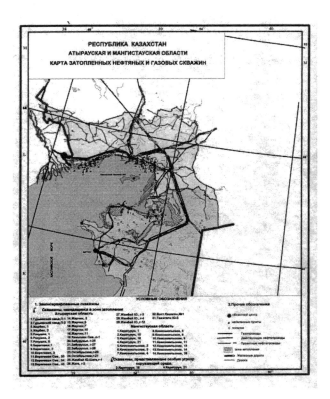

Fig. 6. Republic of Kazakhstan. Styrau and Mangystau oblasts. Map of flooded oil and gas wells.

At these wells, the concentration of hydrocarbons is several times larger than in other regions. The situation is aggravated by the fact that the technical state of most wells is not known and they can cause oil spills at any time.

According to the Kazakh Research Institute for Fisheries' (KRIF) assessment, these processes can cause an increase of MAC from 2-3 (background indicators) to 8 MAC under tide phenomena of average amplitude, and in case of their occurrence with emergencies—up to 400 MAC (the KRIF report, 1995).

The present level rise of the Caspian Sea has been going on for 18 years (1978-1995). During this period, the sea level increased by 2,5 m. In 1995, water level rise slowed down and is presently at minus 27,0 m. Sea level rise has caused flooding and underflooding of most coastal areas with sea water where industrial facilities and engineering communications are located, which resulted in deterioration of soil and

500

vegetative cover, fauna as well as ichthyofauna, water vegetation, sea water pollution and great environmental damage. If there is a possible rise in level to minus 25,0 m, most oil fields, economic facilities, transport and high-voltage electric lines will be flooded and Caspian coastal ecosystems damaged (Fig. 7).

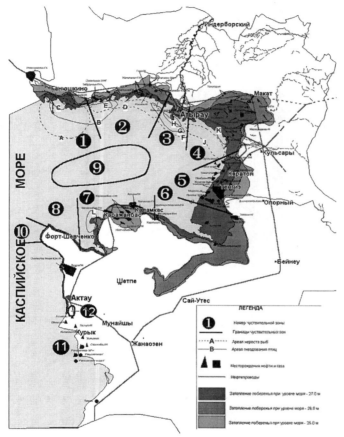

Рис.1.Схема расположения экологически чувствительных зон Казахстанского сектора Каспия

Fig. 7. Scheme of ecologically sensitive areas in the Kazakhstan part of the Caspian Sea.

Effects of long-term sea level rise are complicated with impacts of wind (storm) surges. Inclines of coastal parts are almost extremely small both on land and under sea. Difference in heights for about 2 m at a distance of 50 km is also characteristic. The coastal area may be significantly transformed even if the level changes slightly.

A large surge can cause flooding of 20-30 km of coastal area from permanent shoreline. Surges may result in mass fish mortality, loss of benthos and vegetation at some sites.

Surge waters, extending for scores of kilometers into the coastal zone, flood sites of oil fields with oil-polluted soil and accumulate petroleum products and other contaminants.

Soil pollution with oil products are characteristic for the Caspian region in Kazakhstan.

Development of oil fields and its transportation causes soil pollution as a result of oil spills, discharges of stratal waters and unregulated machinery and transport traffic.

The pipeline transportation system, which is most developed in the region and represented by oil and gas pipelines, also poses serious environmental concerns (Fig. 7).

The Uzen-Atyrau-Samara pipeline, which transports Kazakhstan oil to main export directions of foreign countries, is located on the territory of the Caspian region. The construction of the first part of the Tengiz-Astarakhan-Grozny pipeline is finished. The Kalamkas-Karazhanbas-Aktau and Prorva-Kulsary oil pipelines were put into operation.

The first export pipeline, which should push big oil of the Tengiz oil field and the Caspian offshore towards the world market, is the Atyrau-Novorossiysk oil pipeline. It is constructed within the framework of international Caspian Pipeline Consortium (CPC).

The gas pipeline system on the territory of the Caspian oblasts are represented by transit parts of the Central Asia-Center and Okar (Turkmenistan)-Beyneu gas pipelines as well as the intraregional Uzen-Zhetybai-Aktau gas pipeline.

It is worth noting there are no defectoscopic studies of operating pipelines, although most pipes became obsolete long ago and by means of defectoscopy need to be replaced in damaged sites. In 1999, there were hundreds of breaks in pipelines, which resulted in spills of thousand of tons of oil and formation of four additional oil storages.

At the oil fields of the Mangystau and Atyrau oblasts only, there are hundreds of thousands tons of oil spilled on land, and tens of thousands hectares of soil cover is polluted at depth 30-100 sm, and at older oil fields at 5-10 meters. New types of wastes were discovered: asphalt and tar paraffin sediments formed as a result of tubing clean up and submerged pumps.

The total area of oil-polluted lands in the Kazakhstan part of the Caspian Sea is over 194 th. hectares, and volumes of spilled oil make millions of tons. Soil cover is absorbed with oil at depth from scores of centimeters to 10 meters. Clean up of oil-polluted areas is one of the priority activities to be made in the region.

Negative impact of oil pollution is characterized by:

- Disturbance of natural landscapes and loss of agricultural lands;
- Underground water level rise followed by formation of small saline lakes;
- Mortality of waterfowl and small animals;
- Air pollution as a result evaporation of hydrocarbons at oil fields;
- Economic losses in regard to immobilization of oil at oil storages.

When the ebb takes place, polluted seawater comes back to its natural shore line carrying all pollutions, mainly oil products, to coastal waters. It is well known, tide phenomena of average amplitude can increase pollution of neighboring water area two-three times more, and in combination with emergencies 15 times more.

At the northern coast, near the oilfield Martishy, a month-delayed spike of pollution levels (up to 16 MPC) was recorded after a strong surge (June 1994) at vast expanse (at a range not less than 100 km), followed by a slow decrease to the norm by October. A high concentration of oil products (1,2-1,4 MPC) was recorded after surges, which occurred at the eastern coast in May and November of 1995.

Many days of observation in July and August of 1995, held at the eastern part of the sea, showed that concentrations of oil products in seawater were subject to considerable fluctuations in time and in space, which was due to prevailing oil transportation in form of aggregates of heavy components. The range of fluctuations with 95%-probability was within + 0,031 mg/l.

An increase of sediment pollution by oil products by a factor of 3-15 was recorded in some eastern coastal shallow waters in comparison with 1991. This can be accounted for bioaccumulation of heavy fractions of oil pollution, which settles during calm weather in areas, where waves can't wash the floor. These areas with the depth of 1,5-2,5 m are situated in 25-30 km from eastern coast.

In general, waters of the Northern Caspian are characterized as moderately polluted. Levels of sea pollution by oil have stayed at the level of 0,07-0,21 mg/l (1-4 MPC) during the recent years. The concentration of phenols in north-eastern Caspian amounted to 0,003-0,009 mg/l (3-9 MPC), surfactants - 0,008-0,029 mg/l (1-3 PMC). Chlorine-organic pesticides are always present in water. The estuary of the Ural and an area of immediate impact of the Volga's runoff for the to most polluted areas.

Pollution of the Caspian Sea is a major significant risk factor to biodiversity of this unique reservoir. Meantime, there is permanent environmental pollution (air, land, water) of the Caspian coast due to the intensification of oil and gas complex. In the last years, a steady decrease of pesticides chlorine-organic was observed in water and bottom sediments but an increase of petroleum, phenols and some heavy metals was observed.

Permanent pollution of the Caspian Sea has led to considerable accumulation of toxic compounds in fish tissues and organs.

Pollution analysis of the Caspian fish by petroleum in 1996 pointed out that most chromatograms contained various identified aliphatic hydrocarbon substances, related to biogenic sources (bacteria, pristan), diesel fuel and pollution by crude oil.

Average levels of investigated metals in fish organs and tissues were within the following ranges: in liver: barium 1.5-12.5 mgr/gr; cadmium 0-0.9 mgr/gr; chromium 0-0.7 mgr/gr; copper 12.5-60.6 mgr/gr; iron 193.8-975.0 mgr/gr; mercury 0-0.6 mgr/gr è zinc 77.5-737.5 mgr/gr.; in muscules: barium 0.3-4.6 mgr/gr; chromium 0.8-4.9 mgr/gr; copper 1.3-21.2 mgr/gr; iron 8.9-46.5 mgr/gr; mercury 0.1-1.9 mgr/gr; nickel 0-4.9 mgr/gr; lead 0-0.9 mgr/gr; zinc 19.5-57.3 mgr/gr.; in spawn: barium 0.7-7.6 mgr/gr;

chromium 0.6-3.5 mgr/gr; copper 3.0-11.4 mgr/gr; iron 62.6-166.3 mgr/gr; mercury 0-0.1 mgr/gr; nickel 0-2.0 mgr/gr; lead 0-0.1 mgr/gr.

The prevailing metals in each of the investigated fish patterns were zinc, copper, barium, and iron. Cadmium and vanadium are present in small quantity.

The Caspian region includes habitats of plants and animals, which are very sensitive to deterioration of living conditions. These are wetlands where numerous species of birds, marine vertebrates and invertebrates, spawning places for sturgeon and extensive thickets of reeds are found. Wetlands and reeds–a realm for birds where they nest and breed. These areas are of great value for migratory birds as well. Previous studies marked a decrease of spawning places, and water level rise resulted in flooding and eutrophication of small bays due to overflow of nutrients.

Based on the natural conditions in the Kazakhstan part of the Caspian Sea and data provided by different nature users of the region, there was a zoning of the Kazakhstan part of the Caspian Sea regarding sensitivity to oil spills and locations of potential dangerous oil installations. These factors pose environmental concern in case of large oil spills, environmental and biological impact as a result of oil spills.

The following sensitive areas and sites of high sensitivity were marked in the Kazakhstan part of the Caspian Sea (Fig. 7):

Area 1. Eastern part of the Volga Delta
    Site A. Fish habitats and migration routes;
    Site B. Nesting places of sea-gulls and sea-swallows;
    Site C. Nesting places of cormorants and herons;
Area 2. Volga-Ural interfluve
    Site D. Fish spawning and migratory places;
    Site E. Habitats of sea-gulls and sea-swallows;
Area 3. The Ural Delta
    Site F. Fish spawning and migratory places
    Site H. Nesting places of cormorants and herons;
    Site G. Nesting places of sea-gulls and sea-swallows;
    Site I. Nesting places of purple swamp hen;
Area 4. Ural-Emba interfluve
    Site J. Fish spawning and migratory places;
    Site K. Nesting places of dalmatian pelican;
Area 5. The southern part of the Emba and Prorva Rivers
Area 6. The area of the Komsomolecz bay
Area 7. Buzachi peninsula
    Site L. Concentration places of ornitofauna, including ducks, swans, geese, sea-gulls;

Area 8.   Tuleniy islands
Area 9.   Ural furrow
Area 10.   Mangyshlak peninsula, Bautino region
Area 11.   Caspian shelf from the Mangyshlak peninsula to the Kara-Bogaz-Gol Bay
Area 12.   Karakol lake.

For special sensitive Caspian areas, oil pollution is of particular danger, therefore addressing issues related to oil spill prevention and response, the conservation of ecological system of the Caspian Sea is of great importance.

In this regard, according to the Resolution of the Kazakhstan Government of 29.06.99, 1,876 relevant ministries and departments were to elaborate the National Oil Spill Contingency Action Plan, approved by the Resolution of the Kazakhstan Government of May 6, 2000 # 676.

Oil operations in the Northern Caspian: according to the Law of RK "On Specially Protected Natural Areas," geophysical surveys and explorations and hydrocarbon production, taking into account specific environmental conditions, are allowed in the reserve zone of Northern Caspian. In 1993, "Kazakhstancaspishelf" oil company, which joined the Caspian International Consortium as an operator, and involvement with the largest transnational companies: "Agip", "British Petroleum" in alliance with "Statoil", "British Gas", "Mobil", "Shell" and "Total" was established.

The Ministry of Natural Resources and Environment Protection of RK established "Special Ecological Conditions for Geophysical Surveys in the Kazakhstan part of the Caspian Sea", which are the basis for actions for both industrial organizations and regulatory bodies in the field of environmental protection of the Caspian Sea.

Seismic surveys in the Kazakhstan sector of Caspian sea—the first stage of oil operations—did not cause any serious damage to the Caspian ecosystems and its coast.

All objects for planned drilling activities works are located in the northeastern (at depth 1,8 to 10 m) shallow part of Caspian Sea within the reserve area. At the stage of exploratory drilling, 6 wells are planned to be drilled (Fig. 8).

Projects stipulate a possible threat posed to the Caspian Sea and, in particular, the ecologically sensitive area of the Northern Caspian in case of oil spill emergencies. So, special attention should be paid to expertise of activities related to emergency preparedness, prevention, response. Upper Paleozoic oil of the Caspian sea shelf is a threatening source of possible catastrophic disposals.

Paleozoic oil (Carbon, Devonian) with extreme strata conditions (stratum pressure - up to 1100 atmo., temperature 125-150°C) and aggressive properties (hydrogen sulfide up to 20% etc.) will destructively impact the biological diversity of the Caspian sea.

The gas factor produces 5,683,600 m^3/tons at the Tengiz oil field with large hydrocarbon resources located in Eastern Caspian. The productive capacities are over 1000 m. Staratal conditions are extreme.

Northeast Caspian seabed geomorphology

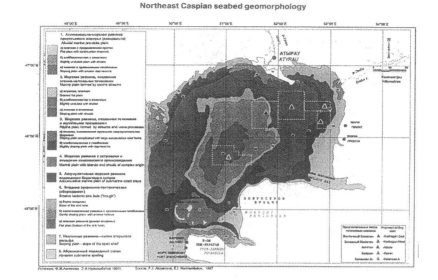

Fig. 8. Northeast Caspian seabed geomorphology.

From June 24, 1985 till July 27, 1986 (398 days) an open oil-gusher was burning. The boring well was gushing from carbonate sediments of carbon. The height of the flame was 180-200m and the flame diameter was about 50m. The temperature of the boring well mouth was 440°C. During gushing 3,5 ml. tons of oil and 1,7 billion cubic m of gas including 516 thousand cubic m of hydrogen sulfide were burnt. Soot was also generated. Sulfur dioxide concentration at the boring well mouth was up to 1100 Limited Allowed Concentration (LAC), at the distance of 500 m - 46 LAC; 45 km - 20-42 LAC; 100 km - 1 LAC. Mathematics modeling has shown that the boring well (N 37) oil-gusher radius reaches 300-350 km.

The rate of illnesses of population in the Jilioysk of Atirau region has increased 2,5 times in the year of boring well (N37) gushing. About 200 thousand people were negatively influenced by this phenomenon. About 200 thousand birds have died.

At present for the first time in the whole history of the Caspian sea, the upper Paleozoic oil (stratum pressure more than 1000 atmos., temperature 110-130°C, power more than 1100 m, with sulfur content in gas - 20-25%) was discovered in the shelf zone of the Caspian in shallow waters (depth 3,5 m) at 70 km to the South-East from Atirau in the structure Vostochniy Kashagan by the Company OKIOK. Analogous types of oil are anticipated in the other shelf structures. These oil deposits represent a gigantic power generation capacity. This fact creates a real, constant (for many years) threat of catastrophic discharges into the sea.

Priority actions for environmental protection at hydrocarbon exploration and production are as follows:

1. Caspian water pollution:
   - Construction of engineering and protective facilities;
   - Inventory and capping flooded oil fields and exploratory wells;
   - Impact of oil pollution on bioresources of Northern Caspian;
   - Establishment of the offshore oil spill prevention system.
2. Degradation of the soil and vegetative cover:
   - Improvement of existing and introduction of modern technologies as well as oil production, transportation and processing techniques;
   - Shutting down existing oil storages;
   - Waste, stratal and borehole water recycling;
   - Capping leaking wells;
   - Response to radioactive pollution of oil fields.
3. Air pollution:
   - Cessation of associated gas flaring at oil fields;
   - Improvement of techniques at oil fields and oil processing enterprises;
   - Application of emission mitigation technologies under exploratory and production drilling.

The first step towards solving common environmental issues of the Caspian Sea and its coastal area is joint efforts of the Caspian littoral states (Azerbaijan, Iran, Kazakhstan, Russian Federation and Turkmenistan) to implement the Caspian Environment Programme (CEP) financed by international organizations (GEF, UNDP, UNEP, World Bank, TACIS).

The Ministry of Natural Resources and Environment Protection of RK performs the functions of a coordinating body in Kazakhstan, with the Minister as a National Focal Point. The Caspian Sea and Oil & Gas Pollution Problems Committee implements current activities under the CEP.

The CEP stipulates support to Priority Investment Project Portfolio (PIPP). In accordance with the Memorandum between the MNREP and U.S. Energy Association (USEA), the Committee takes on joint activities with the USEA in preparation of the Guidelines on Environmental Impact Assessment (EIA) for oil and gas projects, as well as the Guidelines on Establishment of environmental monitoring network for marine oil projects in North-eastern Caspian.

The draft National Oil Spills Contingency Action Plan of the Republic of Kazakhstan, approved by the Kazakhstan Government (Enactment #676 of May 6, 2000), has been developed with the active support of "Kazakhoil" National Oil Company and with the participation of a number of ministries and departments (including the Committee) and the "Hagler Bailly" U.S. Consulting Company. As a part of the Work Group, the Committee submitted a number of proposals to the Contingency Plan, which were further included into the draft.

There is an agreement with the Exxon Mobil company regarding the joint implementation of the institutional project to support thematic directions and the CEP programme coordination. The company has allotted financial resources for the implementation of the given project. The Committee conducts negotiations with oil companies regarding research activities for assessment of the flooded oil wells in the Caspian region, as well as other CEP priority activities.

The Committee participates in the development of normatively legal documents related to the Caspian issues. Amendments and additions were proposed to Special Environmental Requirements to Offshore Oil Exploration in the State Reserved area in Northern Caspian, approved by the Government of the Republic of Kazakhstan.

The efforts on elaboration of the Caspian Biodiversity Conservation Programme within the framework of the CEP have been undertaken for biodiversity conservation of the Caspian region. The Caspian Committee works closely with NATO/CCMS for implementation of environmental projects aimed at environmental protection of the Caspian Sea.

In this connection, the Committee took an active part in the preparation of the pilot project "Review of environmental projects of the Caspian Sea for the planning of future activities," as this project provides encouraging advantages for the future of environmental studies of the Caspian Sea. The project materials have been prepared and presented in Adam (1999). On July 2000, the final version of the project was reviewed at the NATO/CCMS in Brussels.

At present, the NATO/CCMS has submitted for the Committee's consideration the short-term special project on environmental safety of transport of hazardous substances in the Black and Caspian Seas as a prolongation of work under the above project.

As a whole, actions on priority environmental directions related to pollution mitigation and biodiversity conservation in the Caspian region are financed from different sources (loans, grants, technical assistance, budget). Based on the MNREP's initiative, the Governmental Action Plan includes Resolutions to elaborate the State Caspian Biodiversity Conservation Programme as well as implementation of CEP activities. The Caspian region is one of major priorities in the Long-Term Development Strategy of the Republic of Kazakhstan. Addressing environmental issues is within the framework of cooperation between the Republic of Kazakhstan and UNDP–Institutional Strengthening for Sustainable Development Programme for 2001-2004.

POLLUTION IMPACT TO FISH FROM NORTH CASPIAN SEA

YELENA N. ZHIMBEY
Kazakh State University, Almaty, Kazakhstan

IGOR V. MITROFANOV
NGO "Tethys", Tethys Research, Almaty, Kazakhstan, igor@miv.almaty.kz

Histopathological changes of fish can be used as biomarkers of pollution impact in different water bodies. Histological changes and lesions were analyzed in gobies and other species in the Caspian Sea for several years. All investigations were done in collaboration with Woods Hole Oceanographic Institution and budgeted by CRDF (award #KB1-2017).

Sampling organs were gills, heart, liver, spleen, pancreas, intestine, and kidney. For histology analyses, organs were fixed in the field in 10% buffered formalin. All histological processing was routine: embedding in paraffin, cutting slides, staining by hematoxylin-eosin.

Most of the investigations were made on the monkey goby (*N. fluviatilis*) one of the most abundant and wide distributed species in the Caspian Sea. Histological changes findings for this species cover almost the whole specter of changes in Caspian fish. Also round goby (*N. melanostomus*), syrman goby (*N. syrman*), gorlap goby (*N. gorlap*), Caspian sprat (*Clupeonella delicatula*), Caspian sand-smelt (*Atherina boyeri*), Caspian roach (*Rutilus rutilus*) and other species were investigated.

GOBY SPECIES

- Monkey goby–*Neogobius fluviatilis pallasi* (Berg, 1916)–53 specimens
- Round goby–*Neogobius melanostomus affinis* (Eichwald, 1831)–14 specimens
- Syrman goby–*Neogobius syrman eurystomus* (Kessler, 1877)–5 specimens
- Gorlap goby–*Neogobius gorlap* Iljin, 1927 = N. kessleri gorlap–8 specimens
- Caspian goby–*Neogobius caspicus* (Eichwald, 1831)–2 specimens
- Deepwater goby–*Neogobius bathybius* (Kessler, 1877)–9 specimens
- Blackspot goby–*Mesogobius nigronotatus* (Kessler, 1877)–3 specimens

Liver

Fatty degeneration. Fatty degeneration was found for all goby species. The big lipid vacuole appeared in the hepatocytes and driven cytoplasm and nucleus to the walls. Very often it lead to atrophy of the nucleus.

Collangioma (cirrhosis): proliferation of connective tissue inside the liver.

Necrosis. Cytoplasm become colorless, nuclei decrease in size and are destroyed. Destroying of cell wall begins.

Basophilic foci. Cytoplasm of hepatocytes become dark blue. Nuclei increase in size, and have intensive coloration. No vacuoles at all.

Spleen

Macrophage aggregations. Macrophage aggregations were found in almost all investigated fishes.

Atrophy. Erythrocytes disappear. Cytoplasm of some cells become colorless, nuclei have pale coloration.

Proliferation of connective tissue. External tunica of the spleen becomes thicker. Additional connective cords appear inside the organ.

Necrosis. Necrosis was found in one monkey goby, simultaneously with parasite.

Cell vacuolation. Cell vacuolation and tumors were found several times.

Pancreas

Degradation. Degradation of acinar cells around intestine. Acinar cells lose granules of zymogen, become pale, cytoplasm become homogeneous, nuclei don't appear well. This is found in many goby specimens.

Cell vacuolation. Vacuolation of acinar cells detected only in parts of pancreas inside the liver.

Intestine

Degradation. Degradation of intestine walls.

Cell vacuolation. Vacuolation of epithelial cells very often accompanies inflammation of intestine walls.

Kidney

Canaliculus. Destruction of canaliculus were one of the most frequent changes. Cells of canaliculus become round and lose connections with each other.

Cell vacuolation. Vacuolation of canaliculus. Hydropic vacuole appears in single cells. Cytoplasm and nucleus are driven to the wall.

Connection. Connection between interstitial cells destroyed.

Necrosis. Necrosis of canaliculus and destruction of glomerulus. External tunica of glomerulus destroyed, cells lose connection with each other and distance between them become more than a half of its diameter.

Parasites. Parasites were detected twice inside kidney.

Heart

Single changes were found only in monkey goby. All types of changes were observed only once. These were *cytoplasm granulation, fibrosis of muscle cells, atrophy.*

Gills

Hyperplasia. Hyperplasia of epithelial cells is one of the most common lesions. Lamellar and interlamellar epithelium proliferates and packs the space between secondary lamellas.

Lamellas. Aneurysm of secondary lamellas. Dilatation of lamellas and stagnation of blood in it.

Parasites. Many specimens have different type of parasites encapsulated in secondary and/ or primary lamellas. Identification of parasites wasn't made.

CASPIAN ROACH

Rutilus rutilus caspicus (Jakowlew, 1870)
    Only 13 specimens not far from the Ural delta were investigated.

Liver

Macrophage aggregations were found in all investigated fish and once hydropic vacuolation of hepatocytes was found.

Spleen

Macrophage aggregations were found in all investigated fishes.

Kidney
Vacuolation of canaliculus were found out in 8 fishes and macrophages were found out in 7 fishes.

Intestine
Only once were macrophages inside intestine submucosa found.

Pancreas
In all fishes acinar cells around intestine were displaced by adipose tissue. No changes in pancreas inside liver were observed.

Heart
Only once were macrophages were found. In the same fish macrophages were found in the intestine.

Gills
Once proliferation of interlamellar epithelia and twice aneurysm of secondary lamellas were found.

CASPIAN SPRAT

Clupeonella delicatula caspia (Nordmann, 1840)
Only 11 specimens from North and Middle Caspian Sea were investigated.

Liver
There were no changes in the liver of fishes from the North Caspian Sea. In fishes from the Middle Caspian Sea, several lesions were observed. Macrophage aggregations, blood stagnation, necrosis of hepatocytes, granulation of cytoplasm, and parasite were found.

Spleen
In all investigated fishes, macrophage aggregations were found.

Kidney
Macrophage aggregation and necrosis were found in fishes from the Middle Caspian Sea.

Intestine
No changes were found.

Pancreas
In the fishes from the Middle Caspian Sea, macrophage aggregation and replacement of acinar cells by adipose tissue were found.

Heart
No changes were found.

Gills
Only once were aneurysm and parasite detected in the fish from the Middle Caspian Sea.

Gonads
Proliferation of connective tissue was found in one male.

CASPIAN SAND-SMELT

Atherina boyeri caspia (Eichwald, 1831) = Atherina mochon pontica natio caspia
Only six specimens from two locations in the North Caspian Sea and seven specimens from one location in the Middle Caspian Sea were investigated.

Liver
Fatty degradation, macrophage aggregation, and granulation of cytoplasm were found both in North and Middle Caspian locations.

Spleen
Only two fishes were examined. Both have macrophage aggregations and one has collangioma.

Kidney
One fish from North Caspian Sea has parasites and degradation of canaliculus. One fish from Middle Caspian Sea has vacuolation of canalicullus.

Intestine
No changes were found.

Pancreas
No changes were found in fishes from North Caspian Sea. All fishes from Middle Caspian Sea have replacement of acinar cells by adipose tissue. One fish has macrophages between acinar cells.

Heart
No changes were found.

Gills
No changes were found.

Gonads
No changes were found.

DISCUSSION

Compared to the goby species, roach has fewer lesions and changes of inner organs. Nevertheless, there are some abnormalities detected only for roach and never observed in other species. There were hydropic vacuolation of hepatocytes, macrophages between muscle cells in the heart, focal tumors and necrosis in testes. We never observed in the roach lesions such as necrosis, degeneration, and atrophy of pancreas. In general there was more macrophage aggregation in roach than in other species. So, roach has its own specter of pathologies not identical to the goby species even though they share a habitat. There was more pathology in all organs, except the kidney, in goby species compared to roach. However, roach has much more pathology in the kidney than the most sensitive goby species–syrman goby.

Sprat and sand-smelt are pelagic fishes with feeding habits different from gobies and roach. Fat accumulation in sprat occurs around the intestine as in roach. These two species never have fatty degeneration of hepatocytes. In sand-smelt, fat accumulates in the liver as in goby species and all these species have fatty degeneration of hepatocytes. Sand-smelt and sprat have granulation of cytoplasm in hepatocytes, which was never observed in goby species and roach. Both pelagic species were investigated simultaneously on several locations. In all cases there were more pathologies in all organs in sprat, than in sand-smelt. It means that sprat is more sensitive to inconvenient environment than sand-smelt. There were more pathologies in sprat than in goby species on the same location. The number of pathologies in the inner organs of sand-smelt is almost equal to deepwater goby from the same location.

According to the number of different lesions, sprat are more sensitive than all the other investigated species. Goby species were divided into two groups. For the first group (*N. melanostomus; N. bathybius*), more pathological changes were found in the liver and spleen, and for the second group (*N. fluviatilis, N. caspicus, N. syrman*), more changes were found in the kidney and gills. Gorlap goby is more resistant, than other gobies. Most pathological changes in the North Caspian Sea were found in fishes from the mouth of the Ural delta. The environment is better for the fishes. The better environment was in the open sea not far from the Ural delta. Ecological conditions in the mouth of Ural delta are not stable during the year and depend on the amount of pollutants coming in with river influx.

Only two specimens of gobies were investigated from the south of the Middle Caspian Sea near the Azerbaijan cost. Both specimens have significant lesions in liver, kidney, and pancreas. Hepatocytes contain small amount of lipid vacuoles. The biological condition of both fishes is critical. On a location near to Baku city and an oil drilling field, many goby specimens have necrosis, tumors, and blood stagnation in the liver, proliferation of connective tissue in the spleen. All fishes had heavy lesions of gills. Biological condition of all fishes is critical. In the north of the Middle Caspian Sea near to the border with North Caspian Sea, gobies, sprat, and sand-smelt were investigated. In the liver of the goby species, accumulation of glycogen in the liver was from small to medium. Some gobies had blood stagnation in the liver. Some fishes had necrosis and proliferation of connective tissue in different organs. Many fish had lesions in the gills. The biological condition of sprat is dangerous. There were not so much pathologies in the sand-smelt from the same location. The biological condition of gobies and sand-smelt is satisfactory.

CONCLUSION

- For future monitoring sprat and a complex of gobies species (*N. fluviatilis–N. melanostomus–N. gorlap*) are the most suitable.
- The biological condition of fishes in locations remote from deltas is good. Biological condition of fishes in the mouth of the Ural Delta is not stable and changes from good to critical during the year. According to the number of pathological changes, the zone around the mouth of the Ural delta is one of the most polluted sites in the eastern part of the North Caspian Sea.
- The biological condition of all fish species is better in the Middle Caspian Sea (Western coast line) compared to the North Caspian Sea. It is especially true for polluted sites around oil drilling fields.

MONITORING OF MARINE POLLUTION IN THE NORTHERN CASPIAN SEA

ALEXANDER KORSHENKO

Marine Pollution Monitoring, State Oceanographic Institute, Moscow, Russia

Monitoring system of Caspian Sea of former Soviet Union formed during 1970th and consists of net of different type of stations covered all investigated area. For measuring of basic hydrometeorological parameters, like temperature, salinity, density, air conditions, insolation, standard hydrochemistry etc., stations at so-called "century transects" mainly were used. Eight transects crossed the Volga delta (4) and Central Caspian area (4) several time a year. For estimation of pollution level in the sea waters the OGSN stations were served. In the northern shallow waters the position of these stations usually are the same with points at century transects, in contrary with the Central Caspian where they mainly are not coincide. Also many OGSN stations placed in coastal waters in vicinity of large cities or estuarine regions where pollution from the land is most visible. The additional information could be obtained from the net of sea-level points and coastal stations in estuaries and riverbanks.

Monitoring of pollution in the Volga estuarine zone beside standard observation usually includes ammonium, Fe, Cu, detergents, phenols, dissolved organic matter, petroleum hydrocarbons. Heavy metals and petroleum hydrocarbons usually were measured in sea water and sediments. During the additional marine expeditions and other research investigations in the waters influenced by Volga the number of measured parameters and spatial position of station could be varied in a wide range.

In Dagestan coastal waters and in Central Caspian usually during ship expeditions at the monitoring stations the ammonium, phenols, organic contaminants and petroleum hydrocarbons were estimated for times per year during last three decades. The temporal variations of all measured pollutants, seasonal and interannual, and spatial distributions in 8 coastal and estuarine areas of Dagestan were presented as figures and tables. This data allow concluding that the level of pollution in the coastal waters slightly decreased in all investigated areas and strongly depended mainly from Volga output and less from the local sources.

16. SEMINAR PARTICIPANTS

SEMINAR PARTICIPANTS

Professor Adriano Aguzzi	Institute of Neuropathology University Hospital of Zürich Zürich, **Switzerland**
Professor Rashid Ajalov	Caspian Environmental Programme Baku, **Azerbaijan**
H. E. Mario Alessi	Italian Administration for International Organization Rome, **Italy**
Professor H. Alper	Department of Chemistry University of Ottawa Ottawa, **Canada**
Dr. Paul Bartel	USGS Washington, **USA**
Professor William A. Barletta	Accelerator and Fusion Research Division Lawrence Berkeley National Laboratory Berkeley, **USA**
Professor Gunnel Biberfeld	Swedish Institute for Infectious Disease Control & Karolinska Institute Solna, **Sweden**
Dr. Donald F. Boesch	Center for Environmental Science University of Maryland Cambridge, **USA**
Professor J. M. Borthagaray	Instituto Superior de Urbanismo University of Buenos Aires Buenos Aires, **Argentina**
Dr. Olivia Bosch	IISS London, **UK**
Professor David Brookshire	Department of Economics University of New Mexico Albuquerque, **USA**

Professor Herbert Budka	Institute of Neurology University of Vienna Vienna, **Austria**
Professor Franco Buonaguro	Fondazione Pascale Istituto Nazionale dei Tumori Naples, **Italy**
Dr. Gregory Canavan	Physics Division Los Alamos National Laboratory Los Alamos, **USA**
Dr. Larisa Cervenakova	Plasma Derivatives Department American Red Cross Holland Laboratory Rockville, **USA**
Dr. Nathalie Charpak	Kangaroo Mothers Foundation Bogota, **Colombia**
Professor Dimitry Chereshkin	Institute of Systems Analysis Russian Academy of Sciences Moscow, **Russia**
Professor Robert Clark	Hydrology and Water Resources University of Arizona Tucson, **USA**
Professor Guy de Thé	Unité d'Epidémiologie des Virus Oncogènes Institut Pasteur Paris, **France**
Dr. M'Bareck Diop	Technical Advisor to the President of Senegal Dakar, **Senegal**
Dr. Dimitru Dorogan	Ministry of Waters and Environmental Protection Bucharest, **Romania**
Dr. Peter Douglas	California Coastal Commission San Francisco, **USA**
Professor Chris D. Ellis	Landscaping Architecture & Urban Planning Texas A&M University College Station, **USA**

Professor Lorne Everett	The IT Group Santa Barbara, **USA**
Dr. Marina Ferreira Rea	Maternal and Child Health Division Instituto de Saude São Paulo, **Brazil**
Dr. Bertil Galland	24 Heures Buxy, **France**
Dr. Hamid Ghaffarzadeh	Caspian Environmental Programme Baku, **Azerbaijan**
Professor Bernardino Ghetti	Pathology & Laboratory Medicine Department Indiana University Indianapolis, **USA**
Dr. Thomas J. Gilmartin	Global Security Research Lawrence Livermore National Laboratory Livermore, **USA**
Professor Michel Gourdin	Physique Théorique et Hautes Energies Université Pierre et Marie Curie Paris, **France**
Professor Glenda Gray	Perinatal HIV Research Unit Chris Hani Baragwanath Hospital Soweto, **South Africa**
Professor J. Mayo Greenberg	University of Leiden Leiden, **The Netherlands**
Dr. Kevin Gurney	Department of Atmospheric Science Colorado State University Fort Collins, **USA**
Dr. Dagmar Heim	Swiss Federal Veterinary Office Liebefeld, **Switzerland**
Professor Reiner K. Huber	Universität der Bundeswehr München Neubiberg, **Germany**

Dr. Walter F. Huebner	Southwest Research Institute San Antonio, **USA**
Professor Yuping Huo	College of Physics & Engineering Zheng Zhou University Zheng Zhou, **China**
Dr. Christiane Huraux	Monther-Infant HIV Transmission Consultant Paris, **France**
Professor Namik T. Ibrahimov	Caspian Environmental Programme Baku, **Azerbaijan**
Professor Leonardas Kairiukstis	Laboratory of Ecology and Forestry Kaunas-Girlonys, **Lithuania**
Professor Eugenij Klushin	Strategic Studies Center Russian Academy of Sciences Moscow, **Russia**
Dr. Gerhard Knies	DESY Hamburg, **Germany**
Dr. Alexander Korshenko	Marine Pollution Monitoring Laboratory State Oceanographic Institute Moscow, **Russia**
Dr. Andrei Krutskih	Department of Science & Techology Russian Foreign Ministry Moscow, **Russia**
Professor Valery Kukhar	Institute for Bio-organic Chemistry Academy of Sciences Kiev, **Ukraine**
Professor Tsung-Dao Lee	Department of Physics Columbia University New York, **USA**
Professor Axel Lehmann	Universität der Bundeswehr München Neubiberg, **Germany**

Professor Alf A. Lindberg	Campus Mérieux Marcy l'Etoile, **France**
Dr. Gregg Marland	Environmental Sciences Division Oak Ridge National Laboratory Oak Ridge, **USA**
Professor Sergio Martellucci	Faculty of Engineering Università di Roma "Tor Vergata" Rome, **Italy**
Professor Arthur I. Miller	Department of Science & Technology Studies University College London London, **UK**
Dr. Vladimir Mirianashvili	Georgian Academy of Sciences Tbilisi, **Georgia**
Professor Igor V. Mitrofanov	Institute of Zoology MS & HE RK Almaty, **Kazakhstan**
Dr. David Mulenex	Counselor for Environment, S & T U.S. Embassy Rome, **Italy**
Dr. Barry Myers	Accu Weather, Inc. State College, **USA**
Professor Lennart Olsson	Centre for Environmental Studies Lund University Lund, **Sweden**
Professor Jay Orear	Floyd R. Newman Laboratory Cornell University Ithaca, **USA**
Professor Emin Özsoy	Institute of Marine Sciences Middle East Technical University Erdemli, **Turkey**
Professor Donato Palumbo	World Laboratory Fusion Centre Brussels, **Belgium**

Professor Margaret Petersen	Hydrology & Water Resources University of Arizona Tucson, **USA**
Professor Andrei Piontkovsky	Strategic Studies Centre Moscow, **Russia**
Professor Juras Pozela	ICSC World LaboratoryBranch Vilnius, **Lithuania**
Professor Richard Ragaini	LLNL Department of Environmental Protection University of California Livermore, **USA**
Professor Vittorio Ragaini	University of Milan Milan, **Italy**
Professor Karl Rebane	ICSC World Laboratory Branch Estonia Tallinn, **Estonia**
Professor Bob Redfield	Institute of Human Virology University of Maryland Baltimore, **USA**
Dr. Maura Ricketts	Department of Surveillance & Response World Health Organization Geneva, **Switzerland**
Professor Zenonas Rudzikas	Theoretical Physics & Astronomy Institute Lithuanian Academy of Sciences Vilnius, **Lithuania**
Professor Ilkay Salihoglu	Institute of Marine Sciences Middle East Technical University Erdemli, **Turkey**
Professor N. M. Samuel	Experimental Medicine & AIDS Resource Center The Tamil Nadu Medical University Guindy-Chennai, **India**
Professor Gabriella Scarlatti	Laboratory of Virology DIBIT San Raffaele Scientific Institute

Milan, **Italy**

Dr. Don Scavia

National Centers for Coastal Ocean Service
National Oceanic & Atmospheric Administration
Silver Spring, **USA**

Dr. Hiltmar Schubert

Fraunhofer-Institut für Chemische Technologie
Pfinztal, **Germany**

Professor Geraldo Gomes Serra

NUTAU
Sao Paulo State University
Sao Paulo, **Brazil**

Mrs. Ludmilla Shabanova

National Environmental Center for Sustainable
Development
Kokshetau City, **Kazakhstan**

Professor Kai M. B. Siegbahn

Institute of Physicss
University of Uppsala
Uppsala, **Sweden**

Professor K. Sivaramakrishnan

Centre for Policy Research
New Delhi, **India**

Dr. Andrew W. Smith

Infection Research Group
Glasgow Dental Hospital & School
Glasgow, **UK**

Professor Soroosh Sorooshian

Hydrology and Water Resources
University of Arizona
Tucson, **USA**

Professor William A. Sprigg

Institute for the Study of Planet Earth
University of Arizona
Tucson, **USA**

Dr. V. J. Sundaram

National Design & Research Forum
The Institute of Engineers (India)
Bangalore, **India**

Professor Albert Tavkhelidze

National Academy of Sciences
Tbilisi, **Georgia**

Dr. Timothy L. Thomas	U.S. Department of Defense Fort Leavenworth, **USA**
Mrs. Kay Thompson	U.S. Department of Energy Washington, D.C., **USA**
Dr. Larry Tieszen	International Programs EROS Data Center, USGS Sioux Falls, **USA**
Professor Sheila Tlou	University of Botswana Gaborone, **Botswana**
Dr. Petra Tschakert	University of Arizona Tucson, **Arizona**
Professor Vitali Tsygichko	Institute for System Studies Russian Academy of Sciences Moscow, **Russia**
Dr. Teja Valencic	Ministry of Education, Science & Sport Ljubljana, **Slovenia**
Professor Philip Van de Perre	Laboratory of Bacteriology & Virology CHU-Arnaud de Villeneuve Montpellier, **France**
Professor Anatoly Vasiliev	State Oceanographic Institute Moscow, **Russia**
Professor Marcel Vivargent	CERN Geneva, **Switzerland**
Professor François Waelbroeck	Juelich Fusion Centre, St. Amandsberg, **Belgium**
Dr. Andrew W. Warren	Department of Geography University of London London, **UK**
Dr. Henning Wegener	German Ambassador to Spain (former) Madrid, **Spain**

Professor Robert G. Will	Western General Hospital National CJD Surveillance Unit Edinburgh, **UK**
Professor Aaron T. Wolf	Department of Geosciences Oregon State University Corvallis, **USA**
Dr. Zeng Yi	Institute of Virology Academy of Preventive Medicine Beijing, **China**
Dr. Rolf Zetterstösm	Acta Paediatria Stockholm, **Sweden**
Professor Antonino Zichichi	CERN & University of Bologna Geneva, **Switzerland**